Dyadic Green Functions in Electromagnetic Theory

IEEE Series on Electromagnetic Waves

The IEEE PRESS Series on Electromagnetic Waves consists of new titles as well as reprints and revisions of recognized classics that maintain long-term archival significance in electromagnetic waves and applications.

Series Books Published

Collin, R. E., *Field Theory of Guided Waves*, 2d. rev. ed., 1991
Tai, C. T., *Generalized Vector and Dyadic Analysis: Applied Mathematics in Field Theory*, 1991
Elliott, R. S., *Electromagnetics: History, Theory, and Applications*, 1993
Harrington, R. F., *Field Computation by Moment Methods*, 1993
Tai, C. T., *Dyadic Green Functions in Electromagnetic Theory*, 2nd ed., 1993

Future Series Title

Dudley, D. G., *Mathematical Foundations of Electromagnetic Theory*

Dyadic Green Functions in Electromagnetic Theory

Second Edition

Chen-To Tai

Professor Emeritus
Radiation Laboratory
Department of Electrical Engineering
and Computer Science
University of Michigan

IEEE PRESS Series on
Electromagnetic Waves
Donald G. Dudley, Series Editor

IEEE
PRESS

IEEE Antennas and Propagation Society and
IEEE Microwave Theory and Techniques Society, *Co-sponsors*

The Institute of Electrical and Electronics Engineers, Inc., New York

IEEE PRESS
445 Hoes Lane, PO Box 1331
Piscataway, NJ 08855-1331

Technical Reviewers

Nicolaos G. Alexopoulos
University of California at Los Angeles

Edmund K. Miller
Los Alamos National Laboratory

Kai Chang
Texas A & M University

Robert E. Collin
Case Western Reserve University

Printed in the United States of America
10 9 8 7 6 5 4 3 2 1

ISBN 0-7803-0449-7
IEEE Order Number: PC0348-3

Library of Congress Cataloging-in-Publication Data

Tai Chen-To (date)
Dyadic green functions in electromagnetic theory
by Chen-to Tai.—2nd ed.
p. cm.
Sponsors : IEEE Antennas and Propagation Society and IEEE Microwave Theory and Techniques Society.
Includes Bibliographical references and index.
ISBN 0-7803-0449-7
1. Electromagnetic theory—Mathematics. 2. Green's functions. 3. Boundary value problems. I. IEEE Antennas and Propagation Society. II. IEEE Microwave Theory and Techniques Society. III. Title
QC670.T34 1993
530.1′41—dc20

93-24201
CIP

Dedicated
to
Professor Chih Kung Jen
(An Inspiring Teacher of Science and Humanity)

Contents

Preface

The first edition of this book, bearing the same title, was published by Intext Education Publishers in 1971. Since then, several topics in the book have been found to have been improperly treated; in particular, a singular term in the eigenfunction expansion of the electrical dyadic Green function was inadvertently omitted, an oversight that was later amended [Tai, 1973].

In the present edition, some major revisions have been made. First, Maxwell's equations have been cast in a dyadic form to facilitate the introduction of the electric and the magnetic dyadic Green functions. The magnetic dyadic Green function was not introduced in the first edition, but it was found to be a very important entity in the entire theory of dyadic Green functions. Being a solenoidal function, its eigenfunction expansion does not require the use of non-solenoidal vector wave functions or Hansen's L-functions [Stratton, 1941]. With the aid of Maxwell-Ampère equation in dyadic form, one can find the eigenfunction expansion of the electrical dyadic Green function, including the previously missing singular term. This method is used extensively in the present edition.

Several other new features are found in this edition. For example, the integral solutions of Maxwell's equations are now derived with the aid of the vector-dyadic Green's theorem instead of by the vector Green's theorem as in the old treatment. By doing so, many intermediate steps can be omitted. In reviewing Maxwell's theory we have emphasized the necessity of adopting one of two alternative postulates in stating the boundary conditions. The implication is that the boundary conditions cannot be derived from Maxwell's differential equations without a postulate. Reciprocity theorems in electromagnetic theory are discussed in detail. In addition to the classical theorems due to Rayleigh, Carson, and Helmholtz, two complementary reciprocity theorems have been formulated to uncover the symmetrical relations of the magnetic dyadic Green functions not derivable from the Rayleigh-Carson theorem.

Various dyadic Green functions for problems involving plain layered media have been derived, including a two-dimensional Fourier-integral representation of these functions. In the area of moving media, the problem of transient radiation is formulated with the aid of an affine transformation which enables us to solve the Maxwell-Minkowski equation in a relatively simple manner.

Many new exercises have been added to this edition to help the reader better understand the materials covered in the book. Answers for some exercises are given, and sufficient hints are provided for many others so that the book may be used not only as a reference but also as a text for a graduate course in electromagnetic theory.

Acknowledgments

I am very grateful to Professor Per-Olof Brundell of the University of Lund, Sweden, who, in 1972, called my attention to the incompleteness of the eigenfunction expansion of the electric dyadic Green function in the original edition of this book. My discussion with Dr. Olov Einarsson, then a faculty member of the same institution, on the dependence of the integral of the electric dyadic Green function on the shape of the cell in the source region was very valuable, particularly, on the aspect ratio of a cylindrical cell. The works of Prof. Robert E. Collin consolidate our understanding of the singularity behavior of the dyadic Green functions. His many communications with me on this subject were very valuable prior to the publication of a book in this field by Prof. J. Van Bladel [1991]. I am also very grateful to Prof. Donald G. Dudley and Dr. William A. Johnson for their very careful review of my original manuscript. Section 5-8 of Chapter 5 was written as a result of their thoughtful comments.

During the preparation of this manuscript I received the most valuable help from Ms. Bonnie Kidd. Her expertise in typing this manuscript was invaluable. The assistance of Dr. Leland Pierce and Ms. Patricia Wolfe are also very much appreciated.

I would also like to express my sincere thanks to Prof. Fawwaz T. Ulaby, Director of the Radiation Laboratory at the University of Michigan, for his constant encouragement by providing me with the technical support necessary to complete this manuscript. Mr. Dudley Kay, Director of Book Publishing, and Ms. Karen Miller, Production Editor of IEEE Press, have proved to be most efficient and helpful during all stages of the production of this book.

Chen-To Tai
Ann Arbor, Michigan

Dyadic Green Functions in Electromagnetic Theory

1

General Theorems and Formulas

In this chapter we review some of the important theorems and formulas needed in the subsequent chapters. It is assumed that the reader has had an adequate course in advanced calculus, including vector analysis, Fourier series and integrals, and the theory of complex variables. Our review will contain sufficient material so that references to other books will be kept to a minimum. We sacrifice to some extent the mathematic rigor that may be required in a more thorough treatment. For example, we use quite freely the integral representation of the delta function, assuming that an exponential function with imaginary argument is Fourier transformable. Whenever necessary, adequate references[1] will be given to strengthen any plausible statement or to remove possible ambiguity.

1-1 VECTOR NOTATIONS AND THE COORDINATE SYSTEMS

A vector quantity or a vector function will be denoted by $\overline{F}$. A letter with a hat, such as $\hat{x}$, is used to denote a unit vector in the direction of the covered letter. In most cases, these letters correspond to the variables in a coordinate system. The scalar product of two vectors is denoted by $\overline{A} \cdot \overline{B}$ and the vector product by $\overline{A} \times \overline{B}$. The three commonly used systems in this book are

1. Rectangular, or Cartesian, x, y, z
2. Circular cylindrical or simply cylindrical, r, ϕ, z
3. Spherical, R, θ, ϕ

[1]In the citations in the text, the author's name is used as the identification. If it is a book, either the section number or the pages will be cited, if necessary.

The spatial variables associated with these systems are shown in Fig. 1-1. It should be pointed out that the same ϕ-variable is used for both the cylindrical and the spherical systems. The unit vectors belonging to these systems are displayed in Fig. 1-2 in two cross-sectional views. The relation between these unit vectors is summarized in Tables 1-1 and 1-2.

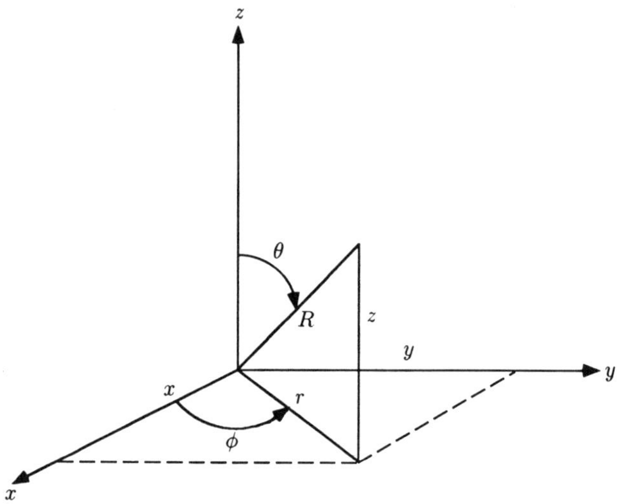

Fig. 1-1 Three commonly used coordinate systems

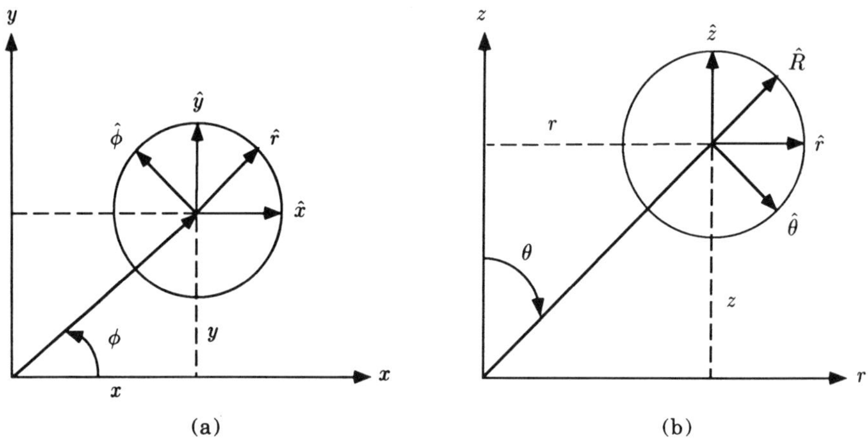

Fig. 1-2 The unit vectors in three commonly used coordinate systems

To express unit vector $\hat{x}$ in terms of the unit vectors in the spherical system, one uses the coefficients in the first column of Table 1-2, which gives

$$\hat{x} = \sin\theta\cos\phi\hat{R} + \cos\theta\cos\phi\hat{\theta} - \sin\phi\hat{\phi}. \tag{1.1}$$

TABLE 1-1 Relations Between the Unit Vectors in the Rectangular and the Cylindrical Coordinate Systems

	$\hat{x}$	$\hat{y}$	$\hat{z}$
$\hat{r}$	$\cos\phi$	$\sin\phi$	0
$\hat{\phi}$	$-\sin\phi$	$\cos\phi$	0
$\hat{z}$	0	0	1

TABLE 1-2 Relations Between the Unit Vectors in the Rectangular and the Spherical Coordinate Systems

	$\hat{x}$	$\hat{y}$	$\hat{z}$
$\hat{R}$	$\sin\theta\cos\phi$	$\sin\theta\sin\phi$	$\cos\theta$
$\hat{\theta}$	$\cos\theta\cos\phi$	$\cos\theta\sin\phi$	$-\sin\theta$
$\hat{\phi}$	$-\sin\phi$	$\cos\phi$	0

Likewise, the second row gives

$$\hat{\theta} = \cos\theta\cos\phi\hat{x} + \cos\theta\sin\phi\hat{y} - \sin\theta\hat{z}. \tag{1.2}$$

The reader can verify for himself or herself that these tables apply equally well to the transform of the components of a vector; for example,

$$A_\theta = \cos\theta\cos\phi A_x + \cos\theta\sin\phi A_y - \sin\theta A_z. \tag{1.3}$$

Another coordinate system used in this book deals with an elliptical cylinder. One set of variables that can be used in this system is designated by (u, v, z). A cross-sectional view of a plane perpendicular to the z-axis is shown in Fig. 1-3. The constant u contours and the constant v contours correspond, respectively, to a family of confocal ellipses and a family of confocal hyperbolas. The relations between (x, y) and (u, v) are

$$x = c\cosh u\cos v \tag{1.4}$$

$$y = c\sinh u\sin v, \tag{1.5}$$

where $\infty > u \geq 0, 2\pi \geq v \geq 0$. Two alternate variables which are used sometimes in place of (u, v) are defined by

$$\xi = \cosh u \tag{1.6}$$

$$\eta = \cos v, \tag{1.7}$$

where $\infty > \xi \geq 0, 1 \geq \eta \geq -1$. Table 1-3 contains the transformation coefficients between the unit vectors of the rectangular system and the elliptical cylinder system.

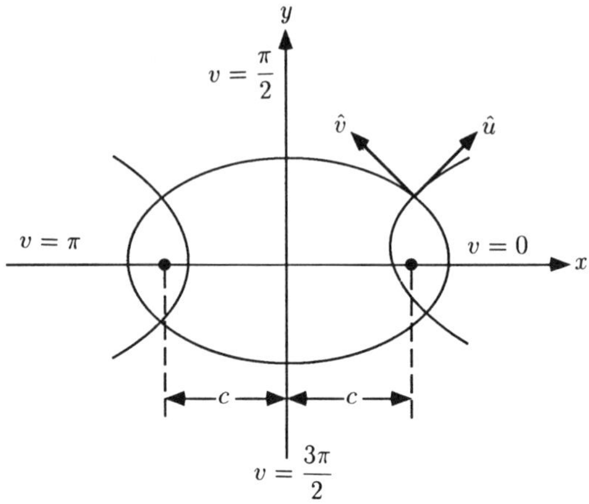

Fig. 1-3 A cross-sectional view of the elliptical coordinate system

TABLE 1-3 Relations Between the Unit Vectors in the Rectangular System and the Elliptical Cylinder System

	$\hat{x}$	$\hat{y}$	$\hat{z}$
$\hat{u}$	$\frac{1}{\Delta}\sinh u \cos v$	$\frac{1}{\Delta}\cosh u \sin v$	0
$\hat{v}$	$-\frac{1}{\Delta}\cosh u \sin v$	$\frac{1}{\Delta}\sinh u \cos v$	0
$\hat{z}$	0	0	1

$$\Delta = \left(\cosh^2 u - \cos^2 v\right)^{\frac{1}{2}}$$

1-2 VECTOR ANALYSIS

The entire subject of vector analysis consists of three definitions, namely, the gradient, the divergence, and the curl; a number of identities; and two theorems named after Gauss and Stokes. For convenient reference some of the identities and formulas are listed in Appendix A. We will not review here the elementary aspects of vector analysis but, rather, will outline the two theorems and a number of useful lemmas that can be derived from these theorems.

Gauss theorem states that for any vector function of position $\overline{F}$ with continuous first derivatives throughout a volume V and over the enclosing surface S,

$$\iiint_V \nabla \cdot \overline{F} dV = \oiint_S \overline{F} \cdot d\overline{S} \text{ (Gauss theorem).} \tag{1.8}$$

The ring around a surface integral is to emphasize the fact that the surface is a closed one. The same notation will be applied to a closed line integral.

Stokes theorem states that for any continuous vector function of position with continuous first derivatives on an open surface S bounded by a contour c:

$$\oiint_S (\nabla \times \overline{F}) \cdot d\overline{S} = \oint_c \overline{F} \cdot d\overline{\ell} \text{ (Stokes theorem).} \tag{1.9}$$

It is understood that the direction of the line integral and the direction of $d\overline{S}$ follows the right-hand screw rule.

In addition to these two important theorems, there are several more theorems in vector analysis, namely,

$$\iiint_V \nabla f dV = \oiint_S f d\overline{S} \text{ (gradient theorem)} \tag{1.10}$$

$$\iiint_V \nabla \times \overline{F} dV = \oiint_S \hat{n} \times \overline{F} dS \text{ (curl theorem),} \tag{1.11}$$

where $\hat{n}$ denotes an outward unit normal to the surface S enclosing the volume V.

If we let

$$\overline{F} = \Phi \nabla \psi - \psi \nabla \Phi, \tag{1.12}$$

where Φ and ψ are two scalar functions of position, then in view of identities (A.11) and (A.16) of Appendix A,

$$\nabla \cdot \overline{F} = \Phi \nabla^2 \psi - \psi \nabla^2 \Phi, \tag{1.13}$$

where $\nabla^2 \psi$ and $\nabla^2 \Phi$ denote, respectively, the Laplacian of ψ and Φ. It follows from Gauss theorem that

$$\iiint_V (\Phi \nabla^2 \psi - \psi \nabla^2 \Phi)\, dV = \oiint_S (\Phi \nabla \psi - \psi \nabla \Phi) \cdot d\overline{S}, \tag{1.14}$$

which is designated as the scalar Green theorem of the second kind.

If we let

$$\overline{F} = \overline{P} \times \nabla \times \overline{Q}, \tag{1.15}$$

where $\overline{P}$ and $\overline{Q}$ are two vector functions, then according to the vector identity (A.13) of Appendix A,

$$\nabla \cdot \overline{F} = (\nabla \times \overline{P}) \cdot (\nabla \times \overline{Q}) - \overline{P} \cdot \nabla \times \nabla \times \overline{Q}. \tag{1.16}$$

Upon substituting it into Gauss theorem, we obtain the vector Green theorem of the first kind

$$\iiint_V [(\nabla \times \overline{P}) \cdot (\nabla \times \overline{Q}) - \overline{P} \cdot \nabla \times \nabla \times \overline{Q}]\, dV$$
$$= \oiint_S (\overline{P} \times \nabla \times \overline{Q}) \cdot d\overline{S} = \oiint_S \hat{n} \cdot (\overline{P} \times \nabla \times \overline{Q})\, dS, \tag{1.17}$$

where $\hat{n}$ denotes the outward unit normal to the surface S.

By interchanging the roles of $\overline{P}$ and $\overline{Q}$ in (1.17) and taking the difference of the two resultant equations we obtain the vector Green theorem of the second kind

$$\begin{aligned}\iiint_V (\overline{P}\cdot\nabla\times\nabla\times\overline{Q}-\overline{Q}\cdot\nabla\times\nabla\times\overline{P})\,dV \\ = \oiint_S (\overline{Q}\times\nabla\times\overline{P}-\overline{P}\times\nabla\times\overline{Q})\cdot d\overline{S} \\ = \oiint_S \hat{n}\cdot(\overline{Q}\times\nabla\times\overline{P}-\overline{P}\times\nabla\times\overline{Q})\,dS. \end{aligned} \tag{1.18}$$

The derivation of these theorems and the relations between them are treated in detail in this author's book on vector and dyadic analysis [Tai, 1992].

1-3 DYADIC ANALYSIS

This section will introduce some essential formulas in dyadic analysis, which is an extension of vector analysis to a higher level.

A vector function or a vector $\overline{F}$ expressed in a Cartesian system is defined by

$$\overline{F}=\sum_{i=1}^{3}F_i\hat{x}_i, \tag{1.19}$$

where F_i with $i=(1,2,3)$ denotes the three scalar components of $\overline{F}$ and $\hat{x}_i$ denotes the three unit vectors in the direction of $\overline{x}_i$. We use x_i in this section to denote the Cartesian variables (x,y,z), so the summation sign can be applied to $\overline{F}$ as in (1.19). From now on, it is understood that the summation index always runs from 1 to 3 unless specified otherwise.

Now we consider three distinct vector functions denoted by

$$\overline{F}_j=\sum_i F_{ij}\hat{x}_i,\ j=1,2,3; \tag{1.20}$$

then a dyadic function or a *dyadic*, denoted by $\overline{\overline{F}}$, can be formed and is defined by

$$\overline{\overline{F}}=\sum_j \overline{F}_j\hat{x}_j, \tag{1.21}$$

where $\overline{F}_j$ with $j=(1,2,3)$ are designated as the three vector components of $\overline{\overline{F}}$. In (1.21) the positioning of $\overline{F}_j$ and $\hat{x}_j$ must be kept in that order. By substituting (1.20) into (1.21) we can write $\overline{\overline{F}}$ in the form

$$\overline{\overline{F}}=\sum_i\sum_j F_{ij}\hat{x}_i\hat{x}_j, \tag{1.22}$$

where F_{ij} are designated as the nine scalar components of $\overline{\overline{F}}$ and the doublet $\hat{x}_i\hat{x}_j$ as the nine unit dyadics or *dyads*, each being formed by a pair of unit vectors in that order, which are not commutative; that is,

$$\hat{x}_i\hat{x}_j \neq \hat{x}_j\hat{x}_i. \tag{1.23}$$

The *transpose* of a dyadic $\overline{\overline{F}}$ expressed by (1.21) will be denoted by $\left(\overline{\overline{F}}\right)^T$ and is defined by

$$\left(\overline{\overline{F}}\right)^T = \sum_j \hat{x}_j\overline{F}_j = \sum_i\sum_j F_{ij}\hat{x}_j\hat{x}_i = \sum_j\sum_i F_{ji}\hat{x}_i\hat{x}_j. \tag{1.24}$$

Comparing (1.24) with (1.21) and (1.22) we see that the positions of $\overline{F}_j$ and $\hat{x}_j$ in $\overline{\overline{F}}$ has been interchanged, or the scalar component F_{ij} in $\overline{\overline{F}}$ has been replaced by F_{ji} in $\left(\overline{\overline{F}}\right)^T$; hence the nomenclature "transpose."

A *symmetrical dyadic*, denoted by $\overline{\overline{F}}_s$, is characterized by $F_{ji} = F_{ij}$; hence

$$\left(\overline{\overline{F}}_s\right)^T = \overline{\overline{F}}_s. \tag{1.25}$$

A symmetrical dyadic therefore has only six distinct scalar components, although it still has nine terms or nine dyadic components.

An *antisymmetric dyadic*, denoted by $\overline{\overline{F}}_a$, is characterized by $F_{ij} = -F_{ji}$; hence $F_{ii} = 0$ for $i = 1, 2, 3$ and

$$\left(\overline{\overline{F}}_a\right)^T = -\overline{\overline{F}}_a. \tag{1.26}$$

An antisymmetric dyadic, therefore, has only three distinct scalar components if we do not consider the negative sign as being distinct, and it has six nonvanishing dyadic components.

One special case of a symmetric dyadic is described by

$$F_{ij} = 1, \quad i = j$$
$$F_{ij} = 0, \quad i \neq j$$

or

$$F_{ij} = \delta_{ij}, \tag{1.27}$$

where δ_{ij} denotes the Kronecker delta function. This dyadic is denoted by $\overline{\overline{I}}$, and it is called an *idem factor*. Its explicit expression is

$$\overline{\overline{I}} = \sum_i \hat{x}_i\hat{x}_i. \tag{1.28}$$

A dyadic by itself, like a matrix, has no algebraic property. It plays the role of an operator when certain products are formed. In particular, we can define two scalar products between a vector $\overline{a}$ and a dyadic $\overline{\overline{F}}$. The *anterior scalar product*, denoted by $\overline{a} \cdot \overline{\overline{F}}$, is defined by

$$\overline{a} \cdot \overline{\overline{F}} = \sum_j \left(\overline{a} \cdot \overline{F}_j\right) \hat{x}_j = \sum_i \sum_j a_i F_{ij} \hat{x}_j, \tag{1.29}$$

which is a vector. The *posterior scalar product*, denoted by $\overline{\overline{F}} \cdot \overline{a}$, is defined by

$$\begin{aligned} \overline{\overline{F}} \cdot \overline{a} &= \sum_j \overline{F}_j \left(\hat{x}_j \cdot \overline{a}\right) = \sum_i \sum_j a_j F_{ij} \hat{x}_i \\ &= \sum_i \sum_j a_i F_{ji} \hat{x}_j, \end{aligned} \tag{1.30}$$

which is also a vector. In general, the two scalar products are not equal unless $\overline{\overline{F}}$ is a symmetrical dyadic. For any dyadic we have the relation

$$\overline{a} \cdot \left(\overline{\overline{F}}\right)^T = \overline{\overline{F}} \cdot \overline{a}. \tag{1.31}$$

This is an important identity in dyadic analysis. As a result of (1.25) and (1.26), one finds

$$\overline{a} \cdot \overline{\overline{F}}_s = \overline{\overline{F}}_s \cdot \overline{a} \tag{1.32}$$

$$\overline{a} \cdot \overline{\overline{F}}_a = -\overline{\overline{F}}_a \cdot \overline{a}. \tag{1.33}$$

If $\overline{\overline{F}}_s = \overline{\overline{I}}$, the idem factor, then

$$\overline{a} \cdot \overline{\overline{I}} = \overline{\overline{I}} \cdot \overline{a} = \overline{a}. \tag{1.34}$$

This is the reason why $\overline{\overline{I}}$ is designated as the idem factor.

We have also two vector products between a vector $\overline{a}$ and a dyadic $\overline{\overline{F}}$. The *anterior vector product*, denoted by $\overline{a} \times \overline{\overline{F}}$, is defined by

$$\overline{a} \times \overline{\overline{F}} = \sum_{j=1}^{3} \left(\overline{a} \times \overline{F}_j\right) \hat{x}_j; \tag{1.35}$$

the *posterior vector product*, denoted by $\overline{\overline{F}} \times \overline{a}$, is defined by

$$\overline{\overline{F}} \times \overline{a} = \sum_{j=1}^{3} \overline{F}_j \left(\overline{x}_j \times \overline{a}\right). \tag{1.36}$$

These vector products are both dyadics, and there is no relation similar to (1.31) for these two products.

In vector analysis we have the following identities involving three vectors:

$$\bar{a} \cdot \left(\bar{b} \times \bar{c}\right) = \bar{b} \cdot \left(\bar{c} \times \bar{a}\right) = \bar{c} \cdot \left(\bar{a} \times \bar{b}\right) . \tag{1.37}$$

These identities can be generalized to involve dyadics. We consider three distinct sets of triple products with three different vector functions $\bar{c}_j$; that is,

$$\bar{a} \cdot \left(\bar{b} \times \bar{c}_j\right) = -\bar{b} \cdot \left(\bar{a} \times \bar{c}_j\right) = \left(\bar{a} \times \bar{b}\right) \cdot \bar{c}_j, \tag{1.38}$$

with $j = (1, 2, 3)$. We purposely place the function $\bar{c}_j$ at the posterior position in order to derive the desired dyadic identities. Now we juxtapose a unit vector $\hat{x}_j$ at the posterior position of each term in (1.38) and sum the resultant equations with respect to j to obtain

$$\bar{a} \cdot \left(\bar{b} \times \bar{\bar{c}}\right) = -\bar{b} \cdot \left(\bar{a} \times \bar{\bar{c}}\right) = \left(\bar{a} \times \bar{b}\right) \cdot \bar{\bar{c}}; \tag{1.39}$$

thus we have elevated the vector triple products to a higher level involving one dyadic and two vectors while each term in (1.38) is a scalar and the corresponding terms in (1.39) are vectors. We can elevate the vector function $\bar{b}$ in the last two terms of (1.39) to a dyadic by considering three distinct equations of the form

$$-\left(\bar{a} \times \bar{\bar{c}}\right)^T \cdot \bar{b}_j = \left(\bar{\bar{c}}\right)^T \cdot \left(\bar{a} \times \bar{b}_j\right), \tag{1.40}$$

with $j = 1, 2, 3$. By juxtaposing a unit vector $\hat{x}_j$ at the posterior position of the two terms in (1.40) and summing the resultant equations with respect to j, we obtain

$$-\left(\bar{a} \times \bar{\bar{c}}\right)^T \cdot \bar{\bar{b}} = \left(\bar{\bar{c}}\right)^T \cdot \left(\bar{a} \times \bar{\bar{b}}\right). \tag{1.41}$$

Each term is the scalar product of two dyadics, and the result gives an identity of two dyadics.

The previous material deals mainly with dyadic algebra. In the following we introduce some definitions and formulas involving the differentiation and the integration of dyadic functions.

The *divergence of a dyadic function*, denoted by $\nabla \cdot \bar{\bar{F}}$, is defined by

$$\nabla \cdot \bar{\bar{F}} = \sum_j \left(\nabla \cdot \bar{F}_j\right) \hat{x}_j = \sum_i \sum_j \frac{\partial F_{ij}}{\partial x_i} \hat{x}_j, \tag{1.42}$$

which is a vector function. The *curl of a dyadic function*, denoted by $\nabla \times \bar{\bar{F}}$, is defined by

$$\nabla \times \bar{\bar{F}} = \sum_j \left(\nabla \times \bar{F}_j\right) \hat{x}_j = \sum_i \sum_j \left(\nabla F_{ij} \times \hat{x}_i\right) \hat{x}_j, \tag{1.43}$$

where we have used the vector identity

$$\nabla \times \left(F_{ij} \hat{x}_j\right) = \nabla F_{ij} \times \hat{x}_j \tag{1.44}$$

to derive (1.43), which is a dyadic function. In addition to these two functions, we will encounter the *gradient of a vector function*, denoted by $\nabla \overline{F}$, which is defined by

$$\nabla \overline{F} = \sum_j (\nabla F_j)\, \hat{x}_j = \sum_i \sum_j \frac{\partial F_j}{\partial x_i} \hat{x}_i \hat{x}_j, \tag{1.45}$$

which is a dyadic.

When a dyadic function is constructed with an idem factor $\overline{\overline{I}}$ and a scalar function f in the form

$$\overline{\overline{F}} = f\overline{\overline{I}},$$

then

$$\begin{aligned} \nabla \cdot \overline{\overline{F}} &= \nabla \cdot \left(f\overline{\overline{I}}\right) = \sum_i \nabla \cdot (f\hat{x}_i)\, \hat{x}_i = \sum_i \frac{\partial f}{\partial x_i} \hat{x}_i \\ &= \nabla f \end{aligned} \tag{1.46}$$

and

$$\begin{aligned} \nabla \times \overline{\overline{F}} &= \nabla \times \left(f\overline{\overline{I}}\right) = \sum_i \nabla \times (f\hat{x}_i)\, \hat{x}_i \\ &= \sum_i (\nabla f \times \hat{x}_i)\, \hat{x}_i = \nabla f \times \overline{\overline{I}}, \end{aligned} \tag{1.47}$$

which is a dyadic.

Having introduced the divergence and the curl of a dyadic, we can elevate several vector Green theorems reviewed in Sec. 1-2 to the dyadic form. We consider three distinct sets of the vector Green theorem of the first kind stated by (1.17)

$$\begin{aligned} &\iiint_V \left[(\nabla \times \overline{P}) \cdot (\nabla \times \overline{Q}_j) - \overline{P} \cdot \nabla \times \nabla \times \overline{Q}_j\right] dV \\ &= \oiint_S \hat{n} \cdot \left(\overline{P} \times \nabla \times \overline{Q}_j\right) dS, \quad j = 1, 2, 3. \end{aligned} \tag{1.48}$$

By juxtaposing a unit vector $\hat{x}_j$ to the posterior position of (1.48) and summing the three resultant equations, we obtain the vector dyadic Green theorem of the first kind

$$\begin{aligned} &\iiint_V \left[(\nabla \times \overline{P}) \cdot \left(\nabla \times \overline{\overline{Q}}\right) - \overline{P} \cdot \nabla \times \nabla \times \overline{\overline{Q}}\right] dV \\ &= \oiint_S \hat{n} \cdot \left(\overline{P} \times \nabla \times \overline{\overline{Q}}\right) dS. \end{aligned} \tag{1.49}$$

To elevate the vector Green theorem of the second kind to the vector-dyadic form, we consider three sets of that theorem written in the form

$$\iiint_V \left[\overline{P} \cdot \nabla \times \nabla \times \overline{Q}_j - (\nabla \times \nabla \times \overline{P}) \cdot \overline{Q}_j\right] dV$$
$$= -\oiint_S \hat{n} \cdot \left[(\nabla \times \overline{P}) \times \overline{Q}_j + \overline{P} \times \nabla \times \overline{Q}_j\right] dS$$
$$= -\iint_S \left[(\hat{n} \times \nabla \times \overline{P}) \cdot \overline{Q}_j + (\hat{n} \times \overline{P}) \cdot \nabla \times \overline{Q}_j\right] dS, \quad j = 1, 2, 3. \tag{1.50}$$

It is observed that we purposely put the function $\overline{Q}_j$ at the posterior position in order to do the elevating. By juxtaposing a unit vector $\hat{x}_j$ at the posterior position of (1.50) and summing the resultant three equations, we obtain the vector-dyadic Green theorem of the second kind; namely,

$$\iiint_V \left[\overline{P} \cdot \nabla \times \nabla \times \overline{\overline{Q}} - (\nabla \times \nabla \times \overline{P}) \cdot \overline{\overline{Q}}\right] dV$$
$$= -\oiint_S \left[(\hat{n} \times \nabla \times \overline{P}) \cdot \overline{\overline{Q}} + (\hat{n} \times \overline{P}) \cdot \nabla \times \overline{\overline{Q}}\right] dS. \tag{1.51}$$

The vector function $\overline{P}$ in (1.49) and (1.51) can now be elevated to a dyadic. Thus we write (1.49) in the form

$$\iiint_V \left[\left(\nabla \times \overline{\overline{Q}}\right)^T \cdot \nabla \times \overline{P} - \left(\nabla \times \nabla \times \overline{\overline{Q}}\right)^T \cdot \overline{P}\right] dV$$
$$= \oiint_S (\hat{n} \times \overline{P}) \cdot \nabla \times \overline{\overline{Q}}\, dS = \oiint_S \left(\nabla \times \overline{\overline{Q}}\right)^T \cdot (\hat{n} \times \overline{P})\, dS.$$

By elevating $\overline{P}$ to a dyadic level, we obtain the dyadic-dyadic Green theorem of the first kind in the form

$$\iiint_V \left[\left(\nabla \times \overline{\overline{Q}}\right)^T \cdot \nabla \times \overline{\overline{P}} - \left(\nabla \times \nabla \times \overline{\overline{Q}}\right)^T \cdot \overline{\overline{P}}\right] dV$$
$$= \oiint_S \left(\nabla \times \overline{\overline{Q}}\right)^T \cdot \left(\hat{n} \times \overline{\overline{P}}\right) dS. \tag{1.52}$$

By following the same procedure for (1.51) we can obtain the dyadic-dyadic Green theorem of the second kind; namely,

$$\iiint_V \left[\left(\nabla \times \nabla \times \overline{\overline{Q}}\right)^T \cdot \overline{\overline{P}} - \left(\overline{\overline{Q}}\right)^T \cdot \nabla \times \nabla \times \overline{\overline{P}}\right] dV$$
$$= -\oiint_S \left[\left(\overline{\overline{Q}}\right)^T \cdot \left(\hat{n} \times \nabla \times \overline{\overline{P}}\right) + \left(\nabla \times \overline{\overline{Q}}\right)^T \cdot \left(\hat{n} \times \overline{\overline{P}}\right)\right] dS. \tag{1.53}$$

These theorems are needed later to integrate Maxwell's equations using dyadic Green functions and to prove the symmetrical properties of dyadic Green functions. The concept of the dyadic Green functions and their precise forms are the main topics of this book which will be discussed shortly.

1-4 FOURIER TRANSFORM AND HANKEL TRANSFORM

In this section we will review the basic formulas in the theory of the Fourier transform and in the theory of the Hankel transform or Fourier-Bessel transform. At the end of this section we will derive the integral representation of the delta functions weighted according to the dimension in which these functions are used. The meaning of a weighted delta function will be explained later.

The Fourier transform of a piecewise continuous function $f(\xi)$ is defined by

$$g(h) = \int_{-\infty}^{\infty} f(\xi)\, e^{-ih\xi} d\xi. \tag{1.54}$$

The existence of $g(h)$ requires that $\int_{-\infty}^{\infty} \mid f(\xi) \mid d\xi$ be bounded. The inverse of (1.54) is given by

$$f\,(x) = \frac{1}{2\pi} \int_{-\infty}^{\infty} g\,(h)\, e^{ihx} dh. \tag{1.55}$$

The Fourier transform can be extended to functions of many variables. In particular, for functions of two variables, we have the following two-dimensional Fourier transform pair

$$g\,(h_1, h_2) = \int_{-\infty}^{\infty} \int_{-\infty}^{\infty} f\,(\xi_1, \xi_2)\, e^{-i(h_1\xi_1 + h_2\xi_2)}\, d\xi_1\, d\xi_2 \tag{1.56}$$

$$f\,(x_1, x_2) = \frac{1}{(2\pi)^2} \int_{-\infty}^{\infty} \int_{-\infty}^{\infty} g\,(h_1, h_2)\, e^{i(h_1x_1 + h_2x_2)}\, dh_1\, dh_2. \tag{1.57}$$

The Hankel transform or Fourier-Bessel transform can be considered as a special case of the two-dimensional Fourier transform. It deals with a class of functions whereby $f\,(x_1, x_2)$ is a function of r and ϕ when expressed in the radial cylindrical variables. To derive the Hankel transform pair from this point of view, we make the following changes of variables:

$$x_1 = r\cos\phi \qquad x_2 = r\sin\phi$$
$$\xi_1 = \rho\cos\beta \qquad \xi_2 = \rho\sin\beta$$
$$h_1 = \lambda\cos\alpha \qquad h_2 = \lambda\sin\alpha.$$

Equations (1.56) and (1.57) can then be written in the form

$$g\,(\lambda, \alpha) = \int_0^{\infty} \rho d\rho \int_0^{2\pi} d\beta f\,(\rho, \beta)\, e^{-i\lambda\rho\cos(\beta-\alpha)} \tag{1.58}$$

$$f\,(r, \phi) = \frac{1}{(2\pi)^2} \int_0^{\infty} \lambda d\lambda \int_0^{2\pi} d\alpha g\,(\lambda, \alpha)\, e^{i\lambda r\cos(\alpha-\phi)}. \tag{1.59}$$

Upon combining (1.58) and (1.59), we obtain

$$f(r,\phi) = \frac{1}{(2\pi)^2}\int_0^\infty \rho\,d\rho \int_0^\infty \lambda\,d\lambda \int_0^{2\pi} d\beta \int_0^{2\pi} d\alpha \cdot f(\rho,\beta)\, e^{i[\lambda r\cos(\alpha-\phi)-\lambda\rho\cos(\beta-\alpha)]}. \tag{1.60}$$

Now let $f(r,\phi)$ be a function in the form $F(r)\,e^{in\phi}$, where n is assumed to be a positive real constant not necessarily integer. Equation (1.60) becomes

$$F(r)\,e^{in\phi} = \frac{1}{(2\pi)^2}\int_0^\infty \rho\,d\rho \int_0^\infty \lambda\,d\lambda \int_0^{2\pi} d\beta \int_0^{2\pi} d\alpha \cdot F(\rho)\,e^{in\beta} e^{i[\lambda r\cos(\alpha-\phi)-\lambda\rho\cos(\beta-\alpha)]}. \tag{1.61}$$

Dividing the entire equation by $e^{in\phi}$ and rearranging the terms in the exponential functions, with the anticipation that the integrals with respect to α and β have the appearance of the integral representations of Bessel functions, one can write the resultant equation in the form

$$F(r) = \frac{1}{(2\pi)^2}\int_0^\infty \rho\,d\rho \int_0^\infty \lambda\,d\lambda \int_0^{2\pi} d\beta \int_0^{2\pi} d\alpha \cdot F(\rho)\,e^{i\left[\lambda r\cos(\alpha-\phi)+n\left(\alpha-\phi-\frac{\pi}{2}\right)-\lambda\rho\cos(\beta-\alpha)+n\left(\beta-\alpha+\frac{\pi}{2}\right)\right]}. \tag{1.62}$$

Let us now consider the integral with respect to β first or, more specifically, the integral

$$\frac{1}{2\pi}\int_0^{2\pi} e^{i\left[-\lambda\rho\cos(\beta-\alpha)+n\left(\beta-\alpha+\frac{\pi}{2}\right)\right]}\,d\beta. \tag{1.63}$$

By changing the variable of integration to ω defined by $\omega = \beta - \alpha$, this becomes

$$\frac{1}{2\pi}\int_{-\alpha}^{2\pi-\alpha} e^{i\left[-\lambda\rho\cos\omega+n\left(\omega+\frac{\pi}{2}\right)\right]}\,d\omega. \tag{1.64}$$

If we judiciously choose the limit of integration such that the contour follows the path from $-\frac{3\pi}{2}+i\infty$ to $\frac{\pi}{2}+i\infty$, then the integral becomes the integral representation of the Bessel function of order n which is assumed to be positive and real but not confined to integers; that is,

$$J_n(\lambda\rho) = \frac{1}{2\pi}\int_{-\frac{3\pi}{2}+i\infty}^{\frac{\pi}{2}+i\infty} e^{-i\left[\lambda\rho\cos\omega-n\left(\omega+\frac{\pi}{2}\right)\right]}\,d\omega. \tag{1.65}$$

The remaining integration with respect to α is now evaluated in a similar manner. We change the variable of integration to $\omega = \alpha - \phi$ and choose the path of integration from $-\frac{\pi}{2}+i\infty$ to $\frac{3\pi}{2}+i\infty$, which yields

$$J_n(\lambda r) = \frac{1}{2\pi}\int_{-\frac{\pi}{2}+i\infty}^{\frac{3\pi}{2}+i\infty} e^{i\left[\lambda r\cos\omega+n\left(\omega-\frac{\pi}{2}\right)\right]}\,d\omega. \tag{1.66}$$

The final form for (1.62) after this reduction of the angular integrals becomes

$$F(r) = \int_0^\infty \lambda\, d\lambda \int_0^\infty \rho\, d\rho F(\rho)\, J_n(\lambda\rho)\, J_n(\lambda r)\,. \tag{1.67}$$

This can be separated in a pair of identities by letting

$$G(\lambda) = \int_0^\infty F(\rho)\, J_n(\lambda\rho)\, \rho\, d\rho; \tag{1.68}$$

then

$$F(r) = \int_0^\infty G(\lambda)\, J_n(\lambda r)\, \lambda\, d\lambda. \tag{1.69}$$

Equations (1.68) and (1.69) constitute the pair of Hankel transforms which are valid for Bessel functions of any order. It should be remarked that the derivation which we have presented here follows very closely the one described by Sommerfeld [1949, pp. 109–111]. However, he derived these expressions under the condition that n is an integer. Later, he applied these formulas to nonintegral values of n without further elaboration [p. 211]. In fact, when the Hankel transform is applied to spherical problems, the Bessel functions involved are of half-integer order or, more precisely, the spherical Bessel functions. For this reason it is more convenient to modify (1.68) and (1.69) so that they would contain these functions directly. To obtain these desired expressions, we let $n = m + \frac{1}{2}$ and change the notation r to R in order to conform to spherical nomenclature. Now, the spherical Bessel function is defined by

$$j_m(\xi) = \left(\frac{\pi}{2\xi}\right)^{\frac{1}{2}} J_{m+\frac{1}{2}}(\xi)\,. \tag{1.70}$$

Equations (1.68) and (1.69) can then be written in the form

$$G(\lambda) = \int_0^\infty \left(\frac{2\lambda\rho}{\pi}\right)^{\frac{1}{2}} F(\rho)\, j_m(\lambda\rho)\, \rho\, d\rho \tag{1.71}$$

$$F(R) = \int_0^\infty \left(\frac{2\lambda R}{\pi}\right)^{\frac{1}{2}} G(\lambda)\, j_m(\lambda R)\, \lambda\, d\lambda. \tag{1.72}$$

To recast (1.71) and (1.72) in a symmetrical form, we introduce two new functions $f(R)$ and $g(\lambda)$ defined by

$$G(\lambda) = \lambda^{\frac{1}{2}} g(\lambda)$$

and

$$F(R) = R^{\frac{1}{2}} f(R)\,. \tag{1.73}$$

The pair of Hankel transforms for $f(R)$ and $g(\lambda)$ in terms of the spherical Bessel function then has the form

$$g(\lambda) = \left(\frac{2}{\pi}\right)^{\frac{1}{2}} \int_0^\infty f(\rho)\, j_m(\lambda\rho)\, \rho^2\, d\rho \tag{1.74}$$

$$f(R) = \left(\frac{2}{\pi}\right)^{\frac{1}{2}} \int_0^\infty g(\lambda)\, j_m(\lambda R)\, \lambda^2\, d\lambda. \tag{1.75}$$

These two expressions have previously been derived by Stratton [1941, pp. 411–412] using a different technique. The present derivation appears to be less formal, but perhaps simpler.

We will now apply the Fourier transform pair (1.54) and (1.55) and the Hankel transform pairs (1.68) and (1.69) and (1.74) and (1.75) to derive the integral representation of the delta function weighted according to the dimension in which the function is used. These weighted delta functions are defined as follows:

$$\text{one dimensional:} \quad \delta(x - x') \tag{1.76}$$

$$\text{two dimensional:} \quad \frac{\delta(r - r')}{r} \tag{1.77}$$

$$\text{three dimensional:} \quad \frac{\delta(R - R')}{R^2}. \tag{1.78}$$

The integral property of a delta function implies that

$$\int_{-\infty}^{\infty} f(x)\, \delta(x - x')\, dx = f(x'). \tag{1.79}$$

Hence, for the weighted delta function in the two-dimensional case,

$$\int_0^\infty f(r) \left[\frac{\delta(r - r')}{r}\right] r\, dr = f(r'); \tag{1.80}$$

likewise, for the three-dimensional case,

$$\int_0^\infty f(R) \left[\frac{\delta(R - R')}{R^2}\right] R^2\, dR = f(R'). \tag{1.81}$$

Upon substituting $f(x) = \delta(x - x')$ or $f(\xi) = \delta(\xi - x')$ into (1.54) and (1.55), one finds

$$\delta(x - x') = \frac{1}{2\pi} \int_{-\infty}^{\infty} e^{ih(x - x')}\, dh. \tag{1.82}$$

Similarly, by letting $F(r) = \delta(r - r')/r$ or $F(\rho) = \delta(\rho - r')/\rho$ in (1.68) and (1.69), we obtain

$$\frac{\delta(r - r')}{r} = \int_0^\infty J_n(\lambda r)\, J_n(\lambda r')\, \lambda\, d\lambda. \tag{1.83}$$

Finally, by applying (1.74) and (1.75) to the weighted delta function in the three-dimensional case, we obtain

$$\frac{\delta(R - R')}{R^2} = \frac{2}{\pi} \int_0^\infty j_n(\lambda R)\, j_n(\lambda R')\, \lambda^2\, d\lambda. \tag{1.84}$$

Expressions (1.82)–(1.84) will be used later very often in solving the vector wave equation by the method of continuous eigenfunction expansion.

1-5 SADDLE-POINT METHOD OF INTEGRATION AND SEMI-INFINITE INTEGRALS OF THE PRODUCT OF BESSEL FUNCTIONS

Complex integrals of the type

$$F(\rho) = \int_{-\infty}^{\infty} f(h)\, e^{i\rho\phi(h)}\, dh \tag{1.85}$$

will occur frequently in our work. When certain conditions are met, the integral can be evaluated approximately by the method of saddle-point integration. The key conditions are that ρ be a large number compared to unity and $\phi(h)$, whose magnitude is of the order of unity, have an extreme value at a certain point h_0, so that $\phi'(h_0) = 0$. The function $f(h)$ is assumed to be a slowly varying function in the neighborhood of h_0. We consider $\phi(h)$ to be an analytic function of the complex variable $h = \xi + i\eta$ so that

$$\phi(h) = u(\xi, \eta) + iv(\xi, \eta)\,; \tag{1.86}$$

then u and v satisfy the Cauchy-Riemann relations

$$\frac{\partial v}{\partial \xi} = \frac{\partial v}{\partial \eta}, \qquad \frac{\partial u}{\partial \eta} = -\frac{\partial v}{\partial \xi}. \tag{1.87}$$

A three-dimensional plot of the surface $z = v(\xi, \eta)$ shows that in the neighborhood of the point $h = h_0$ or $\xi = \xi_0$ and $\eta = \eta_0$, the surface has the shape of a saddle because

$$\left[\frac{\partial^2 v}{\partial \xi^2} \cdot \frac{\partial^2 v}{\partial \eta^2} - \left(\frac{\partial^2 v}{\partial \xi \partial \eta}\right)^2\right]_{\xi=\xi_0, \eta=\eta_0} < 0 \tag{1.88}$$

as a result of the Cauchy-Reimann relations [Courant, Vol. II, p. 205]. The family of curves described by

$$v(\xi, \eta) = c \tag{1.89}$$

for different values of c has the appearance shown in Fig. 1-4 in the neighborhood of the saddle point, where $v_0 = v(\xi_0, \eta_0)$.

The above description also applies to the function $u(\xi, \eta)$ and the family of curves corresponding to $u(\xi, \eta)$ = constant are orthogonal to those shown in Fig. 1-4. We now change the path of integration such that the contour would pass through h_0 and follow a path corresponding to

$$u(\xi, \eta) = u_0 \tag{1.90}$$

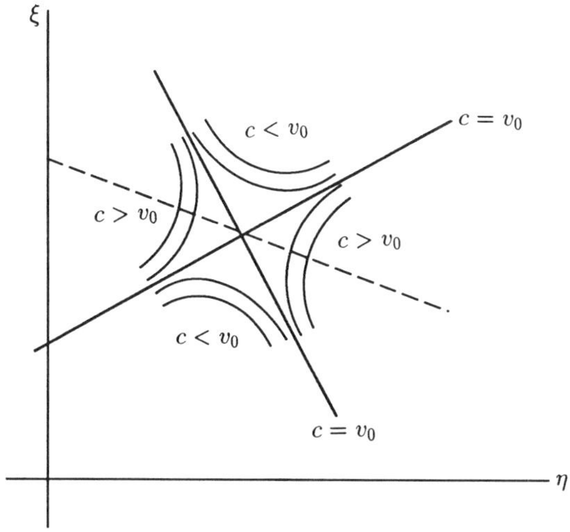

Fig. 1-4 The family of curves described by $v(\xi, \eta) = c$

in the neighborhood of the saddle point. A section of this path is shown in Fig. 1-4 by the dotted line. Under that condition the original integral can be written in the form

$$F(\rho) = \int_c f(h)\, e^{i\rho(u_0+iv)} dh = e^{i\rho u_0} \int_c f(h)\, e^{-\rho v} dh. \tag{1.91}$$

The function $e^{-\rho v}$ has a significant value only along a small segment of the path near the saddle point. Thus we may approximate the function $\phi(h)$ by a series expansion in the neighborhood of h_0 by retaining only the first three terms; that is,

$$\begin{aligned}\phi(h) &= \phi(h_0) + \phi'(h_0)(h - h_0) + \frac{1}{2}\phi''(h_0)(h - h_0)^2 \\ &= \phi(h_0) + \frac{1}{2}\phi''(h_0)(h - h_0)^2.\end{aligned} \tag{1.92}$$

For the slowly varying function $f(h)$ it can be replaced by $f(h_0)$. The original integral, then, can be approximated by

$$F(\rho) = e^{i\rho\phi(h_0)} f(h_0) \int_{h_0-\Delta h}^{h_0+\Delta h} e^{i\frac{1}{2}\rho\phi''(h_0)(h-h_0)^2} dh. \tag{1.93}$$

This can be simplified if we let

$$\begin{aligned}h - h_0 &= s e^{i\alpha} \\ \phi''(h_0) &= |\phi''(h_0)|\, e^{i\beta};\end{aligned}$$

then

$$i\frac{1}{2}\rho\phi''(h_0)(h - h_0)^2 = \frac{1}{2}\rho\, |\phi''(h_0)|\, s^2 e^{i\left(2\alpha+\beta+\frac{\pi}{2}\right)}. \tag{1.94}$$

In order to confine the path of integration along the contour $u = u_0$, the angle α must be so chosen that

$$\alpha = \frac{1}{2}\left(\frac{\pi}{2} - \beta\right).$$

Under this condition, (1.93) reduces to

$$F(\rho) = e^{i\left[\rho\phi(h_0) - \frac{\beta}{2} + \frac{\pi}{4}\right]} f(h_0) \int_{-\tau}^{\tau} e^{-\frac{1}{2}\rho\phi''(h_0)s^2} ds, \tag{1.95}$$

where $\tau = |\Delta h|$. The last integral can be evaluated by a change of variable from s to t

$$t = \left[\frac{\rho\,|\phi''(h_0)|}{2}\right]^{\frac{1}{2}} s \tag{1.96}$$

and in the limit as

$$\left[\frac{\rho\,|\phi''(h_0)|}{2}\right]^{\frac{1}{2}} \tau \tag{1.97}$$

becomes very large we obtain

$$F(\rho) = \left[\frac{2\pi}{\rho\,|\phi''(h_0)|}\right]^{\frac{1}{2}} f(h_0)\, e^{i\left[\rho\phi(h_0) - \frac{\beta}{2} + \frac{\pi}{4}\right]}. \tag{1.98}$$

This is the asymptotic formula for (1.85) under the conditions we have stated.

As an example, let us consider an integral of the form

$$F(r, z) = \int_{-\infty}^{\infty} \frac{G(h)}{\sqrt{2\pi\lambda r}} e^{i(hz + \lambda r)} dh, \tag{1.99}$$

where

$$\lambda = \sqrt{k^2 - h^2} \tag{1.100}$$

and r and z denote two of the variables in the cylindrical coordinate system, k being a constant. Thus we have

$$\begin{aligned}
\rho\phi(h) &= hz + \sqrt{k^2 - h^2}\,r \\
h_0 &= \frac{kz}{\sqrt{r^2 + z^2}} = k\cos\theta \\
\rho\phi(h_0) &= k\sqrt{k^2 + z^2} = kR \\
\rho\phi''(h_0) &= \frac{-k^2 r}{(k^2 - h_0^2)^{\frac{3}{2}}} = -\frac{R}{k\sin^2\theta},
\end{aligned} \tag{1.101}$$

where R and θ denote two of the variables defined in the corresponding spherical coordinate system. When kR is very large, the original integral is well approximated by

$$F(R,\theta) = \frac{G(k\cos\theta)}{R} e^{i\left(kR-\frac{\pi}{4}\right)}. \tag{1.102}$$

This approximation will be used later in finding the asymptotic expressions for various dyadic Green functions.

Another integral which will be encountered often in our work is of the form

$$f(r,r') = \int_0^\infty g(\lambda) \frac{J_\nu(\lambda r) J_\nu(\lambda r')}{\lambda^2 - k^2} d\lambda, \tag{1.103}$$

where $J_\nu(\lambda r)$ and $J_\nu(\lambda r')$ denote two Bessel functions of order ν, not necessarily integers, and $g(\lambda)$ is an odd function of λ with no poles in the complex λ-plane. The evaluation of this type of integral has been described by Sommerfeld [1949, p. 197] for the case where the product is made of two spherical Bessel functions. We shall adopt Sommerfeld's method, but put no restriction on the values of ν. Using the well-known relation between a Bessel function and the two Hankel functions, we can write (1.103) in the form

$$\begin{aligned} f(r,r') &= \frac{1}{2}\int_0^\infty \frac{g(\lambda) J_\nu(\lambda r)}{\lambda^2 - k^2} \left[H_\nu^{(1)}(\lambda r') + H_\nu^{(2)}(\lambda r')\right] d\lambda \\ &= f_1 + f_2. \end{aligned} \tag{1.104}$$

By changing the variable of integration from λ to $\lambda e^{-i\pi}$, we have

$$f_2 = \frac{1}{2}\int_0^{-\infty} \frac{g\left(\lambda e^{-i\pi}\right)}{\lambda^2 - k^2} J_\nu\left(\lambda r e^{-i\pi}\right) H_\nu^{(2)}\left(\lambda r' e^{-i\pi}\right)(-d\lambda). \tag{1.105}$$

According to the circulation relations of the Bessel functions [Sommerfeld, 1949, p. 314], we have

$$\begin{aligned} J_\nu\left(\lambda r e^{-i\pi}\right) &= e^{-iv\pi} J_\nu(\lambda r) \\ H_\nu^{(2)}\left(\lambda r e^{-i\pi}\right) &= -e^{iv\pi} H_\nu^{(1)}(\lambda r). \end{aligned} \tag{1.106}$$

Since $g(\lambda) = -g(-\lambda)$, we obtain

$$f_2 = \frac{1}{2}\int_{-\infty}^0 \frac{g(\lambda)}{\lambda^2 - k^2} J_\nu(\lambda r) H_\nu^{(1)}(\lambda r')\, d\lambda. \tag{1.107}$$

The original integral, therefore, can be written in the form

$$f(r,r') = \frac{1}{2}\int_{-\infty}^\infty \frac{g(\lambda)}{\lambda^2 - k^2} J_\nu(\lambda r) H_\nu^{(1)}(\lambda r')\, d\lambda. \tag{1.108}$$

By a similar reasoning, (1.103) can alternatively be expressed in the form

$$f(r, r') = \frac{1}{2} \int_{-\infty}^{\infty} \frac{g(\lambda)}{\lambda^2 - k^2} H_\nu^{(1)}(\lambda r) J_\nu(\lambda r')\, d\lambda. \tag{1.109}$$

These two integrals can be evaluated in a closed form by completing the contour of integration along a semi-infinite circular path in the upper λ-plane. As a result of the residue theorem, we find

$$f(r, r') = \frac{i\pi g(k)}{2k} \begin{cases} J_\nu(kr) H_\nu^{(1)}(kr'), & r < r' \\ H_\nu^{(1)}(kr) J_\nu(kr'), & r > r'. \end{cases} \tag{1.110}$$

In the case of spherical and conical problems, we will encounter integrals of the type

$$F(R, R') = \int_0^{\infty} \frac{G(\lambda) j_\nu(\lambda R) j_\nu(\lambda R')}{\lambda^2 - k^2}\, d\lambda, \tag{1.111}$$

where $j_\nu(\lambda R')$ and $j_\nu(\lambda r')$ denote two spherical Bessel functions and $G(\lambda)$ is an even function of λ. Using the relation that

$$j_\nu(x) = \left(\frac{\pi}{2x}\right)^{\frac{1}{2}} J_{\nu+\frac{1}{2}}(x), \tag{1.112}$$

(1.111) can be written in the form

$$F(R, R') = \frac{\pi}{2(RR')^{\frac{1}{2}}} \int_0^{\infty} \frac{G(\lambda)}{\lambda(\lambda^2 - k^2)} J_{\nu+\frac{1}{2}}(\lambda R) J_{\nu+\frac{1}{2}}(\lambda R')\, d\lambda. \tag{1.113}$$

This is of the same form as (1.103), provided $G(\lambda)/\lambda$ has no poles in the λ-plane. Applying the general result described by (1.110), we obtain

$$F(R, R') = \frac{i\pi^2 G(k)}{4k^2 (RR')^{\frac{1}{2}}} \begin{cases} J_{\nu+\frac{1}{2}}(kR) H_{\nu+\frac{1}{2}}^{(1)}(kR'), & R < R' \\ H_{\nu+\frac{1}{2}}^{(1)}(kR) J_{\nu+\frac{1}{2}}(kR'), & R > R' \end{cases}$$

$$= \frac{i\pi G(k)}{2k} \begin{cases} j_\nu(kR) h_\nu^{(1)}(kR'), & R < R' \\ h_\nu^{(1)}(kR) j_\nu(kR'), & R > R', \end{cases} \tag{1.114}$$

where $h_\nu^{(1)}(kR)$ denotes the spherical Hankel function of the first kind; that is

$$h_\nu^{(1)}(x) = \left(\frac{\pi}{2x}\right)^{\frac{1}{2}} H_\nu^{(1)}(x). \tag{1.115}$$

Equation (1.114) holds true for functions of any order, not necessarily for ν equal to integers. In fact for conical problems ν is in general fractional.

2

Scalar Green Functions

As an introduction to the terminology, the concept, and the method of deriving various types of the dyadic Green function, we shall first review the transmission line theory from the point of view of the Green function technique since the subject matter, presumably, is already familiar to most of the readers. Much of the material covered here finds its analogy later when we deal with the vector wave equation. The Green functions pertaining to the two-dimensional and three-dimensional scalar wave equations will also be introduced, but no detailed treatment will be given.

2-1 SCALAR GREEN FUNCTIONS OF A ONE-DIMENSIONAL WAVE EQUATION—THEORY OF TRANSMISSION LINES

We consider a transmission line excited by a distributed current source, $K(x)$, as sketched in Fig. 2-1. The line may be finite or infinite, and it may be terminated at either end with an impedance or by another line. For a harmonically oscillating current source $K(x)$, the voltage and the current on the line satisfy the following pair of equations:

$$\frac{dV(x)}{dx} = i\omega LI(x) \tag{2.1}$$

$$\frac{dI(x)}{dx} = i\omega CV(x) + K(x) \tag{2.2}$$

$e^{-i\omega t}$ system is used in this work.

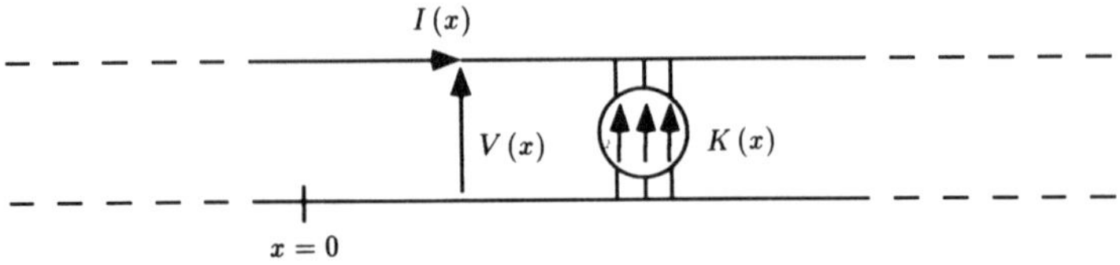

Fig. 2-1 Transmission line excited by a distributed current source, $K(x)$

In (2.1) and (2.2) L and C denote, respectively, the distributed inductance and capacitance of the line. Our problem is to find $V(x)$ and $I(x)$ for certain terminations of the line. Because of (2.1), it is sufficient for us to find $V(x)$ only. By eliminating $I(x)$ between (2.1) and (2.2), we obtain

$$\frac{d^2V(x)}{dx^2} + k^2V(x) = i\omega LK(x), \tag{2.3}$$

where $k = \omega\sqrt{LC}$ denotes the propagation constant of the line. Equation (2.3) has been designated as an inhomogeneous one-dimensional scalar wave equation. The Green function method has been developed particularly to solve this type of equation in a rather elegant way. The method has an analogy in circuit theory whereby the response of a network to any input function can be determined by an integration based on the impulse response of the network. The Green function for a spatial problem plays the same role as the impulse response function in a time-domain problem. For transient field problems, the Green function may be constructed to include the impulsive time characteristics as well. By definition, the Green function pertaining to a one-diminsional scalar wave equation of the form (2.3), denoted by $g(x, x')$, is a solution of the following equation,

$$\frac{d^2g(x, x')}{dx^2} + k^2g(x, x') = -\delta(x - x'), \tag{2.4}$$

where $\delta(x - x')$ represents a delta function already introduced in Sec. 1-4. The physical interpretation of (2.4) is that if we let

$$K(x) = \frac{i}{\omega L}\delta(x - x'), \tag{2.5}$$

then

$$\int_{x'-\epsilon}^{x'+\epsilon} K(x)dx = \frac{i}{\omega L}, \tag{2.6}$$

and (2.3) reduces to (2.4). Equations (2.5) and (2.6) imply that the line is excited by a localized current source of amplitude $i/\omega L$ placed at $x = x'$. It is known from the theory of differential equations that the solution for $g_0(x, x')$ satisfying (2.4) is not completely determined unless we specify the two boundary conditions which the function must satisfy at the extremities of the spatial domain in which the function is defined. The boundary conditions which must be satisfied by $g(x, x')$ are the same as those dictated by the original function which we intend to determine, namely, $V(x)$ in the present case. For this reason, the Green functions are classified according to the boundary conditions which they must obey. Some of the typical ones (for the transmission line) are illustrated in Fig. 2-2. To distinguish various types of functions satisfying different boundary conditions, we use a subscript to identify them. In general, the subscript 0 designates infinite domain so that we have outgoing waves at $x \to \pm\infty$, often called the radiation condition. Subscript 1 means that one of the boundary conditions satisfies the so-called Dirichlet condition while the other satisfies the radiation condition. When one of the boundary conditions satisfies the so-called Neumann condition, we use subscript 2. Subscript 3 is reserved for the mixed type. The same nomenclature will be used later in our classification of the dyadic Green functions. Actually, we should have used a double subscript for two distinct boundary conditions. For example, case (b) of Fig. 2-2 should be denoted by g_{01}, indicating that one radiation condition and one Dirichlet condition are involved. With such an understanding, the simplified notation should be acceptable. In case (d) a superscript becomes necessary because we have two sets of line voltage and current (V_1, I_1) and (V_2, I_2) in this problem, and the Green function also has different forms in the two regions. The first superscript denotes the region where this function is defined, and the second superscript denotes the region where the source is located. Before we give the derivation of the explicit expressions for various types of the Green functions, the main formula showing the application of the Green function should be derived.

Let us consider a single line first and let the domain of x correspond to (x_1, x_2). The function $g(x, x')$ in (2.4) can represent any of the three types, g_0, g_1 and g_2, illustrated in Fig. 2-2. The treatment of case (d) is slightly different, and it will be formulated later. By multiplying (2.3) by $g(x, x')$ and (2.4) by $V(x)$ and taking the difference of the two resultant equations, we obtain

$$V(x)\frac{d^2 g_0(x, x')}{dx^2} - g_0(x, x')\frac{d^2 V(x)}{dx^2} = - V(x)\delta(x - x') - i\omega L g_0(x, x') K(x). \tag{2.7}$$

An integration of (2.7) through the entire domain of x yields

$$\int_{x_1}^{x_2}\left[V(x)\frac{d^2 g_0(x, x')}{dx^2} - g_0(x, x')\frac{d^2 V(x)}{dx^2}\right]dx = -\int_{x_1}^{x_2} V(x)\delta(x - x')dx - i\omega L\int_{x_1}^{x_2} g_0(x, x') K(x)\, dx. \tag{2.8}$$

The first term at the right-hand side of the above equation is simply $V(x')$, and the term at the left-hand side can be simplified by integration by parts, which

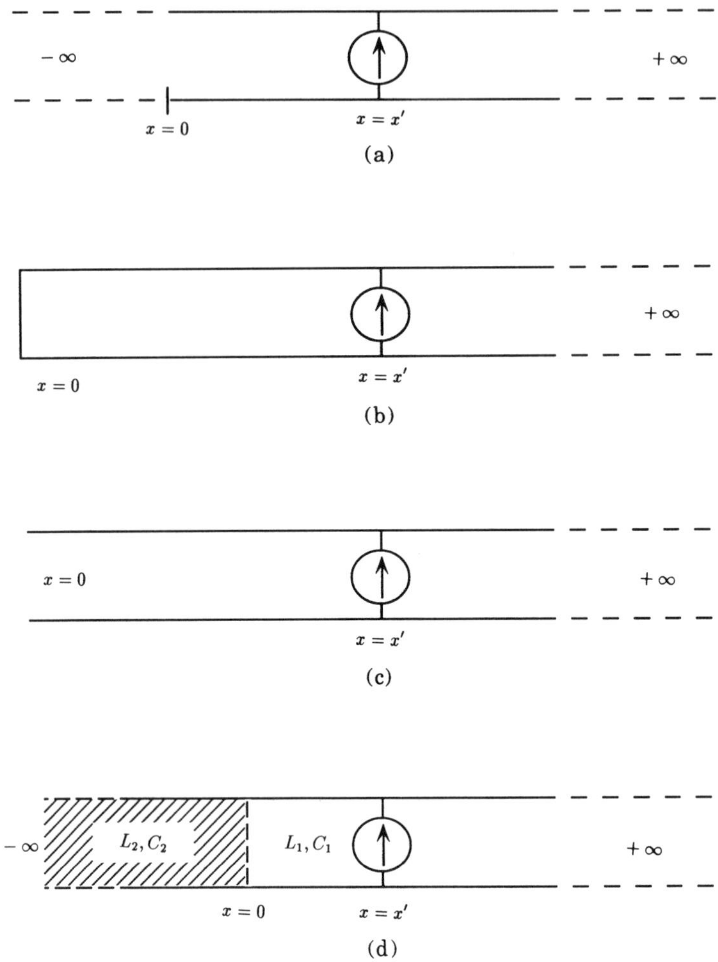

Fig. 2-2 Classification of Green functions according to the boundary conditions

gives

$$V(x') = -i\omega L \int_{x_1}^{x_2} g_0\,(x, x')\,K(x)\,dx. \tag{2.9}$$

If we use the unprimed variable x to denote the position of a field point, as usually is the case, (2.9) can be changed to

$$\begin{aligned} V(x) &= -i\omega L \int_{x_1}^{x_2} g\,(x', x)\,K(x')\,dx' \\ &= -i\omega L \int_{x_1}^{x_2} g_0\,(x, x')\,K(x')\,dx'. \end{aligned} \tag{2.10}$$

The last identity is due to the symmetrical property of the Green function that will be shown in the next section. The shifting of the primed and unprimed variables is often practiced in our work. For this reason, it is important to point out that $g(x', x)$, by definition, satisfies the equation

$$\frac{d^2 g\,(x', x)}{dx'^2} + k^2 g\,(x', x) = -\delta(x' - x). \tag{2.11}$$

A comparison of (2.4) and (2.11) would help us to realize the importance of not arbitrarily altering the positions of x and x' in $g_0\,(x, x')$.

2-2 DERIVATION OF $g_0(x, x')$ BY THE CONVENTIONAL METHOD AND THE OHM-RAYLEIGH METHOD

The expressions for various g's can be derived by the conventional method described in the theory of differential equations. An alternative approach is to apply the method of Ohm-Rayleigh, a terminology introduced by Sommerfeld [1949, p. 179], or the method of eigenfunction expansion. This method is not needed for the problems involving a transmission line. However, it is a necessary tool in our forthcoming treatment of the dyadic Green function. For this reason we introduce it here as a preparation for the future work. In the following, we will derive the various g's by the conventional method first and then apply the Ohm-Rayleigh method to rederive the free-space Green function as a demonstration of the techniques involved in that method.

Case 1. Free-Space Green Function, $g_0\,(x, x')$.

The general solutions for (2.4) in the two regions (see Fig. 2-2 for the layout) are

$$g_0\,(x, x') = Ae^{ikx}, \qquad x > x' \tag{2.12}$$

and

$$g_0\,(x, x') = Be^{-ikx}, \qquad x' > x. \tag{2.13}$$

The choice of the exponential function with the proper sign assures us the satisfaction of boundary conditions at infinity. At $x = x'$, the function must be continuous, and its derivative is discontinuous. The physical interpretation of these two conditions is that the voltage at x' is continuous, but the difference of the line currents at x' must be equal to the source current. Algebraically, if we integrate (2.4) in a small interval around x', we obtain, after identifying $g(x, x')$ as $g_0\,(x, x')$,

$$\int_{x'-\epsilon}^{x'+\epsilon} \frac{d^2 g_0\,(x, x')}{dx^2}\,dx + k^2 \int_{x'-\epsilon}^{x'+\epsilon} g_0\,(x, x')\,dx = -1. \tag{2.14}$$

In the limit as $\epsilon \to 0$, the second term approaches zero, assuming $g_0\,(x, x')$ is finite at $x = x'$, (2.14) then becomes

$$\left[\frac{dg_0(x,x')}{dx}\right]_{x'-0}^{x'+0} = -1. \tag{2.15}$$

The continuity of $g_0(x,x')$ at x' means

$$\left[g_0(x,x')\right]_{x'-0}^{x'+0} = 0. \tag{2.16}$$

Applying these two conditions to (2.12) and (2.13), one finds

$$g_0(x,x') = \begin{cases} \frac{i}{2k}e^{ik(x-x')}, & x \geq x' \\ \frac{i}{2k}e^{-ik(x-x')}, & x' \geq x. \end{cases} \tag{2.17}$$

A more compact expression for (2.17) is

$$g_0(x,x') = \frac{i}{2k}e^{ik|x-x'|}. \tag{2.18}$$

Case 2. Green Function of the First Kind, $g_1(x,x')$.

For this case, we let

$$g_1(x,x') = \begin{cases} Ae^{ikx}, & x \geq x' \\ B\sin kx, & x' \geq x \geq 0. \end{cases} \tag{2.19}$$

The choice of the sine function assures us that a Dirichlet condition is satisfied at $x = 0$. Applying (2.15) and (2.16), with the notation $g_0(x,x')$ changed to $g_1(x,x')$, we have

$$g_1(x,x') = \begin{cases} \frac{1}{k}\sin kx' e^{ikx}, & x \geq x' \\ \frac{1}{k}\sin kx e^{ikx'}, & x' \geq x \geq 0. \end{cases} \tag{2.20}$$

Equation (2.20) may be written in the form

$$g_1(x,x') = \begin{cases} \frac{i}{2k}\left[e^{ik(x-x')} - e^{ik(x+x')}\right], & x \geq x' \\ \frac{i}{2k}\left[e^{-ik(x-x')} - e^{ik(x+x')}\right], & x' \geq x \geq 0. \end{cases} \tag{2.21}$$

In view of (2.17), we may interpret (2.21) as consisting of an incident wave and a scattered wave; that is

$$g_1(x,x') = g_0(x,x') + g_{1s}(x,x'),$$

where

$$g_{1s}(x,x') = \frac{-i}{2k}e^{ik(x+x')}. \tag{2.22}$$

Such a notion is not only physically useful, but mathematically it offers a shortcut to finding a composite Green function. Because once we know $g_0(x, x')$, it is just a matter of finding g_s so that the desired boundary condition is satisfied. It is observed that $g_{1s}(x, x')$ or $g_s(x, x')$ is a solution of the homogeneous differential equation

$$\frac{d^2 g_s}{dx^2} + k^2 g_s = 0 \tag{2.23}$$

and it does not have any discontinuous characteristics as does $g_0(x, x')$. An illustration is given below in determining $g_2(x, x')$ by this shortcut method or the method of scattering superposition.

Case 3. Green Function of the Second Kind, $g_2(x, x')$.

The method of scattering superposition suggests that we can start with

$$g_2(x, x') = g_0(x, x') + Ae^{ikx}. \tag{2.24}$$

To satisfy the Neumann condition at $x = 0$, we require

$$\left[\frac{dg_0(x, x')}{dx} + ikAe^{ikx}\right]_{x=0} = 0 \tag{2.25}$$

or

$$\frac{1}{2}e^{ikx'} + ikA = 0;$$

hence

$$A = \frac{i}{2k}e^{ikx'}. \tag{2.26}$$

The complete expression for $g_2(x, x')$ is therefore given by

$$\begin{aligned} g_2(x, x') &= \frac{i}{2k}\begin{cases} e^{ik(x-x')} + e^{ik(x+x')}, & x \geq x' \\ e^{-ik(x-x')} + e^{ik(x+x')}, & x' \geq x \geq 0 \end{cases} \\ &= \frac{i}{k}\begin{cases} \cos kx' e^{ikx}, & x \geq x' \\ \cos kx e^{ikx'}, & x' \geq x \geq 0. \end{cases} \end{aligned} \tag{2.27}$$

Case 4. Green Function of the Third Kind, $g^{(ij)}(x, x')$.

In this case, we have two differential equations to start with:

$$\frac{d^2 V_1(x)}{dx^2} + k_1^2 V_1(x) = i\omega L_1 K_1(x), \qquad x \geq 0 \tag{2.28}$$

$$\frac{d^2 V_2(x)}{dx^2} + k_2^2 V_2(x) = 0, \qquad 0 \geq x. \tag{2.29}$$

It is assumed that the current source is located in region 1 (see Fig. 2-2). We introduce two Green functions of the third kind, denoted by $g^{(11)}(x, x')$ and

$g^{(21)}(x, x')$. A Green function with double superscript like $g^{(21)}$ signifies that it is a function of the third kind. The first number of the superscript corresponds to the region where the function is defined. The second number corresponds to the region where the source is located; then

$$\frac{d^2 g^{(11)}(x,x')}{dx^2} + k_1^2 g^{(11)}(x,x') = -\delta(x-x'), \qquad x \ge 0 \tag{2.30}$$

and

$$\frac{d^2 g^{(21)}(x,x')}{dx^2} + k_2^2 g^{(21)}(x,x') = 0, \qquad 0 \ge x. \tag{2.31}$$

At the junction corresponding to $x = 0$, $g^{(11)}$ and $g^{(21)}$ satisfy the boundary condition that

$$g^{(11)}(x,x')_{x=0} = g^{(21)}(x,x')_{x=0} \tag{2.32}$$

$$\frac{1}{L_1}\frac{dg^{(11)}(x,x')}{dx}\bigg|_{x=0} = \frac{1}{L_2}\frac{dg^{(21)}(x,x')}{dx}\bigg|_{x=0}. \tag{2.33}$$

The last condition corresponds to the physical requirement that the current at the junction must be continuous. Again, by means of the method of scattering superposition, we write

$$\begin{aligned} g^{(11)}(x,x') &= g_0(x,x') + g_s^{(11)}(x,x') \\ &= \frac{i}{2k_1}\begin{cases} e^{ik_1(x-x')} + Re^{ik_1(x+x')}, & x \ge x' \\ e^{-ik_1(x-x')} + Re^{ik_1(x+x')}, & x' \ge x \ge 0, \end{cases} \end{aligned} \tag{2.34}$$

$$g^{(21)}(x,x') = \frac{i}{2k_1} Te^{-i(k_2 x - k_1 x')}, \qquad 0 \ge x. \tag{2.35}$$

We have also added an extra term, $e^{ik_1 x'}$, with the unknown coefficients R and T to make the final solutions more attractive. Upon applying the boundary conditions (2.32) and (2.33) to these two functions, we find

$$1 + R = T \tag{2.36}$$

$$\frac{k_1}{L_1}(1-R) = \frac{k_2}{L_2} T \tag{2.37}$$

or

$$\frac{1}{Z_1}(1-R) = \frac{1}{Z_2} T, \tag{2.38}$$

where

$$Z_1 = \left(\frac{L_1}{C_1}\right)^{\frac{1}{2}}, \quad Z_2 = \left(\frac{L_2}{C_2}\right)^{\frac{1}{2}}, \tag{2.39}$$

representing, respectively, the characteristic impedance of the lines. By solving for R and T from (2.36) and (2.38), we find

$$R = \frac{Z_2 - Z_1}{Z_2 + Z_1} \tag{2.40}$$

$$T = \frac{2Z_2}{Z_2 + Z_1}. \tag{2.41}$$

These are just the reflection and the transmission coefficient of a wave propagating from line 1 toward line 2. Knowing $g^{(11)}(x, x')$ and $g^{(21)}(x, x')$, we can determine V_1 and V_2 by applying the one-dimensional Green theorem to (2.28) through (2.31) as we did in deriving (2.9). Thus, from (2.28) and (2.30), we obtain

$$V_1(x') = -i\omega L_1 \int_0^\infty g^{(11)}(x, x') K_1(x)\, dx + \left[g^{(11)}(x, x') \frac{dV_1(x)}{dx} - V_1(x) \frac{dg^{(11)}(x, x')}{dx} \right]_0^\infty. \tag{2.42}$$

From (2.29) and (2.31), we have

$$\left[g^{(21)}(x, x') \frac{dV_2(x)}{dx} - V_2(x) \frac{dg^{(21)}(x, x')}{dx} \right]_{-\infty}^0 = 0. \tag{2.43}$$

Because of the radiation condition at $x = -\infty$, and the fact that

$$\frac{dV_2(x)}{dx} = i\omega L_2 I_2(x)$$

$$\frac{dg^{(21)}(x, x')}{dx} = -ik_2 g^{(21)}(x, x')$$

and

$$V_2(x) = Z_2 I_2(x),$$

identity (2.43) is indeed satisfied. Likewise the second term at the right-hand side of (2.42) vanishes for the similar reason. Thus we have

$$V_1(x') = -i\omega L_1 \int_0^\infty g^{(11)}(x, x') K_1(x)\, dx. \tag{2.44}$$

By interchanging the variables x and x', we obtain

$$V_1(x) = -i\omega L_1 \int_0^\infty g^{(11)}(x, x') K_1(x')\, dx', \qquad x \geq 0, \tag{2.45}$$

where we have already made use of the symmetrical property of the Green function, that is, $g^{(11)}(x, x') = g^{(11)}(x', x)$, the proof of which will be found in the next section. To determine $V_2(x)$, we write

$$V_2(x) = V_2(0) e^{-ik_2 x} = V_1(0) e^{-ik_2 x}, \tag{2.46}$$

which is a solution of (2.29) that fulfills the radiation condition at $x = -\infty$. The value of $V_1(0)$ can be found from (2.45) by putting $x = 0$; thus the expression for $V_2(x)$ is given by

$$V_2(x) = \left[-i\omega L_1 \int_0^\infty g^{(11)}(0, x')\, K_1(x')\, dx' \right] e^{-ik_2 x}. \tag{2.47}$$

In view of (2.32) through (2.35), we have

$$g^{(21)}(x, x') = g^{(21)}(0, x')e^{-ik_2 x} = g^{(11)}(0, x')e^{-ik_2 x}; \tag{2.48}$$

thus (2.47) can be written in the form

$$V_2(x) = -i\omega L_1 \int_0^\infty g^{(21)}(x, x') K_1(x')\, dx', \qquad 0 \geq x, \tag{2.49}$$

which has the same appearance as (2.45), except that $g^{(11)}(x, x')$ is replaced by $g^{(21)}(x, x')$ as the domains of the x variables in these two equations are different. This completes our discussion of the derivation and the usage of various kinds of Green function based on the conventional method. As an introduction to the Ohm-Rayleigh method, we will now rederive the expression for $g_0(x, x')$ by this alternative approach. For convenience, the differential equation for $g_0(x, x')$ is repeated here.

$$\frac{d^2 g_0(x, x')}{dx^2} + k^2 g_0(x, x') = -\delta(x - x'), \quad \infty > x > -\infty. \tag{2.50}$$

The key step in the Ohm-Rayleigh method is to expand $\delta(x - x')$ in terms of the eigenfunction of a homogeneous equation of the same type as (2.50). The eigenfunction in this case is e^{ihx}, which is a solution of the equation

$$\frac{d^2 f}{dx^2} + h^2 f = 0, \tag{2.51}$$

where h is an arbitrary constant. Physically, the function e^{ihx} represents a wave that can exist on an infinite line with propagation constant equal to h. Since e^{ihx} is the spectral function used in the Fourier transform, the method of Ohm-Rayleigh in this case is equivalent to the method of Fourier transform in solving (2.50). Before we apply these methods to (2.50), a few words must be said about the question of rigor concerning the existence of the Fourier transform of a delta function. A rigorous treatment of this subject would be based on the theory of generalized functions [Gelfand and Shilov, 1964]. In the present book we merely use the formulas derivable from that theory such as the integral representation of the weighted delta function discussed in the previous chapter. With this understanding in mind, we can solve (2.50) by the method of Fourier transform. Thus we define

$$f(h) = \int_{-\infty}^\infty g_0(x, x')\, e^{-ihx}\, dx; \tag{2.52}$$

then

$$g_0(x, x') = \frac{1}{2\pi} \int_{-\infty}^\infty f(h) e^{ihx}\, dh. \tag{2.53}$$

Applying the Fourier transform to (2.50), and assuming that both g_0 and dg_0/dx approach zero at $x = \pm\infty$, we obtain

$$\int_{-\infty}^{\infty} \left[\frac{d^2 g_0(x, x')}{dx^2} + k^2 g_0(x, x') \right] e^{-ihx}\, dx$$
$$= \left(k^2 - h^2\right) f(h) = -e^{-ihx'} \tag{2.54}$$

or

$$f(h) = \frac{e^{-ihx'}}{h^2 - k^2}. \tag{2.55}$$

Hence

$$g_0(x, x') = \frac{1}{2\pi} \int_{-\infty}^{\infty} \frac{e^{ih(x-x')}}{h^2 - k^2}\, dh. \tag{2.56}$$

The assumption concerning the behavior of g_0 and dg_0/dx at infinity can be justified if we let k be complex with $Im(k) > 0$. Such an artifice is commonly used, and it corresponds to a line with a loss. After the final expression for $g_0(x, x')$ is obtained, we can restore the lossless condition by allowing $Im(k) \to 0$. The procedure also helps us to evaluate the integral representation of $g_0(x, x')$ as expressed by (2.56) in a relatively simple manner. With such an interpretation, the locations of the poles of the integrand in the h-plane are shown in Fig. 2-3 where the contour of integration is assumed to be along the real axis.

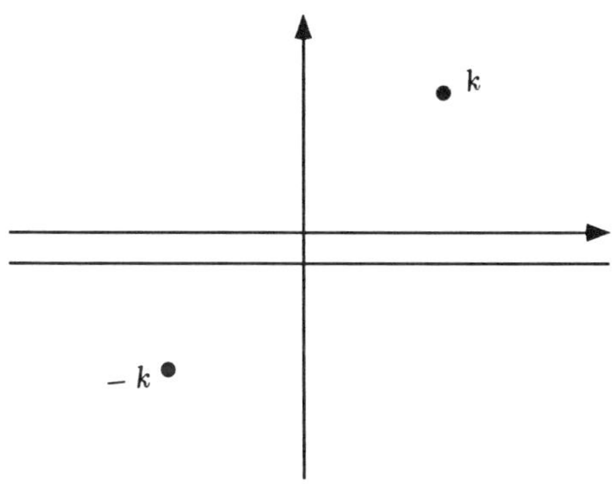

Fig. 2-3 Locations of the poles of the integral representation of $g_0(x, x')$

For $x - x' \geq 0$, the contour can be closed by an infinite path in the upper half-plane without changing its value; for $x - x' \leq 0$, the contour can be closed at the lower half-plane. Applying Cauchy's residue theorem to the closed contour integrals, we obtain

$$g_0(x, x') = \frac{i}{2k} \begin{cases} e^{ik(x-x')}, & x \geq x' \\ e^{-ik(x-x')}, & x' \geq x. \end{cases} \tag{2.57}$$

For real values of k, the same result is obtained by deforming the contour so that it is properly indented at the poles as shown in Fig. 2-4. Equation (2.57), of course, is identical to (2.17). If we had followed precisely the steps involved in the Ohm-Rayleigh method, the procedures are slightly different. First, we let

$$\delta(x - x') = \int_{-\infty}^{\infty} A(h) e^{ihx}\, dh, \tag{2.58}$$

where e^{ihx} is treated as an eigenfunction pertaining to the one-dimensional scalar wave equation for an infinite domain. Multiplying (2.58) by $e^{-ih'x}$ and integrating the resultant equation from $-\infty$ to ∞, we have

$$e^{-ih'x'} = \int_{-\infty}^{\infty} \int_{-\infty}^{\infty} A(h) e^{i(h-h')x}\, dh\, dx, \tag{2.59}$$

where h' denotes an arbitrary constant. In view of (1.82), the integration with respect to x yields $2\pi\delta(h - h')$. Thus

$$e^{-ih'x'} = 2\pi \int_{-\infty}^{\infty} A(h) \delta(h - h')\, dh = 2\pi A(h'); \tag{2.60}$$

hence

$$A(h) = \frac{1}{2\pi} e^{-ihx'}, \tag{2.61}$$

or

$$\delta(x - x') = \frac{1}{2\pi} \int_{-\infty}^{\infty} e^{ih(x-x')}\, dh. \tag{2.62}$$

Now we assume that $g_0(x, x')$ can be represented by a similar integral in terms of the eigenfunction e^{ihx}. More precisely, we let

$$g_0(x, x') = \frac{1}{2\pi} \int_{-\infty}^{\infty} B(h) e^{ih(x-x')}\, dh. \tag{2.63}$$

Upon substituting (2.62) and (2.63) into (2.50), we find that

$$B(h)\left(k^2 - h^2\right) = -1; \tag{2.64}$$

hence

$$g_0(x, x') = \frac{1}{2\pi} \int_{-\infty}^{\infty} \frac{e^{ih(x-x')}}{h^2 - k^2}\, dh. \tag{2.65}$$

Although the direct Fourier transform method and the Ohm-Rayleigh method are equivalent, the concepts behind these two methods are quite different. In the Ohm-Rayleigh method, we emphasize the notion that e^{ihx} is an eigenfunction of the differential equation under consideration, so we treat (2.62) and (2.63) as the eigenfunction expansions of these two functions. The same procedure will be followed later in finding the eigenfunction expansion for the dyadic Green function.

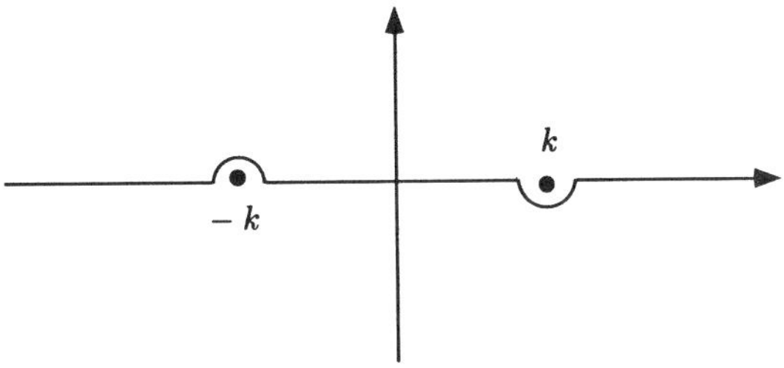

Fig. 2-4 Indentation of contour for real k

2-3 SYMMETRICAL PROPERTIES OF GREEN FUNCTIONS

The symmetrical properties of the Green functions have already been used in deriving (2.10) and (2.45) without a proof. We will now supply the details. We consider two Green functions corresponding to two different source positions, and the domain in which these two functions are defined is the same. Thus, denoting the source positions, respectively, by x_a, and x_b and the domain of x by (x_1, x_2), we have

$$\frac{d^2 g(x, x_a)}{dx^2} + k^2 g(x, x_a) = -\delta(x - x_a) \tag{2.66}$$

$$\frac{d^2 g(x, x_b)}{dx^2} + k^2 g(x, x_b) = -\delta(x - x_b). \tag{2.67}$$

Upon multiplying (2.66) and (2.67), respectively, by $g(x, x_b)$ and $g(x, x_a)$ and taking the difference of two resultant equations and integrating it through the entire domain of x, we obtain

$$g(x_a, x_b) - g(x_b, x_a) = \left[g(x, x_b) \frac{dg(x, x_a)}{dx} - g(x, x_a) \frac{dg(x, x_b)}{dx} \right]_{x_1}^{x_2}. \tag{2.68}$$

If $g(x, x_a)$ and $g(x, x_b)$ represent any of the four kinds, that is, g_0, g_1, g_2, and $g^{(11)}$, they would satisfy the same boundary conditions at the extremities; hence the term at the right-hand side of (2.68) vanishes. We therefore conclude that

$$g(x_a, x_b) = g(x_b, x_a), \tag{2.69}$$

which is the mathematical statement of the symmetrical property of the Green functions g_0, g_1, g_2, and $g^{(11)}$. It is implied that x_a and x_b lie inside the domain where the function is denied. The symmetrical property of $g^{(21)}$ exhibits a slightly different form. In the first place, there are four functions of the third kind, namely, $g^{(11)}, g^{(21)}$, $g^{(22)}$, and $g^{(12)}$. If we let x_a be located in region 1 ($x > 0$) and x_b be located in region 2, the various g's satisfy the following equations:

$$\frac{d^2 g^{(11)}(x, x_a)}{dx^2} + k_1^2 g^{(11)}(x, x_a) = -\delta(x - x_a), \quad x \geq 0 \tag{2.70}$$

$$\frac{d^2 g^{(21)}(x, x_a)}{dx^2} + k_2^2 g^{(21)}(x, x_a) = 0, \quad 0 \geq x \tag{2.71}$$

$$\frac{d^2 g^{(22)}(x, x_b)}{dx^2} + k_2^2 g^{(22)}(x, x_b) = -\delta(x - x_b), \quad 0 \geq x \tag{2.72}$$

$$\frac{d^2 g^{(12)}(x, x_b)}{dx^2} + k_1^2 g^{(12)}(x, x_b) = 0, \quad x \geq 0. \tag{2.73}$$

At $x = +\infty$, both $g^{(11)}$ and $g^{(12)}$ satisfy the same radiation condition specified for region 1, and at $x = -\infty$, both $g^{(22)}$ and $g^{(21)}$ satisfy the identical radiation condition as required for region 2. At $x = 0$, the boundary conditions are

$$\left\{ \begin{aligned} & g^{(11)}(0, x_a) = g^{(21)}(0, x_a) \\ & \frac{1}{L_1} \frac{dg^{(11)}(x, x_a)}{dx}\bigg|_{x=0} = \frac{1}{L_2} \frac{dg^{(21)}(x, x_a)}{dx}\bigg|_{x=0} \\ & g^{(22)}(0, x_b) = g^{(12)}(0, x_b) \\ & \frac{1}{L_2} \frac{dg^{(22)}(x, x_b)}{dx}\bigg|_{x=0} = \frac{1}{L_1} \frac{dg^{(12)}(x, x_b)}{dx}\bigg|_{x=0}. \end{aligned} \right. \tag{2.74}$$

Upon multiplying (2.70) by $g^{(12)}$ and (2.73) by $g^{(11)}$ and integrating the difference of the two resultant equations from $x = 0$ to $x = +\infty$, we obtain

$$g^{(12)}(x_a, x_b)$$

$$= \left[g^{(11)}(x, x_a) \frac{dg^{(12)}(x, x_b)}{dx} - g^{(12)}(x, x_b) \frac{dg^{(11)}(x, x_a)}{dx} \right]_0^\infty \tag{2.75}$$

$$= \left[g^{(12)}(x, x_b) \frac{dg^{(11)}(x, x_a)}{dx} - g^{(11)}(x, x_a) \frac{dg^{(12)}(x, x_b)}{dx} \right]_{x=0}. \tag{2.76}$$

The two terms within the brackets cancel each other at $x = +\infty$ because of the same radiation condition. Repeating a similar calculation for (2.71) and (2.72) in the negative x domain, we have

$$g^{(21)}(x_b, x_a) = \left[g^{(22)}(x, x_b) \frac{dg^{(21)}(x, x_a)}{dx} - g^{(21)}(x, x_a) \frac{dg^{(22)}(x, x_b)}{dx} \right]_{x=0}. \tag{2.77}$$

In view of (2.74), it can be shown readily that

$$L_1 g^{(21)}(x_b, x_a) = L_2 g^{(12)}(x_a, x_b). \tag{2.78}$$

The same result can, of course be obtained from (2.35) by interchanging x with x', k_1 with k_2, and Z_1 with Z_2. However, the present proof does not require the explicit solutions for $g^{(12)}$ or $g^{(21)}$.

Another relationship that has a vector analog later deals with g_1 and g_2. From (2.20) and (2.27) it is obvious that

$$\frac{\partial g_1(x, x')}{\partial x} = -\frac{\partial g_2(x, x')}{\partial x'}. \tag{2.79}$$

This relationship can be derived directly from the differential equations for g_1 and g_2 without resorting to their explicit expressions.

2-4 FREE-SPACE GREEN FUNCTION OF THE THREE-DIMENSIONAL SCALAR WAVE EQUATION

The free-space Green function $G_0(\overline{R}, \overline{R}')$ for a three-dimensional scalar wave equation satisfies the equation

$$\nabla^2 G_0(\overline{R}, \overline{R}') + k^2 G_0(\overline{R}, \overline{R}') = -\delta\left(\overline{R} - \overline{R}'\right). \tag{2.80}$$

The function must satisfy the radiation condition that

$$\lim_{R\to\infty} R\left(\frac{\partial G_0}{\partial R} - ikG_0\right) = 0. \tag{2.81}$$

If we make a change of the variable $\overline{R} - \overline{R}' = \overline{R}_1$ as shown in Fig. 2-5, the problem would have a spherical symmetry with respect to the new origin $0'$; hence the function G_0 would be a function of R_1 only. In terms of the new spherical coordinate system with origin at $0'$, the free-space Green function, henceforth denoted by $G_0\left(\overline{R}_1, 0\right)$, satisfies the equation

$$\frac{1}{R_1^2}\frac{d}{dR_1}\left[R_1^2 \frac{dG_0\left(\overline{R}_1, 0\right)}{dR_1}\right] + k^2 G_0\left(\overline{R}_1, 0\right) = -\frac{\delta(R_1 - 0)}{4\pi R_1^2}. \tag{2.82}$$

The weighting factor $1/4\pi R_1^2$ attached to $\delta(R_1 - 0)$ is due to the fact that

$$\iiint_V \delta\left(\overline{R} - \overline{R}'\right) dV = 1, \tag{2.83}$$

where V encloses $0'$. For $R_1 \neq 0$, the homogeneous equation is the same as the spherical Bessel equation of zeroth order. The appropriate solution for

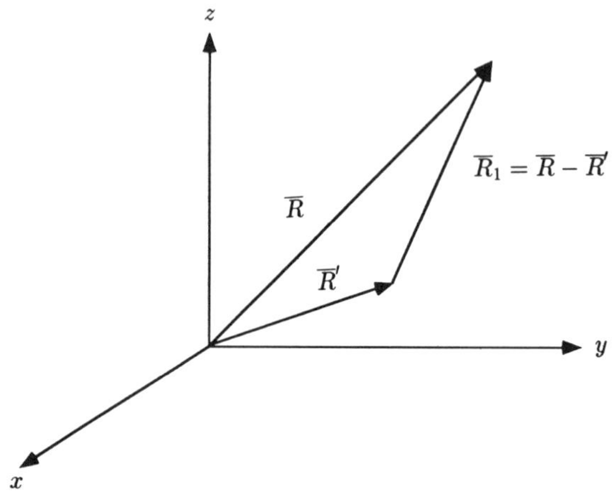

Fig. 2-5 The position vectors in $G_0(\overline{R}, \overline{R}')$

$G_0\left(\overline{R}_1, 0\right)$, which satisfies the radiation condition at infinity, must therefore be proportional to the spherical Hankel function of the first kind of zeroth order [Stratton, 1941, p. 404] or

$$G_0\left(\overline{R}_1, 0\right) = Ah_0^{(1)}\left(kR_1\right) = A\left(\frac{e^{ikR_1}}{ikR_1}\right). \tag{2.84}$$

To determine the constant of proportionality A, we make use of the Gauss theorem that

$$\iiint \nabla^2 \psi dV = \oiint \nabla\psi \cdot \overline{dS}; \tag{2.85}$$

thus a volume of integration of (2.82) through a small spherical region with center at $0'$ gives

$$4\pi R_1^2 \left| \frac{dG_0\left(R_1, 0\right)}{dR_1} \right|_{R_1 = a} + k^2 \int_0^a G_0\left(R_1, 0\right) 4\pi R_1^2 dR_1 = -1. \tag{2.86}$$

Upon substituting (2.84) into (2.86) and letting $a \to 0$, we obtain $A = ik/4\pi$. The complete expression for $G_0\left(\overline{R}_1, 0\right)$ is, therefore, given by

$$G_0\left(\overline{R}_1, 0\right) = \frac{e^{ikR_1}}{4\pi R_1}. \tag{2.87}$$

Transforming back into the original coordinate system with center at 0, we obtain

$$G_0\left(\overline{R}, \overline{R}'\right) = e^{ik\left|\overline{R} - \overline{R}'\right|} / 4\pi \left|\overline{R} - \overline{R}'\right|. \tag{2.88}$$

For a two-dimensional scalar wave equation independent of z in the cylindrical coordinate system, the corresponding free-space Green function is given by

$$G_0(\overline{r}, \overline{r}') = \frac{i}{4} H_0^{(1)}(k|\overline{r} - \overline{r}'|), \tag{2.89}$$

where $\overline{r}$ and $\overline{r}'$ denote the position vector in a cylindrical coordinate system and $H_0^{(1)}$ denotes the Hankel function of zeroth order and of the first kind. The function $G_0(\overline{r}, \overline{r}')$ will not be encountered in the main body of this book.

3

Electromagnetic Theory

Historically, Maxwell's theory of electromagnetism was founded on the basic laws available at his time. His main contribution was to supplement Ampère's law by a now famous term called the *displacement current* to make it compatible with the equation of continuity and with Gauss law. Nowadays it is more practical to present the theory in its entire form without following the historical course. However, it is important to distinguish the dependent equations from the independent ones of the entire set of equations and also to understand the significance of the definite form in contrast to the indefinite form. The meaning of these technical terms will be explained shortly. We also discuss very thoroughly the boundary conditions which have to be postulated in the electromagnetic theory if we consider Maxwell's differential equations as the foundation of his theory.

3-1 THE INDEPENDENT AND DEPENDENT EQUATIONS AND THE INDEFINITE AND DEFINITE FORMS OF MAXWELL'S EQUATIONS

There are three *independent* equations in Maxwell's theory of electromagnetism:

$$\nabla \times \overline{E} = -\frac{\partial \overline{B}}{\partial t} \quad \text{(Faraday's law)} \tag{3.1}$$

$$\nabla \times \overline{H} = \overline{J} + \frac{\partial \overline{D}}{\partial t} \quad \text{(Maxwell-Ampère law)} \tag{3.2}$$

$$\nabla \cdot \overline{J} = -\frac{\partial \rho}{\partial t} \quad \text{(equation of continuity),} \tag{3.3}$$

where

$$\overline{E} = \text{electric field (volt/meter)}$$
$$\overline{D} = \text{electric flux density (coulomb/meter}^2\text{)}$$
$$\overline{H} = \text{magnetic field (ampere/meter)}$$
$$\overline{B} = \text{magnetic flux density (weber/meter}^2\text{)}$$
$$\overline{J} = \text{electric current density (ampere/meter}^2\text{)}$$
$$\rho = \text{electric charge density (coulomb/meter}^3\text{)}.$$

It is understood that all the field quantities, including the current density and the charge density, are functions of position and time. By taking the divergence of (3.1) and setting the constant of integration with respect to time equal to zero, we obtain

$$\nabla \cdot \overline{B} = 0 \quad \text{(Gauss law—magnetic).} \tag{3.4}$$

In a similar manner, upon taking the divergence of (3.2) and eliminating $\overline{J}$ between the resultant equation and (3.3), we have

$$\nabla \cdot \overline{D} = \rho \quad \text{(Gauss law).} \tag{3.5}$$

Since (3.4) and (3.5) are considered to be derivable from (3.1) through (3.3), these two equations should be treated as auxiliary or *dependent* equations in the entire system of equations (3.1) through (3.5). An alternative view is to take (3.1), (3.2), and (3.3) as independent equations while treating (3.3) and (3.4) as dependent equations. Such an alternative choice does not change the basic point of view.

The three independent equations described by (3.1) through (3.3) actually consist of seven scalar differential equations inasmuch as one vector equation is equivalent to three scalar equations. Now each vector function has three components. We have, therefore, 16 unknown scalar functions altogether. It is obvious that the three independent equations are not sufficient to form a complete system of equations to solve for the unknown functions. For clarity, we shall designate (3.1) through (3.3) as Maxwell's equations in the *indefinite form* as long as the constitutive relations between the field quantities are unknown or unspecified. Under such a condition, many alternative forms can be used to describe Maxwell's theory. One common form is to introduce two material field vectors, $\overline{P}$ and $\overline{M}$, which are defined by

$$\overline{D} = \epsilon_0 \overline{E} + \overline{P} \tag{3.6}$$
$$\overline{B} = \mu_0 \left(\overline{H} + \overline{M}\right), \tag{3.7}$$

where

$$\overline{P} = \text{polarization (coulomb/meter}^2\text{)}$$
$$\overline{M} = \text{magnetization (ampere/meter)}$$

$$\begin{aligned}\epsilon_0 &= \text{the electric constant}\\ &= 8.854 \times 10^{-12}(\text{farad/meter})\\ &\simeq \frac{1}{36\pi} \times 10^{-9}(\text{farad/meter})\\ \mu_0 &= \text{the magnetic constant}\\ &= 4\pi \times 10^{-7}(\text{henry/meter}).\end{aligned}$$

When $\overline{E}, \overline{H}, \overline{P}$, and $\overline{M}$ are used, the number of unknowns and the number of equations remain the same, and the essential characteristics of Maxwell's equations are not altered. This is the invariant property of Maxwell's equations. We shall not elaborate here this aspect of the electromagnetic theory since it is irrelevant to the present work. In any event, to make the Maxwell's equation definite we need more information. This additional information is provided by the constitutive relations between the field quantities. For example, in a simple isotropic medium, the field quantities are related as follows:

$$\overline{D} = \epsilon \overline{E} \tag{3.8}$$

$$\overline{B} = \mu \overline{H} \tag{3.9}$$

$$\overline{J} = \sigma \overline{E}, \tag{3.10}$$

where ϵ, μ, and σ denote, respectively, the permittivity, permeability, and conductivity of the medium. Equations (3.8) through (3.10) provide nine more scalar relations that make the number of unknowns and the number of equations compatible. When the constitutive relations between the field quantities are known, Maxwell's equations become *definite*. In many boundary-value problems, the constitutive relations between $\overline{D}, \overline{B}, \overline{E}$, and $\overline{H}$ are usually known while the current density function $\overline{J}$ is treated as a source term. In that case, we are interested in finding the solutions for $\overline{E}$ and $\overline{H}$ in terms of $\overline{J}$ that satisfy certain boundary conditions. Much of the work discussed in this book falls into this case. Thus if the medium under consideration is air, which is practically a vacuum, the definite form of Maxwell's equations becomes

$$\nabla \times \overline{E} = -\mu_0 \frac{\partial \overline{H}}{\partial t} \tag{3.11}$$

$$\nabla \times \overline{H} = \overline{J} + \epsilon_0 \frac{\partial \overline{E}}{\partial t} \tag{3.12}$$

$$\nabla \cdot \overline{J} = -\frac{\partial \rho}{\partial t} \tag{3.13}$$

$$\nabla \cdot \left(\mu_0 \overline{H}\right) = 0 \tag{3.14}$$

$$\nabla \cdot \left(\epsilon_0 \overline{E}\right) = \rho. \tag{3.15}$$

Similar equations for more complex media will be introduced later.

3-2 INTEGRAL FORMS OF MAXWELL'S EQUATIONS

Although the integral forms of Maxwell's equations are not needed in the present work, we will give a brief review here mainly because certain features of these forms are not discussed in standard books on electromagnetic theory, and they are useful in deriving the boundary conditions for the field quantities.

In order to give a complete description of the boundary conditions, we start with the indefinite form of Maxwell's equations. Upon integrating (3.1) through (3.5) through a volume V with an enclosing surface S, we obtain

$$\iiint \nabla \times \overline{E} dV = -\iiint \frac{\partial \overline{B}}{\partial t} dV \tag{3.16}$$

$$\iiint \nabla \times \overline{H} dV = \iiint \left(\overline{J} + \frac{\partial \overline{D}}{\partial t}\right) dV \tag{3.17}$$

$$\iiint \nabla \cdot \overline{J} dV = -\iiint \frac{\partial \rho}{\partial t} dV \tag{3.18}$$

$$\iiint \nabla \cdot \overline{B} dV = 0 \tag{3.19}$$

$$\iiint \nabla \cdot \overline{D} dV = \iiint \rho \, dV. \tag{3.20}$$

Now if the fields and their first derivatives are continuous throughout the region of integration, we can apply the curl theorem and the divergence theorem to obtain

$$\oiint (\hat{n} \times \overline{E}) \, dS = -\iiint \frac{\partial \overline{B}}{\partial t} dV \tag{3.21}$$

$$\oiint (\hat{n} \times \overline{H}) \, dS = \iiint \left(\overline{J} + \frac{\partial \overline{D}}{\partial t}\right) dV \tag{3.22}$$

$$\oiint (\hat{n} \cdot \overline{J}) \, dS = -\iiint \frac{\partial \rho}{\partial t} dV \tag{3.23}$$

$$\oiint (\hat{n} \cdot \overline{B}) \, dS = 0 \tag{3.24}$$

$$\oiint (\hat{n} \cdot \overline{D}) \, dS = \iiint \rho \, dV. \tag{3.25}$$

If we apply an open surface integration to (3.1) and (3.2), we obtain

$$\iint \nabla \times \overline{E} \cdot d\overline{S} = -\iint \frac{\partial \overline{B}}{\partial t} \cdot d\overline{S} \tag{3.26}$$

$$\iint \nabla \times \overline{H} \cdot d\overline{S} = \iint \left(\overline{J} + \frac{\partial \overline{D}}{\partial t} \right) \cdot d\overline{S}. \tag{3.27}$$

Now if the fields and their first derivatives are continuous, we can apply Stokes's theorem to convert (3.26) and (3.27) to the form

$$\oint \overline{E} \cdot d\ell = -\iint \frac{\partial \overline{B}}{\partial t} \cdot d\overline{S} \tag{3.28}$$

$$\oint \overline{H} \cdot d\overline{\ell} = \iint \left(\overline{J} + \frac{\partial \overline{D}}{\partial t} \right) \cdot d\overline{S}. \tag{3.29}$$

It should be emphasized here that (3.21) to (3.25) and (3.28), (3.29) are valid only if the fields and their first derivatives are continuous, a necessary condition to apply the curl theorem, the divergence theorem, and Stokes's theorem to the original integrals.

3-3 BOUNDARY CONDITIONS

Two approaches of presenting the boundary conditions for the electric and the magnetic fields will be discussed. In the first approach, the boundary conditions are postulated that lead to some integral forms of Maxwell's theory which are applicable to discontinuous as well as continuous fields. In the second approach, the process is reversed where the integral forms are postulated first and then the boundary conditions are derived. These two approaches will show clearly that the boundary conditions in electromagnetic theory cannot be derived from the differential equations formulated by Maxwell. A postulate is needed to arrive at these conditions, and the validity of these conditions can be verified only experimentally.

In the first approach, let us consider the application of (3.22) to two adjacent regions shown in Fig. 3-1. It is assumed that the fields $\overline{H}_1, \overline{D}_1$ and their first derivatives are continuous in V_1, and the fields $\overline{H}_2, \overline{D}_2$ and their first derivatives are continuous in V_2. Under these conditions the curl theorem can be applied to (3.16) to yield (3.22). In V_1 we obtain

$$\iint_{S_1} (\hat{n} \times \overline{H}_1)\, dS + \iint_{S_{12}} (-\hat{n}_1 \times \overline{H}_1)\, dS = \iiint_{V_1} \left(\overline{J}_1 + \frac{\partial \overline{D}_1}{\partial t} \right) dV, \tag{3.30}$$

and in V_2, we obtain

$$\iint_{S_2} (\hat{n} \times \overline{H}_2)\, dS + \iint_{S_{12}} (-\hat{n}_2 \times \overline{H}_2)\, dS = \iiint_{V_2} \left(\overline{J}_2 + \frac{\partial \overline{D}_1}{\partial t} \right) dV. \tag{3.31}$$

On S_{12}, the boundary surface between V_1 and V_2, we assume the existence of a surface current with surface current density $\overline{K}$. The total current moment in the entire region, therefore, is given by

$$\iiint_{V_1} \overline{J}_1 dV + \iiint_{V_2} \overline{J}_2 dV + \iint_{S_{12}} \overline{K} dS = \iiint_V \overline{J} dV, \tag{3.32}$$

and it is denoted by $\iiint_V \overline{J} dV$. The sum of (3.30) and (3.31) yields

$$\oiint \hat{n} \times \overline{H} dS - \iint_{S_{12}} \hat{n}_1 \times \left(\overline{H}_1 - \overline{H}_2\right) dS$$
$$= \iiint_V \overline{J} dV - \iint_{S_{12}} \overline{K} dS + \iiint_V \frac{\partial \overline{D}}{\partial t} dV. \tag{3.33}$$

We now postulate the boundary condition that at S_{12}

$$\hat{n}_1 \times \left(\overline{H}_1 - \overline{H}_2\right) = \overline{K} \tag{3.34}$$

then (3.33) reduces to

$$\oiint \hat{n} \times \overline{H} dS = \iiint_V \left(\overline{J} + \frac{\partial \overline{D}}{\partial t}\right) dV. \tag{3.35}$$

Equation (3.35) has the same form as (3.22) except that it is now valid for a discontinuous $\overline{H}$ field, provided that the discontinuity satisfies the boundary condition stated by (3.34). The current moment contained in (3.35) involves both the volume and surface distribution of current inside V. The integral form of the Maxwell-Ampère's law thus derived has, therefore, a broader meaning than its differential form. Most important of all, a boundary condition has been postulated.

By applying the same procedure to Faraday's law we can derive the general integral form of that law

$$\oiint_S \left(\overline{n} \times \overline{E}\right) dS = -\iiint_V \frac{\partial \overline{B}}{\partial t} dV \tag{3.36}$$

under a postulated boundary condition at S_{12}; namely,

$$\hat{n}_1 \times \left(\overline{E}_1 - \overline{E}_2\right) = 0. \tag{3.37}$$

For the Gauss law, we apply (3.25) to the two different regions to obtain

$$\iint_{S_1} \left(\hat{n} \cdot \overline{D}_1\right) dS - \iint_{S_{12}} \left(\hat{n}_1 \cdot \overline{D}_1\right) dS = \iiint_{V_1} \rho_1 dV \tag{3.38}$$

$$\iint_{S_2} \left(\hat{n} \cdot \overline{D}_2\right) dS - \iint_{S_{12}} \left(\hat{n}_2 \cdot \overline{D}_2\right) dS = \iiint_{V_2} \rho_2 dV. \tag{3.39}$$

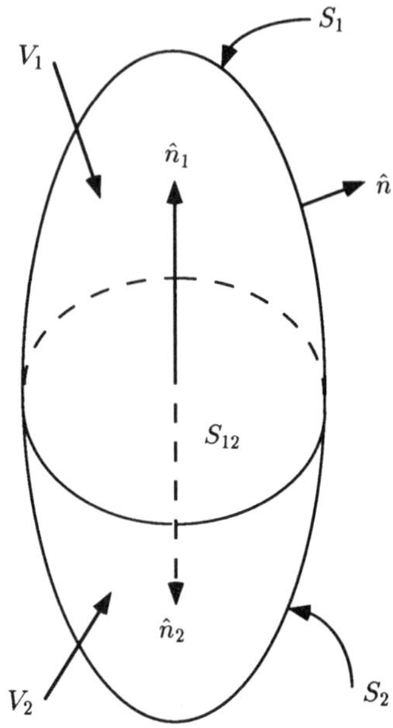

Fig. 3-1 Two adjacent regions with a boundary surface S_{12}

On S_{12}, we assume the existence of a layer of surface charge with surface charge density ρ_s. The total charge inside the entire region is then given by

$$\iiint_{V_1} \rho_1 dV + \iiint_{V_2} \rho_2 dV + \iint_{S_{12}} \rho_s dS = \iiint_V \rho dV. \tag{3.40}$$

The sum of (3.38) and (3.39) yields

$$\begin{aligned} \oiint (\hat{n} \cdot \overline{D})\, dS - \iint_{S_{12}} \hat{n}_1 \cdot (\overline{D}_1 - \overline{D}_2)\, dS \\ = \iiint_V \rho dV - \iint_{S_{12}} \rho_s dS \end{aligned} \tag{3.41}$$

We now postulate the boundary condition that at S_{12}

$$\hat{n}_1 \cdot \left(\overline{D}_1 - \overline{D}_2\right) = \rho_s; \tag{3.42}$$

then

$$\oiint_S (\hat{n} \cdot \overline{D})\, dS = \iiint_V \rho dV. \tag{3.43}$$

Equation (3.43) has the same form as (3.25) except it is now valid for a discontinuous $\overline{D}$ field provided that the discontinuity satisfies the condition stated by (3.42). The total charge represented by the volume integral in (3.43) includes both the volume and the surface distribution of charges inside V, commonly denoted by Q.

By applying the same procedure to (3.24) to the two regions we can derive the general integral form of the magnetic Gauss law; namely,

$$\oiint \left(\hat{n} \cdot \overline{B}\right) dS = 0 \tag{3.44}$$

in a region with a discontinuous B-field provided that the following boundary condition is satisfied

$$\hat{n}_1 \cdot \left(\overline{B}_1 - \overline{B}_2\right) = 0. \tag{3.45}$$

For the law of conservation of charge, we start with (3.23) for the two regions in which we assume $\overline{J}_1, \overline{J}_2$ and their first derivatives are continuous, then

$$\iint_{S_1} \hat{n} \cdot \overline{J}_1 dS - \iint_{S_{12}} \hat{n}_1 \cdot \overline{J}_1 dS = -\iiint_{V_1} \frac{\partial \rho_1}{\partial t} dV \tag{3.46}$$

$$\iint_{S_2} \hat{n} \cdot \overline{J}_2 dS - \iint_{S_{12}} \hat{n}_2 \cdot \overline{J}_2 dS = -\iiint_{V_2} \frac{\partial \rho_2}{\partial t} dV. \tag{3.47}$$

On S_{12}, we assume the existence of a surface current with density $\overline{K}$ and a surface charge with density ρ_s. The total current flowing out the entire surface, denoted by $\oiint \hat{n} \cdot \overline{J} dS$, is then given by

$$\oiint \hat{n} \cdot \overline{J} dS = \iint_{S_1} \hat{n} \cdot \overline{J}_1 dS + \iint_{S_2} \hat{n} \cdot \overline{J}_2 dS + \oint_L \hat{n} \cdot \overline{K} d\ell, \tag{3.48}$$

where L denotes the contour enclosing S_{12}. The total rate of change of the charge inside the volume, denoted by $\iiint \partial\rho/\partial t dV$, is given by

$$\iiint \frac{\partial \rho}{\partial t} dV = \iiint_{V_1} \frac{\partial \rho_1}{\partial t} dV + \iiint_{V_2} \frac{\partial \rho_2}{\partial t} dV + \iint_{S_{12}} \frac{\partial \rho_s}{\partial t} dS. \tag{3.49}$$

The sum of (3.46) and (3.47) can then be written in the form

$$\begin{aligned} &\oiint \hat{n} \cdot \overline{J} dS - \oint_L \hat{n} \cdot \overline{K} d\ell - \iint_{S_{12}} \hat{n}_1 \cdot \left(\overline{J}_1 - \overline{J}_2\right) dS \\ &= -\iiint \frac{\partial \rho}{\partial t} dV + \iint \frac{\partial \rho_s}{\partial t} dS. \end{aligned} \tag{3.50}$$

The surface Gauss theorem in vector analysis [Tai, 1992, p. 87] states that

$$\iint_{S_{12}} \nabla_S \cdot \overline{K} dS = \oint_L \overline{K} \cdot \hat{n} d\ell, \tag{3.51}$$

where $\nabla_S \cdot \overline{K}$ denotes the surface divergence of $\overline{K}$. The line integral in (3.50) therefore can be replaced by a surface integral. Now we postulate the boundary condition

$$\nabla_S \cdot \overline{K} = -\hat{n}_1 \cdot \left(\overline{J}_1 - \overline{J}_2\right) - \frac{\partial \rho_s}{\partial t} \tag{3.52}$$

then (3.50) reduces to

$$\oiint \hat{n} \cdot \overline{J} dS = -\iiint \frac{\partial \rho}{\partial t} dV. \tag{3.53}$$

Equation (3.53) has the same form as (3.23) except it is now applicable to a region containing discontinuous volume current density functions and a surface distribution of charge and a surface current. In conclusion, by postulating the boundary conditions stated by (3.34), (3.37), (3.42), (3.45), and (3.52), we can derive the general integral forms of Maxwell's theory as expressed by (3.35), (3.36), (3.43), (3.44), and (3.53).

The second approach, originally due to Schelkunoff [1972], is to *assume* that the integral form of Maxwell's equations are stated by

$$\oint \overline{E} \cdot d\overline{\ell} = -\iint \frac{\partial \overline{B}}{\partial t} \cdot d\overline{S} \tag{3.54}$$

$$\oint \overline{H} \cdot d\overline{\ell} = \iint \left(\overline{J} + \frac{\partial \overline{D}}{\partial t} \right) \cdot d\overline{S} \tag{3.55}$$

and the three equations described by (3.23) to (3.25). Equations (3.54) and (3.55) represent another integral form of Faraday's law and Ampère-Maxwell's law in contrast to our (3.36) and (3.35). Like our (3.35) and (3.36), (3.54) and (3.55) cannot be derived from (3.1) and (3.2) when $\overline{E}$ and $\overline{H}$ are discontinuous. By postulating these two equations to be valid for any field including discontinuous fields, the boundary conditions for $\overline{E}$ and $\overline{H}$ as stated by (3.34) and (3.37) can, indeed, be derived from these two integral forms. The same is true for (3.23) to (3.25) if they are assumed to be valid for discontinuous as well as continuous fields. When this approach is applied to (3.35), we consider a layer of current confined to a thin region in the form of a thin slab with thickness h and area ΔS. We let

$$\lim_{h \to 0} \overline{J} h = \overline{K}$$

and assume $\overline{D}$ to be finite in the region of integration, then (3.35) yields

$$\left(\hat{n}_1 \cdot \overline{H}_1 + \hat{n}_2 \cdot \overline{H}_2 \right) \Delta S = \overline{K} \Delta S$$

or

$$\hat{n}_1 \cdot \left(\overline{H}_1 - \overline{H}_2 \right) = \overline{K}, \tag{3.56}$$

which is the boundary condition that we have postulated in the first approach. The other conditions can be derived in a similar manner. We emphasize once more that the boundary conditions of the electromagnetic field cannot be deduced from Maxwell's original differential equations.

As far as the boundary conditions for $\overline{E}$ and $\overline{H}$ are concerned, one can start with (3.26) and (3.27) to accomplish the same result. The model based on a volume integral, however, is more convenient, because the same model is applicable to all the equations.

For convenience of reference we tabulate in Table 3-1 the boundary conditions associated with the corresponding differential equations, and the two special cases which are frequently encountered in boundary value problems. The characteristic of an ideal perfect conductor is that it cannot sustain a field inside.

TABLE 3-1 Boundary Conditions

Differential Equations	Boundary Conditions
$\nabla \times \overline{E} = -\frac{\partial \overline{B}}{\partial t}$	1. $\hat{n}_1 \times \left(\overline{E}_1 - \overline{E}_2\right) = 0$ 2. $\hat{n}_1 \times \left(\overline{E}_1 - \overline{E}_2\right) = 0$ 3. $\hat{n}_1 \times \overline{E}_1 = 0$
$\nabla \times \overline{H} = \overline{J} + \frac{\partial \overline{D}}{\partial t}$	1. $\hat{n}_1 \times \left(\overline{H}_1 - \overline{H}_2\right) = \overline{K}$ 2. $\hat{n}_1 \times \left(\overline{H}_1 - \overline{H}_2\right) = 0$ 3. $\hat{n}_1 \times \overline{H}_1 = \overline{K}$
$\nabla \cdot \overline{J} = -\frac{\partial \rho}{\partial t}$	1. $\nabla_S \cdot \overline{K} = -\hat{n}_1 \cdot \left(\overline{J}_1 - \overline{J}_2\right) - \frac{\partial \rho_s}{\partial t}$ 2. $\hat{n}_1 \cdot \left(\overline{J}_1 - \overline{J}_2\right) = -\frac{\partial \rho_s}{\partial t}$ 3. $\nabla_S \cdot \overline{K} = -\hat{n}_1 \cdot \overline{J}_1 - \frac{\partial \rho_s}{\partial t}$
$\nabla \cdot \overline{D} = \rho$	1. $\hat{n}_1 \cdot \left(\overline{D}_1 - \overline{D}_2\right) = \rho_s$ 2. $\hat{n}_1 \cdot \left(\overline{D}_1 - \overline{D}_2\right) = \rho_s$ 3. $\hat{n}_1 \cdot \overline{D}_1 = \rho_s$
$\nabla \cdot \overline{B} = 0$	1. $\hat{n}_1 \cdot \left(\overline{B}_1 - \overline{B}_2\right) = 0$ 2. $\hat{n}_1 \cdot \left(\overline{B}_1 - \overline{B}_2\right) = 0$ 3. $\hat{n}_1 \cdot \overline{B}_1 = 0$

Case 1. General boundary condition
Case 2. Neither of the two adjacent media being a perfect conductor
Case 3. Medium 2 being a perfect conductor

The unit vector $\hat{n}_1$ is pointed from the interface to medium 1.

3-4 MONOCHROMATICALLY OSCILLATING FIELDS IN FREE SPACE

When the field quantities in Maxwell's equations are harmonically oscillating functions with a single angular frequency of oscillation, ω, the system of equations can be simplified considerably. In order to avoid a possible confusion of notation due to the two possible choices of the complex time function, we will give an example. In the first place, we shall use the cosine function to describe the time-varying part. Thus the electric field is represented by

$$\begin{aligned}\overline{E}\,(x,y,z;t) = &E_{x0}\cos\left(\omega t - \Phi_x\right)\hat{x}\\ &+ E_{y0}\cos\left(\omega t - \Phi_y\right)\hat{y}\\ &+ E_{z0}\cos\left(\omega t - \Phi_z\right)\hat{z},\end{aligned} \tag{3.57}$$

where

$$\omega = 2\pi f = \text{angular frequency}.$$

The amplitude functions and the phase functions are, in general, functions of position. We introduce the complex vector function $\overline{E}(x,y,z)$ defined by

$$\overline{E}(x,y,z) = E_{x0}e^{i\Phi_x}\hat{x} + E_{y0}e^{i\Phi_y}\hat{y} + E_{z0}e^{i\Phi_z}\hat{z}; \tag{3.58}$$

then

$$\overline{E}(x,y,z;t) = Re\left[\overline{E}(x,y,z)e^{-i\omega t}\right]. \tag{3.59}$$

We use, therefore, the time function $e^{-i\omega t}$ in our work. In terms of the complex functions, Maxwell's equations in free space with a source function $\overline{J}$ can be described by

$$\nabla \times \overline{E} = i\omega\mu_0\overline{H} \tag{3.60}$$

$$\nabla \times \overline{H} = \overline{J} - i\omega\epsilon_0\overline{E} \tag{3.61}$$

$$\nabla \cdot \overline{J} = i\omega\rho \tag{3.62}$$

$$\nabla \cdot \left(\epsilon_0\overline{E}\right) = \rho \tag{3.63}$$

$$\nabla \cdot \left(\mu_0\overline{H}\right) = 0, \tag{3.64}$$

where the spatial functional dependence (x,y,z) has been omitted for simplicity. By eliminating $\overline{H}$ or $\overline{E}$ between (3.60) and (3.61), we obtain

$$\nabla \times \nabla \times \overline{E} - k^2\overline{E} = i\omega\mu_0\overline{J} \tag{3.65}$$

and

$$\nabla \times \nabla \times \overline{H} - k^2\overline{H} = \nabla \times \overline{J} \tag{3.66}$$

where $k = \omega\sqrt{\mu_0\epsilon_0} = 2\pi/\lambda$ and λ denotes the free-space wavelength. For identification purposes, equations of the type described by (3.65) or (3.66) will be designated as *inhomogeneous vector wave equations*. The entire subject of dyadic Green's technique is developed mainly to find the solutions for this type of equation under the constraint of various boundary conditions. When the domain under consideration is infinite, there are several distinct methods of finding the solution for (3.65) and (3.66). The classical method of potentials is reviewed in the following section. In Chapter 4, we will commence our discussion on dyadic Green function technique based on this classical solution.

3-5 METHOD OF POTENTIALS

In view of (3.64) and the vector identity, $\nabla \cdot \nabla \times \overline{A} = 0$, we can define a vector potential function $\overline{A}$ such that

$$\mu_0 \overline{H} = \nabla \times \overline{A}. \tag{3.67}$$

Upon substituting it into (3.60), and recognizing the identity that $\nabla \times \nabla\psi = 0$, we can introduce a scalar potential function ψ such that

$$\overline{E} = i\omega\overline{A} - \nabla\psi. \tag{3.68}$$

Substituting both (3.67) and (3.68) into (3.61), one finds that

$$\nabla \times \nabla \times \overline{A} = \mu_0 \overline{J} + k^2 \overline{A} + i\omega\mu_0\epsilon_0\nabla\psi. \tag{3.69}$$

Using identity (A.18) of Appendix A, we can change (3.69) into

$$-\nabla^2 \overline{A} + \nabla\nabla \cdot \overline{A} = \mu_0 \overline{J} + k^2 \overline{A} + i\omega\mu_0\epsilon_0\nabla\psi. \tag{3.70}$$

Now we impose the gauge condition that

$$\nabla \cdot \overline{A} = i\omega\mu_0\epsilon_0\psi. \tag{3.71}$$

Equation (3.70) then is reduced to a vector differential equation containing the vector potential function $\overline{A}$ only; that is,

$$\nabla^2 \overline{A} + k^2 \overline{A} = -\mu_0 \overline{J}. \tag{3.72}$$

In contrast to the vector wave equation defined by (3.65), we will designate (3.72) as the inhomogeneous vector Helmholtz equation. The differential equation for ψ can be obtained by taking the divergence of (3.72) and making use of (3.62) and (3.71), in which one obtains the inhomogeneous scalar wave equation

$$\nabla^2\psi + k^2\psi = -\frac{\rho}{\epsilon_0}. \tag{3.73}$$

Solutions for (3.72) and (3.73) that correspond to outgoing waves from the source are given by

$$\overline{A}(\overline{R}) = \mu_0 \iiint \overline{J}(\overline{R}')G_0\left(\overline{R}, \overline{R}'\right) dV' \tag{3.74}$$

and

$$\psi(\overline{R}) = \frac{1}{\epsilon_0} \iiint \rho(\overline{R}')G_0\left(\overline{R}, \overline{R}'\right) dV', \tag{3.75}$$

where

$$G_0\left(\overline{R}, \overline{R}'\right) = \frac{e^{ik|R-R'|}}{4\pi \mid \overline{R} - R' \mid}$$

$$\left|\overline{R} - \overline{R}'\right| = \left[(x-x')^2 + (y-y^2)^2 + (z-z')^2\right]^{\frac{1}{2}}. \tag{3.76}$$

The function $G_0(\overline{R}, \overline{R}')$ is called the free- space Green function for a three-dimensional scalar wave equation, where $\overline{R}'$ denotes the position vector of a source point and $\overline{R}$ denotes the position vector of a field point or that of an observer. A more detailed description of this function is given in Chapter 2, where the origin of this function is shown. Once we know the solution for $\overline{A}(\overline{R})$, the electromagnetic field vector $\overline{E}$ and $\overline{H}$ can be found. As a result of (3.67), (3.68), and (3.71), we have

$$\overline{E} = i\omega\left(\overline{A} + \frac{1}{k^2}\nabla\nabla\cdot\overline{A}\right) \tag{3.77}$$

and

$$\overline{H} = \frac{1}{\mu_0}\nabla\times\overline{A}. \tag{3.78}$$

Before we conclude this section a brief review will be given to the characteristics of the far-zone field. Two conditions are imposed on $\overline{R}$ to derive the expressions for the far-zone field; they are

$$|\overline{R}| \gg |\overline{R}'|$$

and

$$kR \gg 1, \quad \text{or} \quad R \gg \lambda.$$

Under these conditions the function $G_0(\overline{R}, \overline{R}')$ is given approximately by

$$G_0\left(\overline{R}, \overline{R}'\right) = \frac{e^{ik\left(R - \overline{R}\cdot\hat{R}\right)}}{4\pi R}, \tag{3.79}$$

where

$$\overline{R}'\cdot\hat{R} = R'\left[\sin\theta\sin\theta'\cos(\phi-\phi') + \cos\theta\cos\theta'\right]. \tag{3.80}$$

Hence

$$\overline{A}(\overline{R}) = \frac{\mu_0 e^{ikR}}{4\pi R}\iiint\overline{J}(\overline{R}')e^{-ik\overline{R}\cdot\hat{R}}dV' = \frac{\mu_0 e^{ikR}}{4\pi R}\overline{N}. \tag{3.81}$$

The function $\overline{N}$, called the *radiation vector*, is a function of the spherical angular variables (θ, ϕ) only and is not a function of R. Upon substitution of (3.81) into (3.77) and (3.78) and neglecting higher-order terms, one finds that

$$\overline{E} = \frac{i\omega\mu_0 e^{ikR}}{4\pi R}\overline{N}_t = \frac{ikZ_0 e^{ikR}}{4\pi R}\overline{N}_t \tag{3.82}$$

and

$$\overline{H} = \frac{ike^{ikR}}{4\pi R}\hat{R}\times\overline{N}_t = \frac{1}{Z_0}\hat{R}\times\overline{E}, \tag{3.83}$$

where $\overline{N}_t$ denotes the transverse part, with respect to $\hat{R}$, of the radiation vector; that is,

$$\overline{N}_t = \overline{N} - N_R\hat{R}. \tag{3.84}$$

The constant Z_0 denotes the free-space wave impedance, being equal to $(\mu_0/\epsilon_0)^{\frac{1}{2}} \simeq 120\pi$ ohms. The terms which have been neglected are of the order of $1/kR^2$ or higher as compared to $1/R$. Thus the far-zone electromagnetic field due to any current distribution satisfies the condition that

$$\lim_{R\to\infty}\left[\nabla\times\begin{pmatrix}\overline{E}\\\overline{H}\end{pmatrix} - ik\hat{R}\times\begin{pmatrix}\overline{E}\\\overline{H}\end{pmatrix}\right] = 0. \tag{3.85}$$

This is referred to as the *radiation condition* for an electromagnetic field in free space.

Two special cases corresponding to antennas of small size should be mentioned; they are the fields of a small electric dipole and that of a small current loop of arbitrary shape. If we let the origin of the coordinate system to be located inside a current source, and the largest linear dimension of the current source is small compared to a wavelength and also small compared to R, then $k \mid \overline{R}' \mid \ll 1$ and $\mid \overline{R}' \mid \ll R$. Under this condition the approximate expression for the vector potential is given by

$$\overline{A}(\overline{R}) = \frac{\mu_0 e^{ikR}}{4\pi R}\iiint \overline{J}(\overline{R}')dV'. \tag{3.86}$$

Of course, we assume

$$\iiint \overline{J}(\overline{R}')dV' \neq 0. \tag{3.87}$$

The volume integral of $\overline{J}(\overline{R}')$ in (3.86) is designated as the current moment, and it will be denoted by $\overline{c}$; then

$$\overline{A}(\overline{R}) = \frac{\mu_0 \overline{c} e^{ikR}}{4\pi R}, \tag{3.88}$$

with

$$\overline{c} = \iiint \overline{J}(\overline{R}')\, dV'. \tag{3.89}$$

The corresponding electromagnetic fields $\overline{E}$ and $\overline{H}$ can be calculated by means of (3.77) and (3.78). For simplicity, let the direction of $\overline{c}$ be pointed in the z-direction so

$$\overline{c} = c\hat{z} = c\left(\hat{R}\cos\theta - \hat{\theta}\sin\theta\right), \tag{3.90}$$

then we find

$$E_R = \frac{kce^{ikR}}{2\pi\omega\epsilon_0 R^2}\left(1 + \frac{i}{kR}\right)\cos\theta \tag{3.91}$$

$$E_\theta = \frac{-ik^2ce^{ikR}}{4\pi\omega\epsilon_0 R}\left(1 + \frac{i}{kR} - \frac{1}{k^2R^2}\right)\sin\theta \tag{3.92}$$

$$H_\phi = \frac{-ikce^{ikR}}{4\pi R}\left(1 + \frac{i}{kR}\right)\sin\theta. \tag{3.93}$$

When the current element is confined to a filament of length ℓ and of constant amplitude I and pointed in the z-direction, such a model corresponds to a short Hertzian dipole as shown in Fig. 3-2(a). The relationship between the current I and the positive charge q of the dipole is related by $I = -i\omega q$. Thus

$$\overline{c} = I\ell\hat{z} = -i\omega q\ell\hat{z} = -i\omega\overline{p} \tag{3.94}$$

when $\overline{p} = q\ell\hat{z}$ is designated as the dipole moment.

If the current distribution is not uniform, we can still define the dipole moment by writing

$$\overline{c} = \iiint \overline{J}(\overline{R}')dV' = -i\omega\overline{p}. \tag{3.95}$$

For example, if the current is described by a triangular distribution as shown in Fig. 3-2(b), the model is designated as the Abraham dipole, which is a good approximation of a sinusoidal current distribution when ℓ is small compared to a wavelength. In this case, one finds

$$\overline{c} = \frac{I\ell}{2}\hat{z} \quad \text{and} \quad \overline{p} = \frac{iI\ell}{2\omega}.$$

The effective dipole moment of an Abraham dipole is, therefore, equal to one half that a Hertzian dipole of the same length with a constant current I.

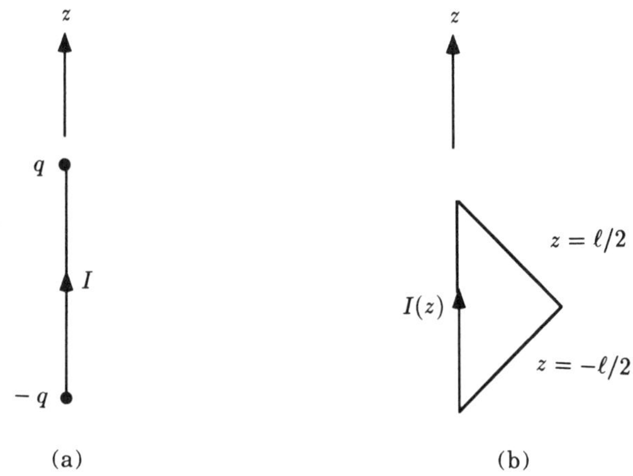

Fig. 3-2 (a) Hertzian dipole; (b) Abraham dipole

If a current distribution has a vanishing current moment, like that of a small loop carrying a circulating current with constant amplitude I, then

$$\iiint \overline{J}(\overline{R}')dV' = \oint I d\overline{\ell} = 0. \tag{3.96}$$

In this case, we need a more refined formula for $\overline{A}(\overline{R})$ to find the corresponding electromagnetic field. We assume that the largest linear dimension of the loop, which may be of arbitrary shape, is small compared to R; then

$$|\overline{R}-\overline{R}'| \doteq R\left(1-\frac{\overline{R}'\cdot\hat{R}}{R}\right).$$

Hence

$$\frac{e^{ik|\overline{R}-\overline{R}'|}}{4\pi|\overline{R}-\overline{R}'|} \doteq \frac{e^{ikR}}{4\pi R}e^{-ik\overline{R}'\cdot\hat{R}}\left(1+\frac{\overline{R}'\cdot\hat{R}}{R}\right).$$

Under this approximation we have

$$\overline{A}(\overline{R}) = \frac{\mu_0 e^{ikR}}{4\pi R}\iiint \overline{J}(\overline{R}')e^{-ik\overline{R}'\cdot\hat{R}}\left(1+\frac{\overline{R}'\cdot\hat{R}}{R}\right)dV'. \tag{3.97}$$

Since we are dealing with an electrically small loop, the exponential function in (3.97) can be approximated by $1-ik\overline{R}'\cdot\hat{R}$, which yields

$$\overline{A}(\overline{R}) = \frac{\mu_0 e^{ikR}}{4\pi R}\iiint \overline{J}(\overline{R}')\left(1-ik\overline{R}'\cdot\hat{R}\right)\left(1+\frac{\overline{R}'\cdot\hat{R}}{R}\right)dV'. \tag{3.98}$$

In view of (3.96), and by neglecting the term $-ik\left(\overline{R}'\cdot\hat{R}\right)^2/R$ in (3.98), we obtain

$$\overline{A}(\overline{R}) = \frac{\mu_0 e^{ikR}}{4\pi R}I\left(-ik+\frac{1}{R}\right)\oint \overline{R}'\cdot\hat{R}\,d\overline{\ell}'. \tag{3.99}$$

Thus the function $\overline{R}'\cdot\hat{R}$ can be converted to

$$\overline{R}'\cdot\hat{R} = \frac{\overline{R}'\cdot\overline{R}}{R} = \frac{xx'+yy'+zz'}{R}$$

and

$$\nabla'\left(\overline{R}'\cdot\hat{R}\right) = \frac{x\hat{x}+y\hat{y}+z\hat{z}}{R} = \frac{\overline{R}}{R} = \hat{R}.$$

Applying the cross-gradient theorem (A.38) of Appendix A,

$$\oint f\,d\overline{\ell}' = \iint \overline{u}_n'\times\nabla' f\,dS' = \iint d\overline{S}'\times\nabla' f,$$

with $f=\overline{R}'\cdot\hat{R}$, (3.99) reduces to

$$\begin{aligned}\overline{A}(\overline{R}) &= \frac{\mu_0 I e^{ikR}}{4\pi R}\left(-ik+\frac{1}{R}\right)\iint d\overline{S}'\times\hat{R}\\ &= \frac{\mu_0 I\overline{S}\times\hat{R}e^{ikR}}{4\pi R}\left(-ik+\frac{1}{R}\right),\end{aligned} \tag{3.100}$$

where $\overline{S}$ denotes the vectorial area of the small loop which may be of arbitrary shape. The quantity $I\overline{S}$ is designated as the magnetic dipole moment, and it will be denoted by $\overline{m}$; hence

$$\overline{A}(\overline{R}) = \frac{-ik\mu_0\overline{m} \times \hat{R}e^{ikR}}{4\pi R}\left(1 + \frac{i}{kR}\right). \tag{3.101}$$

Knowing $\overline{A}(\overline{R})$, we can find the corresponding $\overline{E}$ and $\overline{H}$ by means of (3.77) and (3.78). If we let $\overline{m}$ be pointed in the z-direction so

$$\overline{m} \times \hat{R} = m \sin\theta\hat{\phi}, \tag{3.102}$$

then we obtain

$$H_R = \frac{-ikme^{ikR}}{2\pi R^2}\left(1 + \frac{i}{kR}\right)\cos\theta \tag{3.103}$$

$$H_\theta = \frac{-k^2me^{ikR}}{4\pi R}\left(1 + \frac{i}{kR} - \frac{1}{k^2R^2}\right)\sin\theta \tag{3.104}$$

$$E_\phi = \frac{k^2Z_0me^{ikR}}{4\pi R}\left(1 + \frac{i}{kR}\right)\sin\theta. \tag{3.105}$$

The two sets of fields given by (3.91)–(3.93) and (3.103)–(3.105) demonstrate the so-called duality principle in electromagnetic theory. If we denote the first set by $\overline{E}_e$ and $\overline{H}_e$ with $\overline{p}\,(= i\overline{c}/\omega)$ as the source and the second set by $\overline{E}_m$ and $\overline{H}_m$ with $\overline{m}$ as the source, then $\overline{H}_m = -\overline{E}_e$ and $\overline{E}_m = \overline{H}_e$ with $\overline{p}$ in the first set replaced by $-\mu_0\overline{m}$ for the second set.

4

Dyadic Green Functions

4-1 MAXWELL'S EQUATIONS IN DYADIC FORM AND DYADIC GREEN FUNCTIONS OF THE ELECTRIC AND MAGNETIC TYPE

In order to introduce the concept of dyadic Green functions in electromagnetic theory in a coherent manner, we would like to elevate Maxwell's equations into a dyadic form first. We consider three sets of harmonically oscillating fields with the same frequency and in the same environment which are produced by three distinct current distributions $\bar{J}_j$ with $j = (1, 2, 3)$. Maxwell's equations for these fields can then be written in the form

$$\nabla \times \bar{E}_j = i\omega\mu_0 \bar{H}_j \tag{4.1}$$

$$\nabla \times \bar{H}_j = \bar{J}_j - i\omega\epsilon_0 \bar{E}_j \tag{4.2}$$

$$\nabla \cdot \bar{J}_j = i\omega\rho_j \tag{4.3}$$

$$\nabla \cdot \left(\epsilon_0 \bar{E}_j\right) = \rho_j \tag{4.4}$$

$$\nabla \cdot \left(\mu_0 \bar{H}_j\right) = 0. \tag{4.5}$$

The medium under consideration is assumed to be air. For other isotropic homogeneous media we simply replace the constants μ_0 and ϵ_0 by μ and ϵ. We now change the notation (x, y, z) to (x_1, x_2, x_3). By juxtaposing a unit vector $\hat{x}_j$ at the posterior position of (4.1)–(4.5) and summing the three sets of equations with respect to j, we obtain Maxwell's equations in dyadic form; namely,

$$\nabla \times \bar{\bar{E}} = i\omega\mu_0 \bar{\bar{H}} \tag{4.6}$$

$$\nabla \times \overline{\overline{H}} = \overline{\overline{J}} - i\omega\epsilon_0 \overline{\overline{E}} \tag{4.7}$$

$$\nabla \cdot \overline{\overline{J}} = i\omega\overline{\rho} \tag{4.8}$$

$$\nabla \cdot \left(\epsilon_0 \overline{\overline{E}}\right) = \overline{\rho} \tag{4.9}$$

$$\nabla \cdot \left(\mu_0 \overline{\overline{H}}\right) = 0, \tag{4.10}$$

where

$$\overline{\overline{E}} = \sum_j \overline{E}_j \hat{x}_j = \sum_i \sum_j E_{ij} \hat{x}_i \hat{x}_j \tag{4.11}$$

$$\overline{\overline{H}} = \sum_j \overline{H}_j \hat{x}_j = \sum_i \sum_j H_{ij} \hat{x}_i \hat{x}_j \tag{4.12}$$

$$\overline{\overline{J}} = \sum_j \overline{J}_j \hat{x}_j = \sum_i \sum_j J_{ij} \hat{x}_i \hat{x}_j \tag{4.13}$$

$$\overline{\rho} = \sum_j \rho_j \hat{x}_j. \tag{4.14}$$

According to the nomenclature of dyadic analysis introduced in Sec. 1-3, a dyadic function like $\overline{\overline{E}}$ has three vector components, $\overline{E}_j$ with $j = (1, 2, 3)$, and the vector charge density function $\overline{\rho}$ contains three distinct scalar charge distributions. $\overline{\rho}$ does not have the normal physical meaning of a vector quantity. For example, the magnitude of $\overline{\rho}$ does not have any physical significance.

Let us now consider the three current distributions which correspond to that of three infinitesimal electric dipoles located at $\overline{R} = \overline{R}'$ and oriented in the direction of $\hat{x}, \hat{y}, \hat{z}$ or $\hat{x}_1, \hat{x}_2, \hat{x}_3$; then

$$\overline{J}_j = c_j \delta\left(\overline{R} - \overline{R}'\right) \hat{x}_j, \; j = (1, 2, 3), \tag{4.15}$$

where c_j denotes the current moment of the dipoles; that is,

$$\iiint \overline{J}_j dv = c_j \hat{x}_j. \tag{4.16}$$

We now normalize the current moment such that

$$i\omega\mu_0 c_j = 1; \tag{4.17}$$

then

$$\begin{aligned} i\omega\mu_0 \overline{J}_j &= i\omega\mu_0 c_j \delta\left(\overline{R} - \overline{R}'\right) \hat{x}_j \\ &= \delta\left(\overline{R} - \overline{R}'\right) \hat{x}_j. \end{aligned} \tag{4.18}$$

Under this condition, we introduce a set of new notations for the various dyadic functions. They are

$$\overline{\overline{E}} = \overline{\overline{G}}_e \tag{4.19}$$

$$i\omega\mu_0\overline{\overline{H}} = \overline{\overline{G}}_m \tag{4.20}$$

$$i\omega\mu_0\overline{\overline{J}} = \overline{\overline{I}}\delta\left(\overline{R} - \overline{R}'\right) \tag{4.21}$$

$$\begin{aligned}\overline{\rho} &= \frac{1}{i\omega}\nabla\cdot\overline{\overline{J}} = \frac{-1}{\omega^2\mu_0}\nabla\cdot\left[\overline{\overline{I}}\delta\left(\overline{R} - \overline{R}'\right)\right] \\ &= \frac{-\epsilon_0}{k^2}\nabla\delta\left(\overline{R} - \overline{R}'\right),\end{aligned} \tag{4.22}$$

where

$$k = \omega\left(\mu_0\epsilon_0\right)^{\frac{1}{2}} = \frac{\omega}{c}$$

$$c = \left(\mu_0\epsilon_0\right)^{\frac{1}{2}} = \text{velocity of light in air.}$$

The expression for $\overline{\rho}$ in the form of the gradient of a delta function is a consequence of (1.46). With this change of notation, (4.6), (4.7), (4.9), and (4.10) can be written in the form

$$\nabla\times\overline{\overline{G}}_e = \overline{\overline{G}}_m \tag{4.23}$$

$$\nabla\times\overline{\overline{G}}_m = \overline{\overline{I}}\delta\left(\overline{R} - \overline{R}'\right) + k^2\overline{\overline{G}}_e \tag{4.24}$$

$$\nabla\cdot\overline{\overline{G}}_e = -\frac{1}{k^2}\nabla\delta\left(\overline{R} - \overline{R}'\right) \tag{4.25}$$

$$\nabla\cdot\overline{\overline{G}}_m = 0. \tag{4.26}$$

The relation between $\overline{\overline{J}}$ and $\overline{\rho}$ is described by (4.8). The function of $\overline{\overline{G}}_e$ so defined is designated as the dyadic Green function of the electric type or the electric dyadic Green function, and the function $\overline{\overline{G}}_m$ is designated as the dyadic Green function of the magnetic type or the magnetic dyadic Green function. If we write these two functions in the form

$$\overline{\overline{G}}_e = \sum_j \overline{G}_{ej}\hat{x}_j \tag{4.27}$$

$$\overline{\overline{G}}_m = \sum_j \overline{G}_{mj}\hat{x}_j, \tag{4.28}$$

then $\overline{G}_{ej}$ and $\overline{G}_{mj}$ denote, respectively, the vector Green function of the electric type and the vector Green function of the magnetic type. Physically, $\overline{G}_{ej}$ represents the electric field due to an infinitesimal electric dipole oriented in the direction of $\hat{x}_j$ and located at $\overline{R} = \overline{R}'$, that is,

$$\overline{\overline{G}}_e = \overline{\overline{G}}_e\left(\overline{R}, \overline{R}'\right) \tag{4.29}$$

$$\overline{\overline{G}}_m = \overline{\overline{G}}_m\left(\overline{R}, \overline{R}'\right), \tag{4.30}$$

where $\overline{R}$ denotes the position vector of the field point and $\overline{R}'$ that of the point source. Intuitively, if one knows the electromagnetic fields of three orthogonal infinitesimal electric dipoles, it is conceivable that the field due to any current distribution can be found by a quadrature. The technique of dyadic Green function is based on this premise. The physical meaning of the three vector Green functions $\overline{G}_{ej}$ is illustrated in Fig. 4-1.

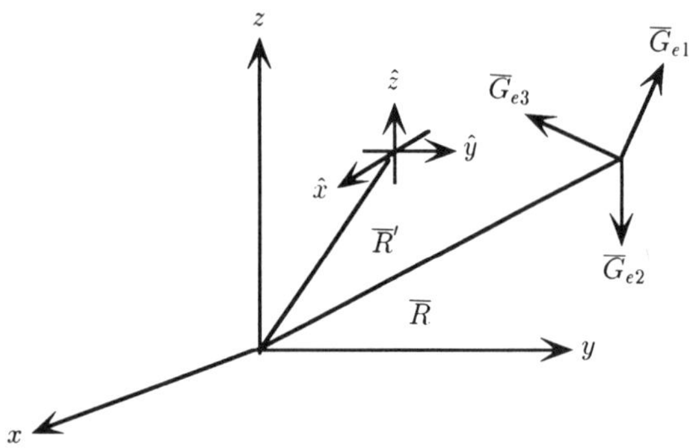

Fig. 4-1 The electric field due to three infinitesimal electric dipoles located at $\overline{R}'$ in the directions of x, y, and z

In addition to Maxwell's equations, the boundary conditions stated by (3.37) and (3.56) can also be cast into dyadic form. In particular, the boundary conditions for the tangential electric and magnetic fields will be written in the form

$$\hat{n} \times \left(\overline{E}^{+} - \overline{E}^{-}\right) = 0 \tag{4.31}$$

$$\hat{n} \times \left(\overline{H}^{+} - \overline{H}^{-}\right) = \overline{J}_s, \tag{4.32}$$

where $\hat{n}$ denotes the unit normal vector pointed from an interface to the positive side of that surface and $\overline{J}_s$ denotes the surface current density. By considering three sets of electric fields due to three orthogonal infinitesimal electric dipoles we can elevate (4.31) into a dyadic form; that is,

$$\hat{n} \times \left(\overline{\overline{G}}_e^{+} - \overline{\overline{G}}_e^{-}\right) = 0. \tag{4.33}$$

When the surface current density function $\overline{J}_s$ in (4.32) corresponds to two tangential infinitesimal electric dipoles, we can define a dyadic surface current density $\overline{\overline{J}}_s$ in the form

$$i\omega\mu_0 \overline{\overline{J}}_s = \overline{\overline{I}}_s \delta\left(\overline{r} - \overline{r}'\right), \tag{4.34}$$

where $\overline{\overline{I}}_s$ denotes the two-dimensional idem factor defined by

$$\overline{\overline{I}}_s = \overline{\overline{I}} - \hat{n}\hat{n} \tag{4.35}$$

and $\delta\left(\overline{r} - \overline{r}'\right)$ denotes the two-dimensional delta function such that

$$\iint \delta\left(\overline{r} - \overline{r}'\right) dS = 1$$

or, in general,

$$\iint f(r)\delta(\bar{r}-\bar{r}')\,dS = f(\bar{r}'),$$

where the region of integration includes the point $\bar{r}'$ on the surface. Equation (4.32) can then be elevated into a dyadic form; that is,

$$\hat{n}\times\left(\overline{\overline{G}}_m^+ - \overline{\overline{G}}_m^-\right) = \overline{\overline{I}}_s\delta(\bar{r}-\bar{r}'), \tag{4.36}$$

where $\overline{\overline{G}}_m^+$ and $\overline{\overline{G}}_m^-$ represent, respectively, $i\omega\mu_0\overline{\overline{H}}^+$ and $i\omega\mu_0\overline{\overline{H}}^-$. The two dyadic boundary conditions stated by (4.33) and (4.36) are two key relations which will be used frequently in subsequent chapters.

4-2 FREE-SPACE DYADIC GREEN FUNCTIONS

The dyadic Green function of the electric and the magnetic type satisfy (4.23) and (4.24). By eliminating one of them from these two equations, we obtain

$$\nabla\times\nabla\times\overline{\overline{G}}_e - k^2\overline{\overline{G}}_e = \overline{\overline{I}}\delta(\overline{R}-\overline{R}') \tag{4.37}$$

$$\nabla\times\nabla\times\overline{\overline{G}}_m - k^2\overline{\overline{G}}_m = \nabla\times\left[\overline{\overline{I}}\delta(\overline{R}-\overline{R}')\right]. \tag{4.38}$$

There are several methods to find the solutions for these equations in free space. One of the methods is to take advantage of the solutions of Maxwell's equations in free space by the method of potentials. According to the formulation in Sec. 3-5, when $i\omega\mu_0\overline{J}(\overline{R}) = \delta(\overline{R}-\overline{R}')\hat{x}_1$ that corresponds to the current distribution of an infinitesimal electric dipole pointed in the x_1 direction, we obtain

$$\overline{A}(\overline{R}) = \frac{1}{i\omega}G_0(\overline{R},\overline{R}')\hat{x}_1; \tag{4.39}$$

then

$$\overline{E}_{01}(\overline{R}) = \overline{G}_{e01}(\overline{R},\overline{R}') = \left(1+\frac{1}{k^2}\nabla\nabla\cdot\right)G_0(\overline{R},\overline{R}')\hat{x}_1, \tag{4.40}$$

where $\overline{G}_{e01}$ denotes the free-space vector Green function of the electric type due to a source pointed in the x_1-direction. Similarly, for sources pointed in the x_2- and in the x_3-direction, respectively, we find

$$\overline{E}_{02}(\overline{R}) = \overline{G}_{e02}(\overline{R},\overline{R}') = \left(1+\frac{1}{k^2}\nabla\nabla\cdot\right)G_0(\overline{R},\overline{R}')\hat{x}_2 \tag{4.41}$$

$$\overline{E}_{03}(\overline{R}) = \overline{G}_{e03}(\overline{R},\overline{R}') = \left(1+\frac{1}{k^2}\nabla\nabla\cdot\right)G_0(\overline{R},\overline{R}')\hat{x}_3. \tag{4.42}$$

The free-space dyadic Green function of the electric type, denoted by $\overline{\overline{G}}_{e0}$, can now be constructed by juxtaposing a unit vector $\hat{x}_i$ at the posterior position of

(4.40) to (4.42) with $i = 1, 2, 3$ and summing the three equations, which yields

$$\begin{aligned}\overline{\overline{G}}_{e0}(\overline{R}, \overline{R}') &= \sum_i \overline{G}_{e0i}(\overline{R}, \overline{R}')\hat{x}_i \\ &= \sum_i \left(1 + \frac{1}{k^2}\nabla\nabla\cdot\right) G_0(\overline{R}, \overline{R}')\hat{x}_i\hat{x}_i. \end{aligned} \tag{4.43}$$

It is understood that the summation is from $i = 1$ to 3, and

$$G_0(\overline{R}, \overline{R}') = e^{ik|\overline{R}-\overline{R}'|}/4\pi \mid \overline{R} - \overline{R}' \mid .$$

Since

$$\sum \hat{x}_i\hat{x}_i = \overline{\overline{I}}$$

and according to (1.46)

$$\nabla \cdot \left[G_0(\overline{R}, \overline{R}')\overline{\overline{I}}\right] = \nabla G_0(\overline{R}, \overline{R}'),$$

(4.43) can be written in the form

$$\overline{\overline{G}}_{e0}(\overline{R}, \overline{R}') = \left(\overline{\overline{I}} + \frac{1}{k^2}\nabla\nabla\right) G_0(\overline{R}, \overline{R}'). \tag{4.44}$$

The subscript "0" attached to $\overline{\overline{G}}_{e0}$ and G_0 represents the free-space condition that means the environment does not have any scattering object. In view of (4.23), the free-space magnetic dyadic Green function is given by

$$\begin{aligned}\overline{\overline{G}}_{m0}(\overline{R}, \overline{R}') &= \nabla \times \left[\overline{\overline{I}}G_0(\overline{R}, \overline{R}')\right] \\ &= \left[\nabla G_0(\overline{R}, \overline{R}')\right] \times \overline{\overline{I}}, \end{aligned} \tag{4.45}$$

where we have made use of the identity

$$\nabla \times (a\overline{\overline{b}}) = a\nabla \times \overline{\overline{b}} + (\nabla a) \times \overline{\overline{b}}, \tag{4.46}$$

with $a = G_0(\overline{R}, \overline{R}')$ and $\overline{\overline{b}} = \overline{\overline{I}}$. It can be verified that (4.44) and (4.45) indeed satisfy (4.23)–(4.26).

Another method in deriving the expression for $\overline{\overline{G}}_{e0}$ is due to Levine and Schwinger [1950]. Since their method will be used later to find the dyadic Green functions in a moving medium, we would like to review that method. The differential equation for $\overline{\overline{G}}_{e0}$ stated by (4.37) with $\overline{\overline{G}}_e$ therein replaced by $\overline{\overline{G}}_{e0}$ can be converted into the form

$$-\nabla^2\overline{\overline{G}}_{e0} + \nabla\nabla \cdot \overline{\overline{G}}_{e0} - k^2\overline{\overline{G}}_{e0} = \overline{\overline{I}}\delta(\overline{R} - \overline{R}') \tag{4.47}$$

because of the identity (A.26) of Appendix A, namely,

$$\nabla \times \nabla \times \bar{\bar{G}}_{e0} = -\nabla^2 \bar{\bar{G}}_{e0} + \nabla\nabla \cdot \bar{\bar{G}}_{e0}. \tag{4.48}$$

By taking the divergence of (4.37), with $\bar{\bar{G}}_e$ replaced by $\bar{\bar{G}}_{e0}$, we obtain

$$\nabla \cdot \bar{\bar{G}}_{e0} = -\frac{1}{k^2}\nabla \cdot \left[\bar{\bar{I}}\delta(\bar{R}, \bar{R}')\right] = -\frac{1}{k^2}\nabla\delta(\bar{R} - \bar{R}'). \tag{4.49}$$

Thus (4.47) can be written in the form

$$\left(\nabla^2 + k^2\right)\bar{\bar{G}}_{e0} = -\left(\bar{\bar{I}} + \frac{1}{k^2}\nabla\nabla\right)\delta(\bar{R} - \bar{R}'). \tag{4.50}$$

To find $\bar{\bar{G}}_{e0}$, we let

$$\bar{\bar{G}}_{e0} = \left(\bar{\bar{I}} + \frac{1}{k^2}\nabla\nabla\right)\psi(\bar{R}, \bar{R}'), \tag{4.51}$$

where $\psi(\bar{R}, \bar{R}')$ is a scalar function to be determined. Substituting (4.51) into (4.50) and rearranging the terms, we have

$$\left(\bar{\bar{I}} + \frac{1}{k^2}\nabla\nabla\right)\left[\left(\nabla^2 + k^2\right)\psi(\bar{R}, \bar{R}') = -\delta(\bar{R} - \bar{R}')\right]. \tag{4.52}$$

The above equation can be satisfied if $\psi(\bar{R}, \bar{R}')$ is a solution of the scalar wave equation

$$\left(\nabla^2 + k^2\right)\psi(\bar{R}, \bar{R}') = -\delta(\bar{R} - \bar{R}'). \tag{4.53}$$

In free space, the solution for $\psi(\bar{R}, \bar{R}')$, according to the discussion in Sec. 2-4, is given by

$$\psi(\bar{R}, \bar{R}') = G_0(\bar{R}, \bar{R}'); \tag{4.54}$$

hence

$$\bar{\bar{G}}_{e0}(\bar{R}, \bar{R}') = \left(\bar{\bar{I}} + \frac{1}{k^2}\nabla\nabla\right)G_0(\bar{R}, \bar{R}'), \tag{4.55}$$

which is the same as (4.44). The ingenuity of this method is evident although one has to accept the concept of the generalized functions such as $\nabla\delta(\bar{R} - \bar{R}')$ in addition to $\delta(\bar{R} - \bar{R}')$. By applying the same technique to the equation for $\bar{\bar{G}}_{m0}$, stated by (4.38), with $\bar{\bar{G}}_m$ replaced by $\bar{\bar{G}}_{m0}$ we can obtain the expression

$$\begin{aligned}\bar{\bar{G}}_{m0}(\bar{R}, \bar{R}') &= \nabla \times \left[\bar{\bar{I}}G_0(\bar{R}, \bar{R}')\right] \\ &= \nabla G_0(\bar{R}, \bar{R}') \times \bar{\bar{I}}.\end{aligned} \tag{4.56}$$

4-3 CLASSIFICATION OF DYADIC GREEN FUNCTIONS

The technique of dyadic Green function is introduced mainly to formulate various canonical electromagnetic problems in a systematic manner to avoid treatments of many special cases which can be treated as one general problem. Some typical problems are illustrated in Fig. 4-2 where (a) shows a current source in the presence of a conducting sphere located in air, (b) shows a conducting cylinder with an aperture which is excited by some source inside the cylinder, (c) shows a rectangular waveguide with a current source placed inside the guide, and (d) shows two semi-infinite isotropic media in contact, such as air and "flat" earth with a current source placed in one of the regions. If the current source in these problems has some specific distributions, we have to consider these distributions as special cases. For example, the waveguide can be excited by a transversal electric dipole or a longitudinal dipole or a magnetic dipole. In Fig. 4-2(a), if the sphere is not there, we need only the free-space dyadic Green function to study the field produced by different distributions of the current source in free space. Unless specified otherwise, we assume that for problems involving only one medium such as (a), (b), and (c) the medium is air, then the wave number k is equal to $\omega\left(\mu_0\epsilon_0\right)^{\frac{1}{2}} = 2\pi/\lambda$. The electromagnetic fields in these cases are solutions of the wave equations

$$\nabla \times \nabla \times \bar{E}(\bar{R}) - k^2\bar{E}(\bar{R}) = i\omega\mu_0\bar{J}(\bar{R}) \tag{4.57}$$

$$\nabla \times \nabla \times \bar{H}(\bar{R}) - k^2\bar{H}(\bar{R}) = \nabla \times \bar{J}\bar{R}. \tag{4.58}$$

The fields must satisfy the boundary conditions required by these problems. For example, inside a rectangular waveguide the tangential components of the electric field must be vanishing at the walls of the guide.

For problems involving two isotropic media such as the configuration shown in Fig. 4-2(d), there are two sets of fields. We denote the wave number in these two regions by $k_1 = \omega\left(\mu_1\epsilon_1\right)^{1/2}$ and $k_2 = \omega\left(\mu_2\epsilon_2\right)^{1/2}$. For a current source located in region 1 only, the two sets of wave equations are

$$\nabla \times \nabla \times \bar{E}_1(\bar{R}) - k_1^2\bar{E}_1(\bar{R}) = i\omega\mu_1\bar{J}_1(\bar{R}) \tag{4.59}$$

$$\nabla \times \nabla \times \bar{H}_1(\bar{R}) - k_1^2\bar{H}_1(\bar{R}) = \nabla \times \bar{J}_1(\bar{R}) \tag{4.60}$$

and

$$\nabla \times \nabla \times \bar{E}_2(\bar{R}) - k_2^2\bar{E}_2(\bar{R}) = 0 \tag{4.61}$$

$$\nabla \times \nabla \times \bar{H}_2(\bar{R}) - k_2^2\bar{H}_2(\bar{R}) = 0. \tag{4.62}$$

By means of the dyadic Green functions we can find the integral solutions of (4.57) and (4.58) in a very compact form. In general, we will use the notations $\bar{\bar{G}}_e$ and $\bar{\bar{G}}_m$ to denote, respectively, the electric and the magnetic dyadic Green functions; they are solutions of the dyadic differential equations

$$\nabla \times \nabla \times \bar{\bar{G}}_e(\bar{R}, \bar{R}') - k^2\bar{\bar{G}}_e(\bar{R}, \bar{R}') = \bar{\bar{I}}\delta(\bar{R} - \bar{R}') \tag{4.63}$$

$$\nabla \times \nabla \times \overline{\overline{G}}_m(\overline{R}, \overline{R}') - k^2 \overline{\overline{G}}_m(\overline{R}, \overline{R}') = \nabla \times \left[\overline{\overline{I}}\delta(\overline{R} - \overline{R}')\right]. \tag{4.64}$$

The dyadic Green functions to integrate (4.59)–(4.62) will be introduced later.

To find the integral solutions for (4.57) we can apply the second vector-dyadic Green's theorem introduced in Sec. 1-3; namely,

$$\begin{aligned}&\iiint_V \left[\overline{P} \cdot \nabla \times \nabla \times \overline{\overline{Q}} - (\nabla \times \nabla \times \overline{P}) \cdot \overline{\overline{Q}}\right] dV \\ &= - \oiint_S \hat{n} \cdot \left[\overline{P} \times \nabla \times \overline{\overline{Q}} + (\nabla \times \overline{P}) \times \overline{\overline{Q}}\right] dS.\end{aligned} \tag{4.65}$$

By letting $\overline{P} = \overline{E}(\overline{R})$ and $\overline{\overline{Q}} = \overline{\overline{G}}_e(\overline{R}, \overline{R}')$, we obtain

$$\begin{aligned}&\iiint_V \left\{\overline{E}(\overline{R}) \cdot \nabla \times \nabla \times \overline{\overline{G}}_e(\overline{R}, \overline{R}') - \left[\nabla \times \nabla \times \overline{E}(\overline{R})\right] \cdot \overline{\overline{G}}_e(\overline{R}, \overline{R}')\right\} dV \\ &= - \oiint_S \hat{n} \cdot \left\{\overline{E}(\overline{R}) \times \nabla \times \overline{\overline{G}}_e(\overline{R}, \overline{R}') + \left[\nabla \times \overline{E}(\overline{R})\right] \times \overline{\overline{G}}_e(\overline{R}, \overline{R}')\right\} dS.\end{aligned} \tag{4.66}$$

In view of (4.57) and (4.63), the above equation can be converted to

$$\begin{aligned}&\iiint_V \left\{\left[k^2 \overline{E}(\overline{R}) + i\omega\mu_0 \overline{J}(\overline{R})\right] \cdot \overline{\overline{G}}_e(\overline{R}, \overline{R}') \right. \\ &\left. - \overline{E}(\overline{R}) \cdot \left[k^2 \overline{\overline{G}}_e(\overline{R}, \overline{R}') + \overline{\overline{I}}\delta(\overline{R} - \overline{R}')\right]\right\} dV \\ &= \oiint_S \hat{n} \cdot \left\{\overline{E}(\overline{R}) \times \nabla \times \overline{\overline{G}}_e(\overline{R}, \overline{R}') \right. \\ &\left. + \left[\nabla \times \overline{E}(\overline{R})\right] \times \overline{\overline{G}}_e(\overline{R}, \overline{R}')\right\} dS.\end{aligned} \tag{4.67}$$

Two of the terms in the volume integral of (4.67) cancel each other, and

$$\iiint_V \overline{E}(\overline{R}) \cdot \overline{\overline{I}}\delta(\overline{R} - \overline{R}')\, dV = \iiint_V \overline{E}(\overline{R})\delta(\overline{R} - \overline{R}')\, dV = \overline{E}(\overline{R}'). \tag{4.68}$$

Hence

$$\begin{aligned}&\overline{E}(\overline{R}') - i\omega\mu_0 \iiint_V \overline{J}(\overline{R}) \cdot \overline{\overline{G}}_e(\overline{R}, \overline{R}') dV \\ &= - \oiint_S \hat{n} \cdot \left\{\overline{E}(\overline{R}) \times \nabla \times \overline{\overline{G}}_e(\overline{R}, \overline{R}') \right. \\ &\left. + \left[\nabla \times \overline{E}(\overline{R})\right] \times \overline{\overline{G}}_e(\overline{R}, \overline{R}')\right\} dS.\end{aligned} \tag{4.69}$$

For a Maxwellian field $\nabla \times \overline{E}(\overline{R}) = i\omega\mu_0 \overline{H}(\overline{R})$, and because of the dyadic identity

$$\overline{a} \cdot (\overline{b} \times \overline{\overline{c}}) = \overline{b} \cdot (\overline{a} \times \overline{\overline{c}}) = (\overline{a} \times \overline{b}) \cdot \overline{\overline{c}}, \tag{4.70}$$

the surface integral in (4.69) can be changed to an alternative form, namely,

$$\begin{aligned}\overline{E}(\overline{R}') - i\omega\mu_0 \iiint_V \overline{J}(\overline{R}) \cdot \overline{\overline{G}}_e(\overline{R}, \overline{R}')dV \\ = \oiint_S \Big\{ \big[i\omega\mu_0 \overline{H}(\overline{R})\big] \cdot \Big[\hat{n} \times \overline{\overline{G}}_e(\overline{R}, \overline{R}')\Big] \\ - \big[\hat{n} \times \overline{E}(\overline{R})\big] \cdot \nabla \times \overline{\overline{G}}_e(\overline{R}, \overline{R}')\Big\} dS. \qquad (4.71)\end{aligned}$$

It should be pointed out that in (4.71) $\overline{R}'$ is now the position vector of the field point and $\overline{R}$ that of a source point. Once $\overline{E}(\overline{R}')$ is known, one can readily find $\overline{H}(\overline{R}')$ using one of the Maxwell's equations. However, in order to discuss the classification of the electric dyadic Green functions which satisfy different boundary conditions, we need an integral expression for $\overline{H}(\overline{R}')$, which can be obtained by putting $\overline{P} = \overline{H}(\overline{R})$ and $\overline{\overline{Q}} = \overline{\overline{G}}_e(\overline{R}, \overline{R}')$ in (4.65), which yields

$$\begin{aligned}\iiint_V \Big\{ \overline{H}(\overline{R}) \cdot \nabla \times \nabla \times \overline{\overline{G}}_e(\overline{R}, \overline{R}') - \big[\nabla \times \nabla \times \overline{H}(\overline{R})\big] \cdot \overline{\overline{G}}_e(\overline{R}, \overline{R}')\Big\} dV \\ = - \oiint_S \hat{n} \cdot \Big\{ \overline{H}(\overline{R}) \times \nabla \times \overline{\overline{G}}_e(\overline{R}, \overline{R}') + \big[\nabla \times \overline{H}(\overline{R})\big] \times \overline{\overline{G}}_e(\overline{R}, \overline{R}')\Big\} dS. \quad (4.72)\end{aligned}$$

In view of (4.58), (4.63), and the identity (4.70), the above equation can be written in the form

$$\begin{aligned}\iiint_V \Big\{ \overline{H}(\overline{R}) \cdot \Big[k^2 \overline{\overline{G}}_e(\overline{R}, \overline{R}') + \overline{\overline{I}}\delta(\overline{R} - \overline{R}')\Big] \\ - \big[k^2 \overline{H}(\overline{R}) + \nabla \times \overline{J}(\overline{R})\big] \cdot \overline{\overline{G}}_e(\overline{R}, \overline{R}')\Big\} dV \\ = \oiint_S \Big\{ \overline{H}(\overline{R}) \cdot \Big[\hat{n} \times \nabla \times \overline{\overline{G}}_e(\overline{R}, \overline{R}')\Big] \\ - \big[\hat{n} \times \nabla \times \overline{H}(\overline{R})\big] \cdot \overline{\overline{G}}_e(\overline{R}, \overline{R}')\Big\} dS \qquad (4.73)\end{aligned}$$

or

$$\begin{aligned}\overline{H}(\overline{R}') - \iiint_V \big[\nabla \times \overline{J}(\overline{R})\big] \cdot \overline{\overline{G}}_e(\overline{R}, \overline{R}')\, dV \\ = \oiint_S \Big\{ \overline{H}(\overline{R}) \cdot \Big[\hat{n} \times \nabla \times \overline{\overline{G}}_e(\overline{R}, \overline{R}')\Big] \\ - \big[\hat{n} \times \nabla \times \overline{H}(\overline{R})\big] \cdot \overline{\overline{G}}_e(\overline{R}, \overline{R}')\Big\} dS. \qquad (4.74)\end{aligned}$$

Applying the dyadic identity

$$\nabla \cdot \big(\overline{a} \times \overline{\overline{b}}\big) = (\nabla \times \overline{a}) \cdot \overline{\overline{b}} - \overline{a} \cdot \nabla \times \overline{\overline{b}}, \qquad (4.75)$$

we can split the volume integral in (4.74) to two terms; namely,

$$\begin{aligned}&\iiint_V [\nabla \times \bar{J}(\bar{R})] \cdot \bar{\bar{G}}_e(\bar{R}, \bar{R}')dV \\ &= \iiint_V \left\{\nabla \cdot \left[\bar{J}(\bar{R}) \times \bar{\bar{G}}_e(\bar{R}, \bar{R}')\right] + \bar{J}(\bar{R}) \cdot \nabla \times \bar{\bar{G}}_e(\bar{R}, \bar{R}')\right\} dV. \qquad (4.76)\end{aligned}$$

The volume integral of the divergence of a dyadic function can be changed to a surface integral by means of the dyadic divergence theorem; that is,

$$\begin{aligned}&\iiint_V \nabla \cdot \left[\bar{J}(\bar{R}) \times \bar{\bar{G}}_e(\bar{R}, \bar{R}')\right] dV \\ &= \oiint_S \hat{n} \cdot \left[\bar{J}(\bar{R}) \times \bar{\bar{G}}_e(\bar{R}, \bar{R}')\right] dS \\ &= \oiint_S [\hat{n} \times \bar{J}(\bar{R})] \cdot \bar{\bar{G}}_e(\bar{R}, \bar{R}')dS. \qquad (4.77)\end{aligned}$$

Using these relations, we can write (4.74) in the form

$$\begin{aligned}&\bar{H}(\bar{R}') - \iiint_V \bar{J}(\bar{R}) \cdot \nabla \times \bar{\bar{G}}_e(\bar{R}, \bar{R}')dV \\ &= \oiint_S \left\{\bar{H}(\bar{R}) \cdot \left[\hat{n} \times \nabla \times \bar{\bar{G}}_e(\bar{R}, \bar{R}')\right]\right. \\ &\quad \left. - \left(\hat{n} \times \left[\nabla \times \bar{H}(\bar{R}) - \bar{J}(\bar{R})\right]\right) \cdot \bar{\bar{G}}_e(\bar{R}, \bar{R}')\right\} dS. \qquad (4.78)\end{aligned}$$

For a Maxwellian field

$$\nabla \times \bar{H}(\bar{R}) = \bar{J}(\bar{R}) - i\omega\epsilon_0 \bar{E}(\bar{R});$$

hence

$$\begin{aligned}&\bar{H}(\bar{R}') - \iiint_V \bar{J}(\bar{R}) \cdot \nabla \times \bar{\bar{G}}_e(\bar{R}, \bar{R}')dV \\ &= \oiint_S \left\{\bar{H}(\bar{R}) \cdot \left[\hat{n} \times \nabla \times \bar{\bar{G}}_e(\bar{R}, \bar{R}')\right]\right. \\ &\quad \left. -i\omega\epsilon_0 \left[\hat{n} \times \bar{E}(\bar{R})\right] \cdot \bar{\bar{G}}_e(\bar{R}, \bar{R}')\right\} dS. \qquad (4.79)\end{aligned}$$

Equation (4.79) is a companion equation to (4.71); their relationship will be revealed later.

In both (4.71) and (4.79) we have not yet specified the surface(s) enclosing the volume V. Let us consider the case that the region is bounded interiorly by a surface S_d and exteriorly by a surface S_∞ at infinity. At S_∞, $\bar{E}(\bar{R})$ and $\bar{H}(\bar{R})$ satisfy the radiation condition; namely,

$$\lim_{R\to\infty} R\left[\nabla \times \bar{E}(\bar{R}) - ik\hat{R} \times \bar{E}(\bar{R})\right] = 0 \qquad (4.80)$$

and

$$\lim_{R\to\infty} R\left[\nabla\times\overline{H}(\overline{R}) - ik\hat{R}\times\overline{H}(\overline{R})\right] = 0. \tag{4.81}$$

The electric vector Green function $\overline{G}_{ej}(\overline{R},\overline{R}')$ with $j=1,2,3$ satisfy the same condition as (4.80). By combining the three equations for $\overline{G}_{ej}(\overline{R},\overline{R}')$ to a dyadic form, we obtain the radiation condition for $\overline{\overline{G}}_e(\overline{R},\overline{R}')$; namely,

$$\lim_{R\to\infty}\left[\nabla\times\overline{\overline{G}}_e(\overline{R},\overline{R}') - ik\hat{R}\times\overline{\overline{G}}_e(\overline{R},\overline{R}')\right] = 0. \tag{4.82}$$

The function $\overline{\overline{G}}_m(\overline{R},\overline{R}')$ satisfies the same condition at infinity. As a result of the radiation condition, the surface integrals in (4.71) and (4.79) evaluated at S_∞ are equal to zero; only the contribution from S_d needs to be considered. Thus we replace S therein with S_d.

The classification of the electric dyadic Green function is based on the boundary condition placed on this function at S_d. Electric dyadic Green function of the first kind, denoted by $\overline{\overline{G}}_{e1}(\overline{R},\overline{R}')$, is required to satisfy the dyadic Dirichlet condition on S_d, namely,

$$\hat{n}\times\overline{\overline{G}}_{e1}(\overline{R},\overline{R}') = 0. \tag{4.83}$$

Under this condition (4.71) reduces to

$$\begin{aligned}\overline{E}(\overline{R}') &- i\omega\mu_0\iiint_V \overline{J}(\overline{R})\cdot\overline{\overline{G}}_{e1}(\overline{R},\overline{R}')\,dV\\ &= -\oiint_{S_d}\left[\hat{n}\times\overline{E}(\overline{R})\right]\cdot\nabla\times\overline{\overline{G}}_{e1}(\overline{R},\overline{R}')\,dS.\end{aligned} \tag{4.84}$$

If the surface S_d corresponds to that of a perfectly conducting body like the one shown in Fig. 4-2(a), then $\hat{n}\times\overline{E}(\overline{R}) = 0$ and the surface integral on S_d vanishes completely, we obtain simply

$$\overline{E}(\overline{R}') = i\omega\mu_0\iiint_V \overline{J}(\overline{R})\cdot\overline{\overline{G}}_{e1}(\overline{R},\overline{R}')\,dV. \tag{4.85}$$

Knowing $\overline{\overline{G}}_{e1}(\overline{R},\overline{R}')$, we can find $\overline{E}(\overline{R}')$. Much of the work in this book deals with the finding of this kind of dyadic Green function for bodies of simple geometrical shapes.

For a scattering body which is partly conducting, such as the conducting cylinder with an aperture shown in Fig. 4-2(b) where there is no current source outside of the cylinder, then (4.71) reduces to

$$\overline{E}(\overline{R}') = -\iint_{S_A}\left[\hat{n}\times\overline{E}(\overline{R})\right]\cdot\nabla\times\overline{\overline{G}}_{e1}(\overline{R},\overline{R}')\,dS, \tag{4.86}$$

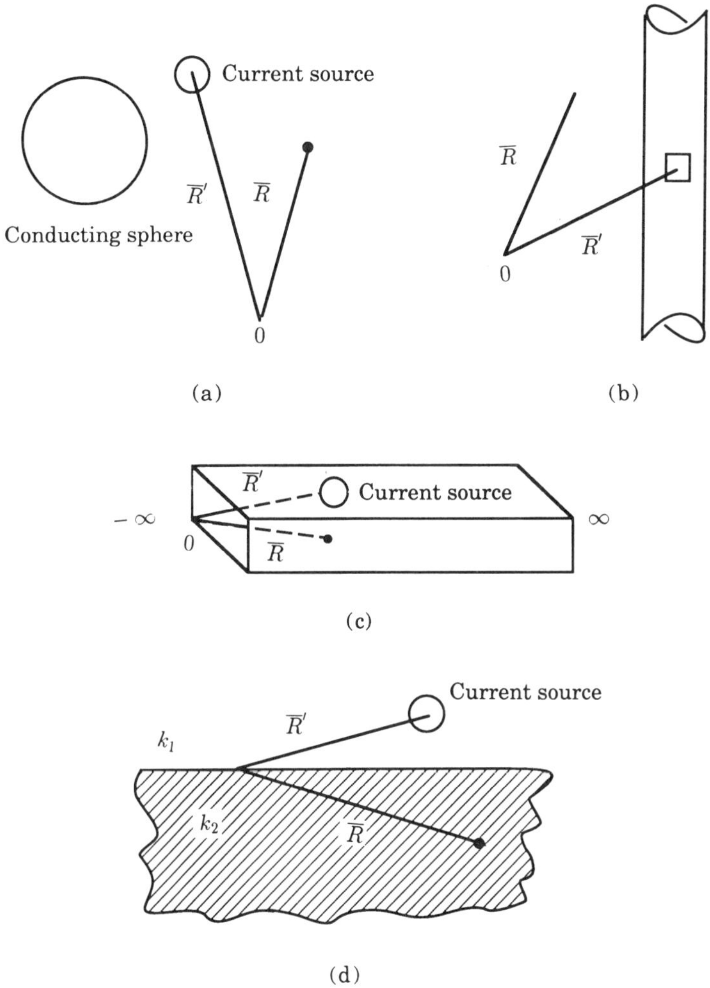

Fig. 4-2 Some typical boundary value problems

where S_A denotes the area occupied by the aperture. Given an aperture field distribution, one can calculate the field outside of the cylinder with the aid of $\overline{\overline{G}}_{e1}$ or more precisely $\nabla \times \overline{\overline{G}}_{e1}(\overline{R}, \overline{R}')$.

When the electric dyadic Green function is required to satisfy the dyadic Neumann boundary condition on S_d, namely,

$$\hat{n} \times \nabla \times \overline{\overline{G}}_{e2}(\overline{R}, \overline{R}') = 0, \tag{4.87}$$

such a function is designated as the electric dyadic Green function of the second kind, and it is denoted by $\overline{\overline{G}}_{e2}(\overline{R}, \overline{R}')$. When $\overline{\overline{G}}_{e2}$ is used in (4.79) with S_d

replacing S, it becomes

$$\overline{H}(\overline{R}') - \iiint \overline{J}(\overline{R}) \cdot \nabla \times \overline{\overline{G}}_{e2}(\overline{R}, \overline{R}')\, dV$$
$$= -i\omega\epsilon_0 \oiint_S [\hat{n} \times \overline{E}(\overline{R})] \cdot \overline{\overline{G}}_{e2}(\overline{R}, \overline{R}')\, dS. \tag{4.88}$$

For a perfectly conducting body, the surface integral vanishes; we obtain

$$\overline{H}(\overline{R}') = \iiint \overline{J}(\overline{R}) \cdot \nabla \times \overline{\overline{G}}_{e2}(\overline{R}, \overline{R}')\, dV. \tag{4.89}$$

For a partly conducting body with an aperture and without a current source outside of the body, (4.88) reduces to

$$\overline{H}(\overline{R}') = -i\omega\epsilon_0 \iint_{S_A} [\hat{n} \times \overline{E}(\overline{R})] \cdot \overline{\overline{G}}_{e2}(\overline{R}, \overline{R}')\, dS. \tag{4.90}$$

We will demonstrate later that in general (4.88) is compatible with (4.84).

The classification of the magnetic dyadic Green function $\overline{\overline{G}}_m$ can be inferred from the relationship between $\overline{\overline{G}}_e$ and $\overline{\overline{G}}_m$; namely,

$$\nabla \times \overline{\overline{G}}_e = \overline{\overline{G}}_m \tag{4.91}$$
$$\nabla \times \overline{\overline{G}}_m = \overline{\overline{I}}\delta(\overline{R} - \overline{R}') + k^2 \overline{\overline{G}}_e. \tag{4.92}$$

The magnetic dyadic Green function of the first kind, denoted by $\overline{\overline{G}}_{m1}$, is required to satisfy the boundary condition

$$\hat{n} \times \overline{\overline{G}}_{m1} = 0 \tag{4.93}$$

on S_d.

In view of (4.91) this condition corresponds to the Neumann condition for $\overline{\overline{G}}_{e2}$; therefore,

$$\nabla \times \overline{\overline{G}}_{e2} = \overline{\overline{G}}_{m1}. \tag{4.94}$$

The magnetic dyadic Green function of the second kind, denoted by $\overline{\overline{G}}_{m2}$, is required to satisfy the boundary condition

$$\hat{n} \times \nabla \times \overline{\overline{G}}_{m2} = 0 \tag{4.95}$$

on S_d. In view of (4.92) this condition corresponds to the Dirichlet condition for $\overline{\overline{G}}_{e1}$; therefore,

$$\nabla \times \overline{\overline{G}}_{m2} = \overline{\overline{I}}\delta(\overline{R} - \overline{R}') + k^2 \overline{\overline{G}}_{e1}. \tag{4.96}$$

Equations (4.94) and (4.96) are important relationships. For example, (4.96) will be used later to find the expression for $\overline{\overline{G}}_{e1}$ from the expression for $\overline{\overline{G}}_{m2}$. It happens that it is simpler to determine $\overline{\overline{G}}_{m2}$ first and then to calculate $\overline{\overline{G}}_{e1}$ through (4.96), mainly because $\overline{\overline{G}}_{m2}$ is a solenoidal dyadic function, that is,

$$\nabla \cdot \overline{\overline{G}}_{m2} = 0, \tag{4.97}$$

but $\overline{\overline{G}}_{e2}$ is nonsolenoidal.

For a problem with two media, normally isotropic, such as the one shown in Fig. 4-2(d), the dyadic Green functions involved are of the third kind, and we need a more elaborate notation for these functions by using a superscript with two numerals. There are four functions for the dyadic Green function of the electric type and another four functions for the magnetic type, denoted, respectively, by $\overline{\overline{G}}_e^{(11)}$, $\overline{\overline{G}}_e^{(12)}$, $\overline{\overline{G}}_e^{(22)}$, and $\overline{\overline{G}}_e^{(21)}$ and $\overline{\overline{G}}_m^{(11)}$, $\overline{\overline{G}}_m^{(12)}$, $\overline{\overline{G}}_m^{(22)}$, and $\overline{\overline{G}}_m^{(21)}$. For more than two media, the functions will be denoted by $\overline{\overline{G}}_e^{(ij)}$ and $\overline{\overline{G}}_m^{(ij)}$, where i and j run from one to the number of the media. Because of the presence of the superscript it is not necessary to attach a subscript "3" to indicate that the function is of the third kind. These functions are used to integrate the field equations in the two regions. If a current source is placed in region 1, the relevant wave equations for the electromagnetic fields are

$$\nabla \times \nabla \times \overline{E}_1(\overline{R}) - k_1^2 \overline{E}_1(\overline{R}) = i\omega\mu_1 \overline{J}_1(\overline{R}) \tag{4.98}$$
$$\nabla \times \nabla \times \overline{E}_2(\overline{R}) - k_2^2 \overline{E}_2(\overline{R}) = 0 \tag{4.99}$$
$$\nabla \times \nabla \times \overline{H}_1(\overline{R}) - k_1^2 \overline{H}_1(\overline{R}) = \nabla \times \overline{J}_1(\overline{R}) \tag{4.100}$$
$$\nabla \times \nabla \times \overline{H}_2(\overline{R}) - k_2^2 \overline{H}_2(\overline{R}) = 0. \tag{4.101}$$

When a current source is placed in region 2, the corresponding wave equations are

$$\nabla \times \nabla \times \overline{E}_1(\overline{R}) - k_1^2 \overline{E}_1(\overline{R}) = 0 \tag{4.102}$$
$$\nabla \times \nabla \times \overline{E}_2(\overline{R}) - k_2^2 \overline{E}_2(\overline{R}) = i\omega\mu_2 \overline{J}_2(\overline{R}) \tag{4.103}$$
$$\nabla \times \nabla \times \overline{H}_1(\overline{R}) - k_1^2 \overline{H}_1(\overline{R}) = 0 \tag{4.104}$$
$$\nabla \times \nabla \times \overline{H}_2(\overline{R}) - k_2^2 \overline{H}_2(\overline{R}) = \nabla \times \overline{J}_2(\overline{R}). \tag{4.105}$$

The wave equations for the electric dyadic Green function to be used to integrate (4.98) and (4.99) are

$$\nabla \times \nabla \times \overline{\overline{G}}_e^{(11)}(\overline{R}, \overline{R}') - k_1^2 \overline{\overline{G}}_e^{(11)}(\overline{R}, \overline{R}') = \overline{\overline{I}}\delta(\overline{R} - \overline{R}') \tag{4.106}$$

and

$$\nabla \times \nabla \times \overline{\overline{G}}_e^{(21)}(\overline{R}, \overline{R}') - k_2^2 \overline{\overline{G}}_e^{(21)}(\overline{R}, \overline{R}') = 0. \tag{4.107}$$

For (4.102) and (4.103), we need $\overline{\overline{G}}_e^{(12)}$ and $\overline{\overline{G}}_e^{(22)}$, which satisfy

$$\nabla \times \nabla \times \overline{\overline{G}}_e^{(12)}(\overline{R}, \overline{R}') - k_1^2 \overline{\overline{G}}_e^{(12)}(\overline{R}, \overline{R}') = 0 \tag{4.108}$$

$$\nabla \times \nabla \times \overline{\overline{G}}_e^{(22)}(\overline{R}, \overline{R}') - k_2^2 \overline{\overline{G}}_e^{(22)}(\overline{R}, \overline{R}') = \overline{\overline{I}}\delta(\overline{R} - \overline{R}'). \tag{4.109}$$

The superscript notation in $\overline{\overline{G}}_e^{(11)}$ means that both the field point and the source point are located in region 1. For $\overline{\overline{G}}_e^{(12)}$, it means that the field point is located in region 1 and the source point is located in region 2. The others are interpreted in a similar way. The magnetic dyadic Green functions of the third kind are related to the electric dyadic Green functions of the third kind by

$$\nabla \times \overline{\overline{G}}_e^{(ii)}(\overline{R}, \overline{R}') = \overline{\overline{G}}_m^{(ii)}(\overline{R}, \overline{R}') \tag{4.110}$$

and

$$\nabla \times \overline{\overline{G}}_m^{(ii)}(\overline{R}, \overline{R}') = \overline{\overline{I}}\delta(\overline{R} - \overline{R}') + k_i^2 \overline{\overline{G}}_e^{(ii)}(\overline{R}, \overline{R}') \tag{4.111}$$

for i = 1, 2 and

$$\nabla \times \overline{\overline{G}}_e^{(ij)}(\overline{R}, \overline{R}') = \overline{\overline{G}}_m^{(ij)}(\overline{R}, \overline{R}') \tag{4.112}$$

$$\nabla \times \overline{\overline{G}}_m^{(ij)}(\overline{R}, \overline{R}') = k_j^2 \overline{\overline{G}}_e^{(ij)}(\overline{R}, \overline{R}') \tag{4.113}$$

for $i, j = 1, 2$ with $i \neq j$. The wave equations for $\overline{\overline{G}}_m^{(ij)}$ can be obtained by taking the curl of (4.106)–(4.109) and making use of (4.110) and (4.112). With the aid of the electric dyadic Green functions of the third kind, we can find the integral solutions for (4.98)–(4.99) and (4.102)–(4.103). We apply now the second vector-dyadic Green theorem, (4.65), to region 1 with

$$\overline{P} = \overline{E}_1(\overline{R})$$
$$\overline{\overline{Q}} = \overline{\overline{G}}_e^{(11)}(\overline{R}, \overline{R}'),$$

where $\overline{E}_1(\overline{R})$ satisfies (4.98) and $\overline{\overline{G}}_e^{(11)}(\overline{R}, \overline{R}')$ satisfies (4.106). After deleting the surface integral at infinity and making use of the differential equations for these two functions, we obtain

$$\begin{aligned} \overline{E}_1(\overline{R}') - i\omega\mu_1 \iiint_{V_1} \overline{J}_1(\overline{R}) \cdot \overline{\overline{G}}_e^{(11)}(\overline{R}, \overline{R}') dV \\ = -\iint_S \Big\{ \left[\hat{n}_1 \times \nabla \times \overline{E}_1(\overline{R})\right] \cdot \overline{\overline{G}}_e^{(11)}(\overline{R}, \overline{R}') \\ + \left[\hat{n}_1 \times \overline{E}_1(\overline{R})\right] \cdot \nabla \times \overline{\overline{G}}_e^{(11)}(\overline{R}, \overline{R}') \Big\} dS, \end{aligned} \tag{4.114}$$

where V_1 denotes the volume in region 1 and S the interface of the two media, $\hat{n}_1$ being the unit normal vector to the interface pointed away from region 1. We are now going to show that the surface integral in (4.114) vanishes. To prove this identity we need an integral expression involving $\overline{E}_2(\overline{R})$ and $\overline{\overline{G}}_e^{(21)}$, which satisfy, respectively, (4.99) and (4.107). By applying (4.65) to region 2 with

$$\overline{P} = \overline{E}_2(\overline{R})$$
$$\overline{\overline{Q}} = \overline{\overline{G}}_e^{(21)}(\overline{R}, \overline{R}')$$

we obtain, after deleting the surface integral at infinity,

$$\iint_S \left\{ \left[\hat{n}_1 \times \nabla \times \overline{E}_2(\overline{R})\right] \cdot \overline{\overline{G}}_e^{(21)}(\overline{R}, \overline{R}') + \left[\hat{n}_1 \times \overline{E}_2(\overline{R})\right] \cdot \nabla \times \overline{\overline{G}}_e^{(21)}(\overline{R}, \overline{R}') \right\} = 0. \tag{4.115}$$

At the interface, the electromagnetic field and the corresponding dyadic Green function satisfy the following boundary conditions:

$$\hat{n}_1 \times \left[\overline{E}_1(\overline{R}) - \overline{E}_2(\overline{R})\right] = 0 \tag{4.116}$$
$$\hat{n}_1 \times \left[\overline{H}_1(\overline{R}) - \overline{H}_2(\overline{R})\right] = 0 \tag{4.117}$$
$$\hat{n}_1 \times \left[\overline{\overline{G}}_e^{(11)}(\overline{R}, \overline{R}') - \overline{\overline{G}}_e^{(21)}(\overline{R}, \overline{R}')\right] = 0 \tag{4.118}$$
$$\hat{n}_1 \times \left[\frac{\nabla \times \overline{\overline{G}}_e^{(11)}(\overline{R}, \overline{R}')}{\mu_1} - \frac{\nabla \times \overline{\overline{G}}_e^{(21)}(\overline{R}, \overline{R}')}{\mu_2}\right] = 0. \tag{4.119}$$

Equation (4.118) is a dyadic version of (4.116) for point sources and (4.119) that of (4.117). The presence of the magnetic constants μ_1 and μ_2 in (4.119) is due to the fact that

$$\nabla \times \overline{E}_1(\overline{R}) = i\omega\mu_1 \overline{H}_1(\overline{R}) \tag{4.120}$$
$$\nabla \times \overline{E}_2(\overline{R}) = i\omega\mu_2 \overline{H}_2(\overline{R}), \tag{4.121}$$

but by definition,

$$\nabla \times \overline{\overline{G}}_e^{(11)}(\overline{R}, \overline{R}') = \overline{\overline{G}}_m^{(11)}(\overline{R}, \overline{R}') \tag{4.122}$$
$$\nabla \times \overline{\overline{G}}_e^{(21)}(\overline{R}, \overline{R}') = \overline{\overline{G}}_m^{(21)}(\overline{R}, \overline{R}'). \tag{4.123}$$

As a result of these boundary conditions, the surface integral in (4.114) is identical to the one in (4.115) except a difference in sign; hence, it also vanishes. Equation (4.114), therefore, reduces to

$$\overline{E}_1(\overline{R}') = i\omega\mu_1 \iiint_{V_1} \overline{J}_1(\overline{R}) \cdot \overline{\overline{G}}_e^{(11)}(\overline{R}, \overline{R}')\, dV, \tag{4.124}$$

where $\overline{R}'$ now denotes the position vector of a field point and $\overline{R}$ that of a point inside the current source.

To determine $\overline{E}_2(\overline{R})$ for a current source placed in region 1 we apply (4.65) to region 2 with $\overline{P} = \overline{E}_2(\overline{R}), \overline{\overline{Q}} = \overline{\overline{G}}_e^{(22)}(\overline{R}, \overline{R}')$, where these two functions satisfy, respectively, (4.99) and (4.109). The result yields

$$\overline{E}_2(\overline{R}') = \iint_S \Big\{ \big[\hat{n}_1 \times \nabla \times \overline{E}_2(\overline{R})\big] \cdot \overline{\overline{G}}_e^{(22)}(\overline{R}, \overline{R}') + \big[\hat{n}_1 \times \overline{E}_2(\overline{R})\big] \cdot \nabla \times \overline{\overline{G}}_e^{(22)}(\overline{R}, \overline{R}') \Big\} \, dS, \tag{4.125}$$

where we have already deleted the surface integral at infinity. In addition to (4.116) and (4.117) the boundary conditions for $\overline{\overline{G}}_e^{(22)}$ and $\nabla \times \overline{\overline{G}}_e^{(22)}$ are

$$\hat{n}_1 \times \Big[\overline{\overline{G}}_e^{(22)}(\overline{R}, \overline{R}') - \overline{\overline{G}}_e^{(12)}(\overline{R}, \overline{R}')\Big] = 0 \tag{4.126}$$

and

$$\hat{n}_1 \times \left[\frac{\nabla \times \overline{\overline{G}}_e^{(22)}(\overline{R}, \overline{R}')}{\mu_2} - \frac{\nabla \times \overline{\overline{G}}_e^{(12)}(\overline{R}, \overline{R}')}{\mu_1} \right] = 0. \tag{4.127}$$

Hence, (4.125) is equivalent to

$$\overline{E}_2(\overline{R}') = \frac{\mu_2}{\mu_1} \iint_S \Big\{ \big[\hat{n}_1 \times \nabla \times \overline{E}_1(\overline{R})\big] \cdot \overline{\overline{G}}_e^{(12)}(\overline{R}, \overline{R}') + \big[\hat{n}_1 \times \overline{E}_1(\overline{R})\big] \cdot \nabla \times \overline{\overline{G}}_e^{(12)}(\overline{R}, \overline{R}') \Big\} \, dS. \tag{4.128}$$

By applying (4.65) to region 1 with $\overline{P} = \overline{E}_1(\overline{R})$ and $\overline{\overline{Q}} = \overline{\overline{G}}_e^{(12)}(\overline{R}, \overline{R}')$ where these two functions satisfy, respectively, (4.98) and (4.108), we obtain

$$\begin{aligned} i\omega\mu_1 \iiint_{V_1} & \overline{J}_1(\overline{R}) \cdot \overline{\overline{G}}_e^{(12)}(\overline{R}, \overline{R}') dV \\ = \iint_S & \Big\{ \big[\hat{n}_1 \times \nabla \times \overline{E}_1(\overline{R})\big] \cdot \overline{\overline{G}}_e^{(12)}(\overline{R}, \overline{R}') \\ & \big[\hat{n}_1 \times \overline{E}_1(\overline{R})\big] \cdot \nabla \times \overline{\overline{G}}_e^{(12)}(\overline{R}, \overline{R}') \Big\} \, dS. \end{aligned} \tag{4.129}$$

The two surface integrals in (4.128) and (4.129) are identical; hence, the ratio of these two equations yields

$$\overline{E}_2(\overline{R}') = i\omega\mu_2 \iiint_{V_1} \overline{J}_1(\overline{R}) \cdot \overline{\overline{G}}_e^{(12)}(\overline{R}, \overline{R}') \, dS. \tag{4.130}$$

The significance of the two electric dyadic Green functions $\overline{\overline{G}}_e^{(11)}$ and $\overline{\overline{G}}_e^{(12)}$ is now very clear by looking at (4.124) and (4.130). The formulation discussed here is analogous to the one for a composite transmission line made of two sections of semi-infinite lines with different line constants. It should be observed that the magnetic constant μ_1 is involved in (4.124) while μ_2 appears in (4.130). When the media under consideration are nonmagnetic $\mu_1 = \mu_2 = \mu_0$.

Since the designation of regions 1 and 2 is quite arbitrary, by interchanging "1" with "2" in (4.124) and (4.130), we can obtain the solutions for (4.102) and (4.103) for a current source placed in region 2. They are

$$\bar{E}_2(\bar{R}') = i\omega\mu_2 \iiint_{V_2} \bar{J}_2(\bar{R}) \cdot \bar{\bar{G}}_e^{(22)}(\bar{R}, \bar{R}')\, dV \tag{4.131}$$

and

$$\bar{E}_1(\bar{R}') = i\omega\mu_1 \iiint_{V_2} \bar{J}_2(\bar{R}) \cdot \bar{\bar{G}}_e^{(21)}(\bar{R}, \bar{R}')\, dV. \tag{4.132}$$

All four electric dyadic Green functions of the third kind have now been used in the formulation.

Although we have derived the integral solutions of various types of problems using the dyadic Green function of the first, the second, and the third kind, the only inconvenience is that in (4.124), (4.130), (4.131), and (4.132) $\bar{R}'$ is being used to denote the position vector of the field point and $\bar{R}$ for a typical point inside the source in the final form of these solutions. This inconvenience is not only a matter of notation, but it involves a deeper implication. For example, if we interchange $\bar{R}'$ with $\bar{R}$ in (4.131), we obtain

$$\bar{E}_2(\bar{R}) = i\omega\mu_2 \iiint_{V_2} \bar{J}_2(\bar{R}') \cdot \bar{\bar{G}}_e^{(22)}(\bar{R}', \bar{R})\, dV'. \tag{4.133}$$

In (4.133), $\bar{R}$ now becomes the position vector of a field point while $\bar{R}'$ becomes that of a source point. This is very desirable; it is a conventional notation. However, the order of the two position vectors in $\bar{\bar{G}}_e^{(22)}(\bar{R}', \bar{R})$ implies that the function satisfies the differential equation

$$\nabla' \times \nabla' \times \bar{\bar{G}}_e^{(22)}(\bar{R}', \bar{R}) - k_2^2 \bar{\bar{G}}_3^{(22)}(\bar{R}' - \bar{R}) = \bar{\bar{I}}\delta(\bar{R}' - \bar{R}). \tag{4.134}$$

In other words it is defined in a primed coordinate system with a source located at $\bar{R}$. In order to use functions defined in the original unprimed system we need the symmetrical relationship between $\bar{\bar{G}}_e^{(22)}(\bar{R}, \bar{R}')$ and $\bar{\bar{G}}_e^{(22)}(\bar{R}', \bar{R})$ so that we can convert (4.133) to a form involving $\bar{\bar{G}}_e^{(22)}(\bar{R}, \bar{R}')$. The symmetrical relationships of the dyadic Green function are not merely mathematical conversions; they are intimately connected with the reciprocity theorems in electromagnetic theory. The subject is discussed in the following two sections.

4-4 SYMMETRICAL PROPERTIES OF DYADIC GREEN FUNCTIONS

The symmetrical relationships of the free-space dyadic Green functions $\overline{\overline{G}}_{e0}$ and $\overline{\overline{G}}_{m0}$ can be derived by using the explicit expressions for these two functions derived in Sec. 4-2 which are repeated here,

$$\overline{\overline{G}}_{e0}(\overline{R}, \overline{R}') = \left(\overline{\overline{I}} + \frac{1}{k^2}\nabla\nabla\right) G_0(\overline{R}, \overline{R}') \tag{4.135}$$

$$\overline{\overline{G}}_{m0}(\overline{R}, \overline{R}') = \nabla \times \left[\overline{\overline{I}} G_0(\overline{R}, \overline{R}')\right], \tag{4.136}$$

where

$$G_0(\overline{R}, \overline{R}') = e^{ik|\overline{R}-\overline{R}'|}/4\pi\left|\overline{R} - \overline{R}'\right|. \tag{4.137}$$

If we denote the gradient operator in the primed variables (x', y', z') by ∇', then $\nabla' G_0 = -\nabla G_0$ and $\nabla'\nabla' G_0 = \nabla\nabla G_0$. By interchanging $\overline{R}$ and $\overline{R}'$ in (4.135), we have

$$\begin{aligned} \overline{\overline{G}}_{e0}(\overline{R}', \overline{R}) &= \left(\overline{\overline{I}} + \frac{1}{k^2}\nabla'\nabla'\right) G_0(\overline{R}', \overline{R}) \\ &= \left(\overline{\overline{I}} + \frac{1}{k^2}\nabla\nabla\right) G_0(\overline{R}, \overline{R}') \\ &= \overline{\overline{G}}_{e0}(\overline{R}, \overline{R}') \end{aligned} \tag{4.138}$$

Since $\overline{\overline{G}}_{e0}(\overline{R}, \overline{R}')$ or $\overline{\overline{G}}_{e0}(\overline{R}', \overline{R})$ is a symmetrical dyadic,

$$\left[\overline{\overline{G}}_{e0}(\overline{R}', \overline{R})\right]^T = \overline{\overline{G}}_{e0}(\overline{R}', \overline{R}); \tag{4.139}$$

hence

$$\left[\overline{\overline{G}}_{e0}(\overline{R}', \overline{R})\right]^T = \overline{\overline{G}}_{e0}(\overline{R}, \overline{R}'). \tag{4.140}$$

This is the symmetrical relationship between $\overline{\overline{G}}_{e0}(\overline{R}', \overline{R})$ and $\overline{\overline{G}}_{e0}(\overline{R}, \overline{R}')$. In the case of $\overline{\overline{G}}_{m0}$, by interchanging $\overline{R}$ and $\overline{R}'$ in (4.136), we have

$$\begin{aligned} \overline{\overline{G}}_{m0}(\overline{R}', \overline{R}) &= \nabla' \times \left[\overline{\overline{I}} G_0(\overline{R}', \overline{R})\right] \\ &= -\nabla \times \left[\overline{\overline{I}} G_0(\overline{R}, \overline{R}')\right] = -\overline{\overline{G}}_{m0}(\overline{R}, \overline{R}'). \end{aligned} \tag{4.141}$$

Now $\overline{\overline{G}}_{m0}(\overline{R}', \overline{R})$ is an antisymmetrical dyadic, so

$$\left[\overline{\overline{G}}_{m0}(\overline{R}', \overline{R})\right]^T = -\overline{\overline{G}}_{m0}(\overline{R}', \overline{R}); \tag{4.142}$$

hence

$$\left[\overline{\overline{G}}_{m0}(\overline{R}', \overline{R})\right]^T = \overline{\overline{G}}_{m0}(\overline{R}, \overline{R}'). \tag{4.143}$$

This is the symmetrical relationship between $\overline{\overline{G}}_{m0}(\overline{R}', \overline{R})$ and $\overline{\overline{G}}_{m0}(\overline{R}, \overline{R}')$.

In the absence of a scattering body, the surface integrals in (4.71) and (4.79) are absent and the functions $\overline{\overline{G}}_e$ and $\overline{\overline{G}}_m$ therein correspond to $\overline{\overline{G}}_{e0}(\overline{R}', \overline{R})$ and $\overline{\overline{G}}_{m0}(\overline{R}', \overline{R})$; hence

$$\overline{E}(\overline{R}') = i\omega\mu_0 \iiint \overline{J}(\overline{R}) \cdot \overline{\overline{G}}_{e0}(\overline{R}, \overline{R}')\, dV \tag{4.144}$$

$$\begin{aligned}\overline{H}(\overline{R}') &= \iiint \overline{J}(\overline{R}) \cdot \nabla \times \overline{\overline{G}}_{e0}(\overline{R}, \overline{R}')\, dV \\ &= \iiint \overline{J}(\overline{R}) \cdot \overline{\overline{G}}_{m0}(\overline{R}, \overline{R}')\, dV.\end{aligned} \tag{4.145}$$

By interchanging $\overline{R}$ and $\overline{R}'$ in the above two expressions and making use of (4.140) and (4.143), we find

$$\begin{aligned}\overline{E}(\overline{R}) &= i\omega\mu_0 \iiint \overline{J}(\overline{R}') \cdot \overline{\overline{G}}_{e0}(\overline{R}', \overline{R})\, dV' \\ &= i\omega\mu_0 \iiint \left[\overline{\overline{G}}_{e0}(\overline{R}', \overline{R})\right]^T \cdot \overline{J}(\overline{R}')\, dV';\end{aligned}$$

thus

$$\overline{E}(\overline{R}) = i\omega\mu_0 \iiint \overline{\overline{G}}_{e0}(\overline{R}, \overline{R}') \cdot \overline{J}(\overline{R}')\, dV'. \tag{4.146}$$

Similarly,

$$\begin{aligned}\overline{H}(\overline{R}) &= \iiint \overline{J}(\overline{R}') \cdot \overline{\overline{G}}_{m0}(\overline{R}', \overline{R})\, dV' \\ &= \iiint \left[\overline{\overline{G}}_{m0}(\overline{R}', \overline{R})\right]^T \cdot \overline{J}(\overline{R}')\, dV' \\ &= \iiint \overline{\overline{G}}_{m0}(\overline{R}, \overline{R}') \cdot \overline{J}(\overline{R}')\, dV',\end{aligned} \tag{4.147}$$

or

$$\overline{H}(\overline{R}) = \iiint \nabla \times \overline{\overline{G}}_{e0}(\overline{R}, \overline{R}') \cdot \overline{J}(\overline{R}')\, dV'. \tag{4.148}$$

Equations (4.146) and (4.148) are the "standard" expressions which can be used to calculate the electromagnetic field in free space for a current distribution $\overline{J}(\overline{R}')$, where $\overline{R}$ is now the position vector commonly adopted for a field point and $\overline{R}'$ is that of a source point. It is also evident that

$$\nabla \times \overline{E}(\overline{R}) = i\omega\mu_0 \overline{H}(\overline{R})$$

and

$$\begin{aligned}\nabla \times \overline{H}(\overline{R}) &= \iiint \nabla \times \nabla \times \overline{\overline{G}}_{e0}(\overline{R}, \overline{R}') \cdot \overline{J}(\overline{R}')\, dV' \\ &= \iiint \left[k^2 \overline{\overline{G}}_{e0}(\overline{R}, \overline{R}') + \overline{\overline{I}}\delta\left(\overline{R} - \overline{R}'\right)\right] \cdot \overline{J}(\overline{R}')\, dV'\end{aligned}$$

$$= \bar{J}(\bar{R}) + k^2 \iiint \bar{\bar{G}}_{e0}(\bar{R}, \bar{R}') \cdot \bar{J}(\bar{R}')\, dV'$$
$$= \bar{J}(\bar{R}) - i\omega\epsilon_0 \bar{E}(\bar{R}).$$

The expressions for $\bar{E}(\bar{R})$ and $\bar{H}(\bar{R})$ given by (4.146) and (4.148), therefore, do satisfy Maxwell's equations.

The symmetrical property of the free-space dyadic Green functions has been shown using the known expressions for these functions. However, if the expressions for the other kinds of functions such as $\bar{\bar{G}}_{e1}$, $\bar{\bar{G}}_{e2}$, and $\bar{\bar{G}}_e^{(ij)}$ are not yet known, it is still possible to derive their symmetrical relationships based on a general method. The tool to be used is the dyadic-dyadic Green theorem of the second kind introduced in Sec. 1-3, namely,

$$\iiint_V \left\{ \left[\nabla \times \nabla \times \bar{\bar{Q}}\right]^T \cdot \bar{\bar{P}} - \left[\bar{\bar{Q}}\right]^T \cdot \nabla \times \nabla \times \bar{\bar{P}} \right\} dV$$
$$= \oiint_S \left\{ \left[\hat{n} \times \bar{\bar{Q}}\right]^T \cdot \nabla \times \bar{\bar{P}} - \left[\nabla \times \bar{\bar{Q}}\right]^T \cdot \left(\hat{n} \times \bar{\bar{P}}\right) \right\} dS. \tag{4.149}$$

Let us apply this theorem to two dyadic functions with

$$\bar{\bar{P}} = \bar{\bar{G}}_{e1}(\bar{R}, \bar{R}_a) \tag{4.150}$$

and

$$\bar{\bar{Q}} = \bar{\bar{G}}_{e1}(\bar{R}, \bar{R}_b), \tag{4.151}$$

where $\bar{R}_a$ and $\bar{R}_b$ denote, respectively, the position vectors of two point sources at different locations and $\bar{\bar{G}}_{e1}$ represents the electric dyadic Green function of the first kind for a certain problem such as the one shown in Fig. 4-1(a) or (b). By definition, the two functions are solutions of the wave equations

$$\nabla \times \nabla \times \bar{\bar{G}}_{e1}(\bar{R}, \bar{R}_a) - k^2 \bar{\bar{G}}_{e1}(\bar{R}, \bar{R}_a) = \bar{\bar{I}}\delta(\bar{R} - \bar{R}_a) \tag{4.152}$$
$$\nabla \times \nabla \times \bar{\bar{G}}_{e1}(\bar{R}, \bar{R}_b) - k^2 \bar{\bar{G}}_{e1}(\bar{R}, \bar{R}_b) = \bar{\bar{I}}\delta(\bar{R} - \bar{R}_b), \tag{4.153}$$

and they both satisfy the radiation condition at infinity and the Dirichlet boundary condition $\hat{n} \times \bar{\bar{G}}_{e1} = 0$ on S_d.

Substituting (4.150) and (4.151) into (4.149) and making use of (4.152) and (4.153), we obtain

$$\iiint_V \left\{ \left[k^2 \left[\bar{\bar{G}}_{e1}(\bar{R}, \bar{R}_b)\right]^T + \bar{\bar{I}}\delta(\bar{R} - \bar{R}_b)\right] \cdot \bar{\bar{G}}_{e1}(\bar{R}, \bar{R}_a) \right.$$
$$\left. - \left[\bar{\bar{G}}_{e1}(\bar{R}, \bar{R}_b)\right]^T \cdot \left[k^2 \bar{\bar{G}}_{e1}(\bar{R}, \bar{R}_a) + \bar{\bar{I}}\delta(\bar{R} - \bar{R}_a)\right] \right\} dV$$
$$= \oiint_S \left\{ \left[\hat{n} \times \bar{\bar{G}}_{e1}(\bar{R}, \bar{R}_b)\right]^T \cdot \nabla \times \bar{\bar{G}}_{e1}(\bar{R}, \bar{R}_a) \right.$$
$$\left. - \left[\nabla \times \bar{\bar{G}}_{e1}(\bar{R}, \bar{R}_b)\right]^T \cdot \left[\hat{n} \times \bar{\bar{G}}_{e1}(\bar{R}, \bar{R}_a)\right] \right\} dS. \tag{4.154}$$

The surface integral in the above equation vanishes because of the radiation condition at S_∞ and the Dirichlet boundary condition on S_d, and the volume integral yields

$$\left[\overline{\overline{G}}_{e1}(\overline{R}_a, \overline{R}_b)\right]^T = \overline{\overline{G}}_{e1}(\overline{R}_b, \overline{R}_a). \tag{4.155}$$

By changing $\overline{R}_a$ and $\overline{R}_b$ to $\overline{R}'$ and $\overline{R}$, respectively, we have

$$\left[\overline{\overline{G}}_{e1}(\overline{R}', \overline{R})\right]^T = \overline{\overline{G}}_{e1}(\overline{R}, \overline{R}'). \tag{4.156}$$

This is the symmetrical relationship for the electric dyadic Green function of the first kind. Since $\overline{\overline{G}}_{e1}$ and $\overline{\overline{G}}_{m2}$ satisfy the relation,

$$\nabla \times \overline{\overline{G}}_{m2}(\overline{R}, \overline{R}') = \overline{\overline{I}}\delta(\overline{R} - \overline{R}') + k^2\overline{\overline{G}}_{e1}(\overline{R}, \overline{R}') \tag{4.157}$$

and

$$\left[\overline{\overline{I}}\right]^T = \overline{\overline{I}}. \tag{4.158}$$

Equation (4.156) implies

$$\left[\nabla' \times \overline{\overline{G}}_{m2}(\overline{R}', \overline{R})\right]^T = \nabla \times \overline{\overline{G}}_{m2}(\overline{R}, \overline{R}'). \tag{4.159}$$

Following a similar procedure it can be shown that

$$\left[\overline{\overline{G}}_{e2}(\overline{R}', \overline{R})\right]^T = \overline{\overline{G}}_{e2}(\overline{R}, \overline{R}') \tag{4.160}$$

or, equivalently,

$$\left[\nabla' \times \overline{\overline{G}}_{m1}(\overline{R}', \overline{R})\right]^T = \nabla \times \overline{\overline{G}}_{m1}(\overline{R}, \overline{R}'). \tag{4.161}$$

There are two more relationships involving $\nabla \times \overline{\overline{G}}_{e1}$ and $\nabla \times \overline{\overline{G}}_{e2}$ or, equivalently, $\overline{\overline{G}}_{m2}$ and $\overline{\overline{G}}_{m1}$ which need to be derived. For that purpose, we let

$$\overline{\overline{P}} = \overline{\overline{G}}_{e1}(\overline{R}, \overline{R}_a) \tag{4.162}$$

$$\overline{\overline{Q}} = \nabla \times \overline{\overline{G}}_{e2}(\overline{R}, \overline{R}_b). \tag{4.163}$$

These two functions are required to satisfy the boundary conditions

$$\hat{n} \times \overline{\overline{G}}_{e1}(\overline{R}, \overline{R}_a) = 0 \tag{4.164}$$

$$\hat{n} \times \nabla \times \overline{\overline{G}}_{e2}(\overline{R}, \overline{R}_b) = 0 \tag{4.165}$$

on the surface of a diffracting body, denoted by S_d. By substituting (4.162) and (4.163) into (4.149), we obtain

$$\begin{aligned}\iiint_V \Big\{ & \left[\nabla \times \nabla \times \nabla \times \overline{\overline{G}}_{e2}(\overline{R}, \overline{R}_b)\right]^T \cdot \overline{\overline{G}}_{e1}(\overline{R}, \overline{R}_a) \\ & - \left[\nabla \times \overline{\overline{G}}_{e2}(\overline{R}, \overline{R}_b)\right]^T \cdot \nabla \times \nabla \times \overline{\overline{G}}_{e1}(\overline{R}, \overline{R}_a) \Big\}\, dV \\ = \oiint_S \Big\{ & \left[\hat{n} \times \nabla \times \overline{\overline{G}}_{e2}(\overline{R}, \overline{R}_b)\right]^T \cdot \nabla \times \overline{\overline{G}}_{e1}(\overline{R}, \overline{R}_a) \\ & - \left[\nabla \times \nabla \times \overline{\overline{G}}_{e2}(\overline{R}, \overline{R}_b)\right]^T \cdot \left[\hat{n} \times \overline{\overline{G}}_{e1}(\overline{R}, \overline{R}_a)\right] \Big\}\, dS. \qquad (4.166)\end{aligned}$$

Since both $\overline{\overline{G}}_{e1}$ and $\overline{\overline{G}}_{e2}$ are solutions of the dyadic wave equation (4.152) and (4.153) with $\overline{\overline{G}}_{e1}(\overline{R}, \overline{R}_b)$ therein replaced by $\overline{\overline{G}}_{e2}(\overline{R}, \overline{R}_b)$, (4.166) can be changed to

$$\begin{aligned}\iiint_V \Big\{ & \left[k^2 \nabla \times \overline{\overline{G}}_{e2}(\overline{R}, \overline{R}_b) + \nabla \times \left[\overline{\overline{I}}\delta(\overline{R} - \overline{R}_b)\right]\right]^T \cdot \overline{\overline{G}}_{e1}(\overline{R}, \overline{R}_a) \\ & - \left[\nabla \times \overline{\overline{G}}_{e2}(\overline{R}, \overline{R}_b)\right]^T \cdot \left[k^2 \overline{\overline{G}}_{e1}(\overline{R}, \overline{R}') + \overline{\overline{I}}\delta(\overline{R} - \overline{R}_a)\right] \Big\}\, dV \\ = \oiint_S \Big\{ & \left[\hat{n} \times \nabla \times \overline{\overline{G}}_{e2}(\overline{R}, \overline{R}')\right]^T \cdot \nabla \times \overline{\overline{G}}_{e1}(\overline{R}, \overline{R}_a) \\ & - \left[k^2 \overline{\overline{G}}_{e2}(\overline{R}, \overline{R}_b) + \overline{\overline{I}}\delta(\overline{R} - \overline{R}_b)\right]^T \cdot \left[\hat{n} \times \overline{\overline{G}}_{e1}(\overline{R}, \overline{R}_a)\right] \Big\}\, dS. \qquad (4.167)\end{aligned}$$

Because of the radiation condition at infinity and the boundary conditions on S_d the surface integral vanishes and (4.167) reduces to

$$\begin{aligned}& \iiint_V \left\{\nabla \times \left[\overline{\overline{I}}\delta(\overline{R} - \overline{R}_b)\right]\right\}^T \cdot \overline{\overline{G}}_{e1}(\overline{R}, \overline{R}_a)\, dV \\ & = \left[\nabla_a \times \overline{\overline{G}}_{e2}(\overline{R}_a, \overline{R}_b)\right]^T. \qquad (4.168)\end{aligned}$$

The volume integral in (4.168) can be evaluated with the aid of the vector-dyadic divergence theorem, (A.16) of Appendix A,

$$\iiint_V \left[(\nabla \times \overline{A}) \cdot \overline{\overline{B}} - \overline{A} \cdot \nabla \times \overline{\overline{B}}\right] dV = \oiint_S \hat{n} \cdot \left(\overline{A} \times \overline{\overline{B}}\right) dS. \qquad (4.169)$$

This theorem can also be written in the form

$$\iiint_V \left[\left(\overline{\overline{B}}\right)^T \cdot \nabla \times \overline{A} - \left(\nabla \times \overline{\overline{B}}\right)^T \cdot \overline{A}\right] dV = \oiint_S \hat{n} \cdot \left(\overline{A} \times \overline{\overline{B}}\right) dS. \qquad (4.170)$$

We now identify $\overline{\overline{B}}$ to be $\overline{\overline{I}}\delta(\overline{R}-\overline{R}_b)$, and we consider three distinct vector functions $\overline{A}_i(\overline{R})$ with $i = 1, 2, 3$, corresponding to the vector components of the function $\overline{\overline{G}}_{e1}(\overline{R}, \overline{R}_a)$; that is,

$$\overline{\overline{G}}_{e1}(\overline{R}, \overline{R}_a) = \sum_i \overline{A}_i(\overline{R})\hat{x}_i = \sum_i \overline{G}_{e1i}(\overline{R}, \overline{R}_a)\hat{x}_i. \tag{4.171}$$

Then (4.170) becomes

$$\begin{aligned}\iiint_V \Big\{ & \left[\overline{\overline{I}}\delta(\overline{R}-\overline{R}_b)\right]^T \cdot \nabla \times \overline{A}_i(\overline{R}) \\ & - \left[\nabla \times \overline{\overline{I}}\delta(\overline{R}-\overline{R}_b)\right]^T \cdot \overline{A}_i(\overline{R}) \Big\}\, dV \\ = \oiint_S & \hat{n} \cdot \left[\overline{A}_i(\overline{R}) \times \overline{\overline{I}}\delta(\overline{R}-\overline{R}_b)\right] dS, \quad i = 1, 2, 3. \end{aligned} \tag{4.172}$$

The surface integral in (4.172) vanishes as the position vector $\overline{R}_b$ is located in a region exterior to the surface S, (4.172) therefore yields

$$\nabla_b \times \overline{A}_i\,(R_b) = \iiint_V \left[\nabla \times \overline{\overline{I}}\delta(\overline{R}-\overline{R}_b)\right]^T \cdot \overline{A}_i(\overline{R})\, dV \tag{4.173}$$

By juxtaposing a unit vector $\hat{x}_i$ at the posterior position of (4.173) and summing the resultant three equations, we obtain

$$\begin{aligned}&\nabla_b \times \overline{\overline{G}}_{e1}(\overline{R}_b, \overline{R}_a) \\ &= \iiint_V \left[\nabla \times \overline{\overline{I}}\delta(\overline{R}-\overline{R}_b)\right]^T \cdot \overline{\overline{G}}_{e1}(\overline{R}, \overline{R}_a)\, dV.\end{aligned} \tag{4.174}$$

The volume integral in (4.174) is identical to the integral of (4.168); thus we obtain the symmetrical relationship

$$\left[\nabla_a \times \overline{\overline{G}}_{e2}(\overline{R}_a, \overline{R}_b)\right]^T = \nabla_b \times \overline{\overline{G}}_{e1}(\overline{R}_b, \overline{R}_a). \tag{4.175}$$

By letting $\overline{R}_a = \overline{R}'$, $\overline{R}_b = \overline{R}$, (4.175) can be written in form

$$\left[\nabla' \times \overline{\overline{G}}_{e2}(\overline{R}', \overline{R})\right]^T = \nabla \times \overline{\overline{G}}_{e1}(\overline{R}, \overline{R}'). \tag{4.176}$$

Since the transpose of a transposed dyadic function is equal to the original dyadic function, (4.176) implies that

$$\left[\nabla' \times \overline{\overline{G}}_{e1}(\overline{R}', \overline{R})\right]^T = \nabla \times \overline{\overline{G}}_{e2}(\overline{R}, \overline{R}'). \tag{4.177}$$

The above two relations are equivalent to

$$\left[\overline{\overline{G}}_{m1}(\overline{R}', \overline{R})\right]^T = \overline{\overline{G}}_{m2}(\overline{R}, \overline{R}') \tag{4.178}$$

and

$$\left[\overline{\overline{G}}_{m2}(\overline{R}', \overline{R})\right]^T = \overline{\overline{G}}_{m1}(\overline{R}, \overline{R}'). \tag{4.179}$$

The symmetrical relationships which we have derived so far are tabulated in Table 4-1 for frequent references.

TABLE 4-1 Symmetrical Relationships of Free-Space Dyadic Green Functions and Functions of the First and the Second Kinds

$$\left[\overline{\overline{G}}_{e0}(\overline{R}', \overline{R})\right]^T = \overline{\overline{G}}_{e0}(\overline{R}, \overline{R}')$$

$$\left[\overline{\overline{G}}_{m0}(\overline{R}', \overline{R})\right]^T = \overline{\overline{G}}_{m0}(\overline{R}, \overline{R}')$$

$$\left[\overline{\overline{G}}_{e1}(\overline{R}', \overline{R})\right]^T = \overline{\overline{G}}_{e1}(\overline{R}, \overline{R}')$$

$$\left[\overline{\overline{G}}_{e2}(\overline{R}', \overline{R})\right]^T = \overline{\overline{G}}_{e2}(\overline{R}, \overline{R}')$$

$$\left[\overline{\overline{G}}_{m1}(\overline{R}', \overline{R})\right]^T = \overline{\overline{G}}_{m2}(\overline{R}, \overline{R}')$$

$$\left[\overline{\overline{G}}_{m2}(\overline{R}', \overline{R})\right]^T = \overline{\overline{G}}_{m1}(\overline{R}, \overline{R}')$$

$$\left[\nabla' \times \overline{\overline{G}}_{e0}(\overline{R}', \overline{R})\right]^T = \nabla \times \overline{\overline{G}}_{e0}(\overline{R}, \overline{R}')$$

$$\left[\nabla' \times \overline{\overline{G}}_{m0}(\overline{R}', \overline{R})\right]^T = \nabla \times \overline{\overline{G}}_{m0}(\overline{R}, \overline{R}')$$

$$\left[\nabla' \times \overline{\overline{G}}_{e1}(\overline{R}', \overline{R})\right]^T = \nabla \times \overline{\overline{G}}_{e2}(\overline{R}, \overline{R}')$$

$$\left[\nabla' \times \overline{\overline{G}}_{e2}(\overline{R}', \overline{R})\right]^T = \nabla \times \overline{\overline{G}}_{e1}(\overline{R}, \overline{R}')$$

$$\left[\nabla' \times \overline{\overline{G}}_{m1}(\overline{R}', \overline{R})\right]^T = \nabla \times \overline{\overline{G}}_{m1}(\overline{R}, \overline{R}')$$

$$\left[\nabla' \times \overline{\overline{G}}_{m2}(\overline{R}', \overline{R})\right]^T = \nabla \times \overline{\overline{G}}_{m2}(\overline{R}, \overline{R}')$$

Once the symmetrical relationships of the functions of the first and second kinds have been found, the integral expressions for the electromagnetic field derived in Sec. 4-3 can be casted into standard form using $\overline{R}$ as the position vector for a field point and $\overline{R}'$ as that for a point inside a source. For example, by interchanging $\overline{R}$ and $\overline{R}'$ in (4.84), we can convert that equation to

$$\begin{aligned}&\overline{E}(\overline{R}) - i\omega\mu_0 \iiint_V \overline{J}(\overline{R}') \cdot \overline{\overline{G}}_{e1}(\overline{R}', \overline{R})\, dV' \\ &= -\oint\!\!\oint_{S_d} [\hat{n}' \times \overline{E}(\overline{R}')] \cdot \nabla' \times \overline{\overline{G}}_{e1}(\overline{R}', \overline{R})\, dS'\end{aligned}$$

or

$$\begin{aligned}\bar{E}(\bar{R}) - i\omega\mu_0 \iiint_V \left[\bar{\bar{G}}_{e1}(\bar{R}', \bar{R})\right]^T \cdot \bar{J}(\bar{R}')\, dV' \\ = - \oiint_{S_d} \left[\nabla' \times \bar{\bar{G}}_{e1}(\bar{R}', \bar{R})\right]^T \cdot \left[\hat{n}' \times \bar{E}(\bar{R}')\right] dS'.\end{aligned} \tag{4.180}$$

In view of (4.155) and (4.177) the above equation can be written in the form

$$\begin{aligned}\bar{E}(\bar{R}) =& i\omega\mu_0 \iiint_V \bar{\bar{G}}_{e1}(\bar{R}, \bar{R}') \cdot \bar{J}(\bar{R}')\, dV' \\ &- \oiint_{S_d} \left[\nabla \times \bar{\bar{G}}_{e2}(\bar{R}, \bar{R}')\right] \cdot \left[\hat{n}' \times \bar{E}(\bar{R}')\right] dS'.\end{aligned} \tag{4.181}$$

When the surface of an otherwise conducting scattering body has an aperture, the surface integral in (4.180) does not vanish. By the same technique other integral solutions represented by (4.85), (4.86), and (4.88)–(4.90) can be converted to the standard form with $\bar{R}$ as the position vector for the field and $\bar{R}'$ that of a source point; they are

$$\bar{E}(\bar{R}) = i\omega\mu_0 \iiint_V \bar{\bar{G}}_{e1}(\bar{R}, \bar{R}') \cdot \bar{J}(\bar{R}')\, dV' \tag{4.182}$$

$$\bar{E}(\bar{R}) = - \iint_{S_A} \left[\nabla \times \bar{\bar{G}}_{e2}(\bar{R}, \bar{R}')\right] \cdot \left[\hat{n}' \times \bar{E}(\bar{R}')\right] dS' \tag{4.183}$$

$$\begin{aligned}\bar{H}(\bar{R}) - \iiint \left[\nabla \times \bar{\bar{G}}_{e1}(\bar{R}, \bar{R}')\right] \cdot \bar{J}(\bar{R}')\, dV' \\ = -i\omega\epsilon_0 \oiint_{S_d} \bar{\bar{G}}_{e2}(\bar{R}, \bar{R}') \cdot \left[\hat{n}' \times \bar{E}(\bar{R}')\right] dS'\end{aligned} \tag{4.184}$$

$$\bar{H}(\bar{R}) = \iiint_V \left[\nabla \times \bar{\bar{G}}_{e1}(\bar{R}, \bar{R}')\right] \cdot \bar{J}(\bar{R}')\, dV' \tag{4.185}$$

$$\bar{H}(\bar{R}) = -i\omega\epsilon_0 \iint_{S_A} \bar{\bar{G}}_{e2}(\bar{R}, \bar{R}') \cdot \left[\hat{n}' \times \bar{E}(\bar{R})\right] dS'. \tag{4.186}$$

The derivation of the symmetrical relationships of the dyadic Green functions of the third kind is more involved. Let us consider first the function $\bar{\bar{G}}_e^{(11)}$ of a problem with the configuration shown in Fig. 4-2(d) where both regions are unbounded at the extreme side of each region. The interface of the two regions is denoted by S. The function $\bar{\bar{G}}_e^{(11)}$ satisfies (4.106) in region 1. Applying the dyadic-dyadic Green theorem (4.149) to that region with

$$\bar{\bar{P}} = \bar{\bar{G}}_e^{(11)}\left(\bar{R}, R_a\right) \tag{4.187}$$

$$\bar{\bar{Q}} = \bar{\bar{G}}_e^{(11)}(\bar{R}, \bar{R}_b), \tag{4.188}$$

where $\bar{R}_a$ and $\bar{R}_b$ denote the locations of two point sources in that region, we obtain

$$\iiint_{V_1} \left\{ \left[k_1^2 \left[\overline{\overline{G}}_e^{(11)}(\overline{R}, \overline{R}_b) \right]^T + \overline{\overline{I}} \delta(\overline{R} - \overline{R}_b) \right] \cdot \overline{\overline{G}}_e^{(11)}(\overline{R}, \overline{R}_a) \right.$$
$$\left. - \left[\overline{\overline{G}}_e^{(11)}(\overline{R}, \overline{R}_b) \right]^T \cdot \left[k_1^2 \overline{\overline{G}}_e(\overline{R}, \overline{R}_a) + \overline{\overline{I}} \delta(\overline{R} - \overline{R}_a) \right] \right\} dV$$
$$= \oiint_S \left\{ \left[\hat{n}_1 \times \overline{\overline{G}}_e^{(11)}(\overline{R}, \overline{R}_b) \right]^T \cdot \nabla \times \overline{\overline{G}}_e^{(11)}(\overline{R}, \overline{R}_a) \right.$$
$$\left. - \left[\nabla \times \overline{\overline{G}}_e^{(11)}(\overline{R}, \overline{R}_b) \right]^T \cdot \left[\hat{n}_1 \times \overline{\overline{G}}_e^{(11)}(\overline{R}, \overline{R}_a) \right] \right\} dS, \qquad (4.189)$$

where $\hat{n}_1$ denotes the unit normal vector at S pointed away from region 1. We have already omitted the surface integral on the semi-infinite surface at the extreme side of that region as a result of the radiation condition. After simplifying the volume integral in (4.189), we obtain

$$\overline{\overline{G}}_e^{(11)}(\overline{R}_b, \overline{R}_a) - \left[\overline{\overline{G}}_e^{(11)}(\overline{R}_a, \overline{R}_b) \right]^T$$
$$= \oiint_S \left\{ \left[\hat{n}_1 \times \overline{\overline{G}}_e^{(11)}(\overline{R}, \overline{R}_b) \right]^T \cdot \nabla \times \overline{\overline{G}}_e^{(11)}(\overline{R}, \overline{R}_a) \right.$$
$$\left. - \left[\nabla \times \overline{\overline{G}}_e^{(11)}(\overline{R}, \overline{R}_b) \right]^T \cdot \left[\hat{n}_1 \times \overline{\overline{G}}_e^{(11)}(\overline{R}, \overline{R}_a) \right] \right\} dS. \qquad (4.190)$$

We are now going to show that the surface integral in (4.189) is equal to zero. To prove this identity let us apply the formula (4.149) to region 2 with

$$\overline{\overline{P}} = \overline{\overline{G}}_e^{(21)}(\overline{R}, \overline{R}_a) \qquad (4.191)$$
$$\overline{\overline{Q}} = \overline{\overline{G}}_e^{(21)}(\overline{R}, \overline{R}_b), \qquad (4.192)$$

where $\overline{\overline{G}}_e^{(21)}$ satisfies (4.105) and $\overline{R}_a$ and $\overline{R}_b$ are the same position vectors used in $\overline{\overline{G}}_e^{(11)}$. Substituting (4.191) and (4.192) into (4.149) and deleting the surface integral at the semi-infinite surface in region 2 as a result of the radiation condition, we obtain

$$\oiint_S \left\{ \left[\hat{n}_1 \times \overline{\overline{G}}_e^{(21)}(\overline{R}, \overline{R}_b) \right]^T \cdot \nabla \times \overline{\overline{G}}_e^{(21)}(\overline{R}, \overline{R}_a) \right.$$
$$\left. - \left[\nabla \times \overline{\overline{G}}_e^{(21)}(\overline{R}, \overline{R}_b) \right]^T \cdot \left[\hat{n}_1 \times \overline{\overline{G}}_e^{(21)}(\overline{R}, \overline{R}_a) \right] \right\} dS = 0. \qquad (4.193)$$

On S, the boundary conditions for the electromagnetic field are

$$\hat{n}_1 \times \left[\overline{E}_1(\overline{R}) - \overline{E}_2(\overline{R}) \right] = 0$$
$$\hat{n}_1 \times \left[\overline{H}_1(\overline{R}) - \overline{H}_2(\overline{R}) \right] = 0.$$

These conditions, after generalized to the dyadic Green functions, yield

$$\hat{n}_1 \times \left[\overline{\overline{G}}_e^{(11)}(\overline{R}, \overline{R}') - \overline{\overline{G}}_e^{(21)}(\overline{R}, \overline{R}') \right] = 0 \qquad (4.194)$$

$$\hat{n}_1 \times \left[\frac{\nabla \times \overline{\overline{G}}_e^{(11)}(\overline{R}, \overline{R}')}{\mu_1} - \frac{\nabla \times \overline{\overline{G}}_e^{(21)}(\overline{R}, \overline{R}')}{\mu_2} \right] = 0, \tag{4.195}$$

where $\overline{R}'$ in (4.194) and (4.195) represents either $\overline{R}_a$ or $\overline{R}_b$. The presence of the magnetic constants μ_1 and μ_2 in (4.195) is due to the fact that we define $\overline{\overline{G}}_m$ as

$$\nabla \times \overline{\overline{G}}_e = \overline{\overline{G}}_m,$$

but

$$\nabla \times \overline{E} = i\omega\mu\overline{H}.$$

As a result of (4.194) and (4.195) we see that the surface integral in (4.190) is equal to the surface integral in (4.193) which is equal to zero; hence

$$\left[\overline{\overline{G}}_e^{(11)}(\overline{R}_a, \overline{R}_b)\right]^T = \overline{\overline{G}}_e^{(11)}(\overline{R}_b, \overline{R}_a)$$

or, equivalently,

$$\left[\overline{\overline{G}}_e^{(11)}(\overline{R}', \overline{R})\right]^T = \overline{\overline{G}}_e^{(11)}(\overline{R}, \overline{R}'). \tag{4.196}$$

Since the designation of regions 1 and 2 is quite arbitrary, it is obvious that

$$\left[\overline{\overline{G}}_e^{(22)}(\overline{R}', \overline{R})\right]^T = \overline{\overline{G}}_e^{(22)}(\overline{R}, \overline{R}'). \tag{4.197}$$

The remaining symmetrical relationship to be discussed is the one dealing with $\overline{\overline{G}}_e^{(12)}$ and $\overline{\overline{G}}_e^{(21)}$. We first apply formula (4.149) to region 1 with

$$\overline{\overline{P}} = \overline{\overline{G}}_e^{(12)}(\overline{R}, \overline{R}_2) \tag{4.198}$$

and

$$\overline{\overline{Q}} = \overline{\overline{G}}_e^{(11)}(\overline{R}, \overline{R}_1). \tag{4.199}$$

These two functions satisfy, respectively, (4.108) and (4.106).

We put into evidence that $\overline{R}_2$ denotes the position vector of a point source placed in region 2 and $\overline{R}_1$ the position vector of a point source in region 1. The superscript notation attached to these two functions already implies the location of the source. Substituting (4.198) and (4.199) into (4.149), we obtain, after deleting the surface integral at the semi-infinite surface in region 1,

$$\begin{aligned}\overline{\overline{G}}_e^{(12)}(\overline{R}_1, \overline{R}_2) = \iint_S \Big\{ &\left[\hat{n}_1 \times \overline{\overline{G}}_e^{(11)}(\overline{R}, \overline{R}_1)\right]^T \cdot \nabla \times \overline{\overline{G}}_e^{(12)}(\overline{R}, \overline{R}_2) \\ - &\left[\nabla \times \overline{\overline{G}}_e^{(11)}(\overline{R}, \overline{R}_1)\right]^T \cdot \left[\hat{n}_1 \times \overline{\overline{G}}_e^{(12)}(\overline{R}, \overline{R}_2)\right] \Big\} \, dS. \end{aligned} \tag{4.200}$$

Formula (4.149) is applied now to region 2 with

$$\overline{\overline{P}} = \overline{\overline{G}}_e^{(22)}(\overline{R}, \overline{R}_2) \tag{4.201}$$

$$\overline{\overline{Q}} = \overline{\overline{G}}_e^{(21)}(\overline{R}, \overline{R}_1), \tag{4.202}$$

which yields, after deleting the surface integral at infinity, the following expression:

$$\left[\overline{\overline{G}}_e^{(21)}(\overline{R}_2, \overline{R}_1)\right]^T = \oiint_S \left\{\left[\hat{n}_1 \times \overline{\overline{G}}_e^{(21)}(\overline{R}, \overline{R}_1)\right]^T \cdot \nabla \times \overline{\overline{G}}_e^{(22)}(\overline{R}, \overline{R}_2)\right.$$
$$\left. - \left[\nabla \times \overline{\overline{G}}_e^{(21)}(\overline{R}, \overline{R}_1)\right]^T \cdot \left[\hat{n}_1 \times \overline{\overline{G}}_e^{(22)}(\overline{R}, \overline{R}_2)\right]\right\} dS. \qquad (4.203)$$

As a result of the boundary conditions stated in (4.194) and (4.195) and two similar conditions for $\overline{\overline{G}}_e^{(22)}$ and $\overline{\overline{G}}_e^{(12)}$, namely,

$$\hat{n}_1 \times \left[\overline{\overline{G}}_e^{(22)}(\overline{R}, \overline{R}') - \overline{\overline{G}}_e^{(12)}(\overline{R}, \overline{R}')\right] = 0$$
$$\hat{n}_1 \times \left[\frac{\nabla \times \overline{\overline{G}}_e^{(22)}(\overline{R}, \overline{R}')}{\mu_2} - \frac{\nabla \times \overline{\overline{G}}_e^{(12)}(\overline{R}, \overline{R}')}{\mu_1}\right] = 0,$$

one finds that the ratio of the two surface integrals in (4.200) and (4.203) is equal to μ_1/μ_2; hence, we obtain the symmetrical relationship

$$\frac{1}{\mu_2}\left[\overline{\overline{G}}_e^{(21)}(\overline{R}_2, \overline{R}_1)\right]^T = \frac{1}{\mu_1}\overline{\overline{G}}_e^{(12)}(\overline{R}_1, \overline{R}_2).$$

By replacing $\overline{R}_2$ by $\overline{R}'$ and $\overline{R}_1$ by $\overline{R}$ in the above equation, we have the relationship expressed in the standard notation, namely,

$$\frac{1}{\mu_2}\left[\overline{\overline{G}}_e^{(21)}(\overline{R}', \overline{R})\right]^T = \frac{1}{\mu_1}\overline{\overline{G}}_e^{(12)}(\overline{R}, \overline{R}'). \qquad (4.204)$$

The derivation of the symmetrical relationship of $\nabla \times \overline{\overline{G}}_e^{(ij)}$ or $\overline{\overline{G}}_m^{(ij)}$ with $i \neq j$ is discussed in Chapter 11, where two complementary models are needed. The result is

$$\frac{1}{k_i^2}\left[\nabla' \times \overline{\overline{G}}_e^{(ij)}(\overline{R}', \overline{R})\right]^T = \frac{1}{k_j^2}\nabla \times \overline{\overline{G}}_e^{(ji)}(\overline{R}, \overline{R}'). \qquad (4.205)$$

It should be emphasized that (4.204) and (4.205) have been derived under the condition that the two regions are unbounded at the extremities. However, they are still valid if one of the regions is bounded, corresponding to the problem of a dielectric body placed in air for example. In that case, there is no radiation condition involved inside the dielectric body and the proof is practically the same. In summary, for the functions of the third kind, the symmetrical relationships are

$$\frac{1}{\mu_i}\left[\overline{\overline{G}}_e^{(ij)}(\overline{R}', \overline{R})\right]^T = \frac{1}{\mu_j}\overline{\overline{G}}_e^{(ji)}(\overline{R}, \overline{R}') \qquad (4.206)$$

and

$$\frac{1}{k_i^2}\left[\nabla' \times \overline{\overline{G}}_e^{(ij)}(\overline{R}', \overline{R})\right]^T = \frac{1}{k_j^2}\nabla \times \overline{\overline{G}}_e^{(ji)}(\overline{R}, \overline{R}'), \qquad (4.207)$$

where i can be equal to or different from j. These relationships can be extended to more than two isotropic media, such as a dielectric cylinder coated by another layer of material with different permittivity.

With these relationships at our disposal, (4.124), (4.130)–(4.132) can be converted to a standard form with $\bar{R}$ as the position vector for the field and $\bar{R}'$ as that of a source point. They are

$$\bar{E}_1(\bar{R}) = i\omega\mu_1 \iiint_{V_1} \bar{\bar{G}}_e^{(11)}(\bar{R}, \bar{R}') \cdot \bar{J}_1(\bar{R}')\, dV' \tag{4.208}$$

$$\bar{E}_2(\bar{R}) = i\omega\mu_2 \iiint_{V_1} \bar{\bar{G}}_e^{(21)}(\bar{R}, \bar{R}') \cdot \bar{J}_1(\bar{R}')\, dV' \tag{4.209}$$

$$\bar{E}_2(\bar{R}) = i\omega\mu_2 \iiint_{V_2} \bar{\bar{G}}_e^{(22)}(\bar{R}, \bar{R}') \cdot \bar{J}_2(\bar{R}')\, dV' \tag{4.210}$$

$$\bar{E}_1(\bar{R}) = i\omega\mu_1 \iiint_{V_2} \bar{\bar{G}}_e^{(12)}(\bar{R}, \bar{R}') \cdot \bar{J}_2(\bar{R}')\, dV'. \tag{4.211}$$

In problems dealing with dielectric media $\mu_1 = \mu_2 = \mu_0$, but, in general, we should observe the different magnetic constants in these expressions.

4-5 RECIPROCITY THEOREMS

The symmetrical relationships derived in the previous section are intimately related to the reciprocity theorems in electromagnetic theory. We will discuss these theorems based on a general approach with the aid of the vector Green theorem; namely,

$$\iiint_V (\bar{P} \cdot \nabla \times \nabla \times \bar{Q} - \bar{Q} \cdot \nabla \times \nabla \times \bar{P})\, dV$$
$$= -\oiint_S \left[(\hat{n} \times \nabla \times \bar{P}) \cdot \bar{Q} + (\hat{n} \times \bar{P}) \cdot \nabla \times \bar{Q}\right] dS. \tag{4.212}$$

We consider two sets of electromagnetic fields with two different current sources. The environment of the problem is assumed to be the same for the two sets, like the one shown in Fig. 4-1. The wave equations for these fields are

$$\nabla \times \nabla \times \bar{E}_a(\bar{R}) - k^2 \bar{E}_a(\bar{R}) = i\omega\mu \bar{J}_a(\bar{R}) \tag{4.213}$$

$$\nabla \times \nabla \times \bar{H}_a(\bar{R}) - k^2 \bar{H}_a(\bar{R}) = \nabla \times \bar{J}_a(\bar{R}) \tag{4.214}$$

and

$$\nabla \times \nabla \times \bar{E}_b(\bar{R}) - k^2 \bar{E}_b(\bar{R}) = i\omega\mu \bar{J}_b(\bar{R}) \tag{4.215}$$

$$\nabla \times \nabla \times \bar{H}_b(\bar{R}) - k^2 \bar{H}_b(\bar{R}) = \nabla \times \bar{J}_b(\bar{R}). \tag{4.216}$$

For a problem with two isotropic media, the above system of equations apply to either region with $k = k_1$ or k_2. To be more specific we consider the problem like the one shown in Fig. 4-2(a). There are two possible choices for the functions $\overline{P}$ and $\overline{Q}$ in (4.212).

Case 1. $\overline{P} = \overline{E}_a(\overline{R}), \quad \overline{Q} = \overline{E}_b(\overline{R});$

then

$$\iiint_V \{\overline{E}_a(\overline{R}) \cdot [k^2\overline{E}_b(\overline{R}) + i\omega\mu\overline{J}_b(\overline{R})] \\ -\overline{E}_b(\overline{R}) \cdot [k^2\overline{E}_a(\overline{R}) + i\omega\mu\overline{J}_a(\overline{R})]\}\, dV \\ = -\oiint_S \{[\hat{n} \times \nabla \times \overline{E}_a(\overline{R})] \cdot \overline{E}_b(\overline{R}) \\ + [\hat{n} \times \overline{E}_a(\overline{R})] \cdot \nabla \times \overline{E}_b(\overline{R})\}\, dS. \tag{4.217}$$

Hence

$$\iiint_V [\overline{J}_b(\overline{R}) \cdot \overline{E}_a(\overline{R}) - \overline{J}_a(\overline{R}) \cdot \overline{E}_b(\overline{R})]\, dV \\ = -\oiint_S \{[\hat{n} \times \overline{H}_a(\overline{R})] \cdot \overline{E}_b(\overline{R}) + [\hat{n} \times \overline{E}_a(\overline{R})] \cdot \overline{H}_b(\overline{R})\}\, dS \\ = \oiint_S \hat{n} \cdot [\overline{E}_b(\overline{R}) \times \overline{H}_a(\overline{R}) - \overline{E}_a(\overline{R}) \times \overline{H}_b(\overline{R})]\, dS, \tag{4.218}$$

where we have used the relationship $\nabla \times \overline{E} = i\omega\mu\overline{H}$ to change (4.217) to (4.218). Let the volume of integration correspond to the unbounded region exterior to the scattering body, assuming it to be perfectly conducting, then there are two surfaces enclosing this volume; one lies at infinity and another corresponds to the surface of the conducting body. At infinity the surface integral vanishes because of the radiation condition and at the surface of the conducting body $\hat{n} \times \overline{E}_a = 0, \hat{n} \times \overline{E}_b = 0$, thus, the entire surface integral of (4.128) vanishes, and the result is

$$\iiint_{V_a} \overline{J}_a(\overline{R}) \cdot \overline{E}_b(\overline{R})\, dV = \iiint_{V_b} \overline{J}_b(\overline{R}) \cdot \overline{E}_a(\overline{R})\, dV. \tag{4.219}$$

Equation (4.219) represents the well-known reciprocity theorem of Rayleigh-Carson. The theorem, of course, is valid when the scattering body is absent. In that case, there is only a surface at infinity to be considered. In (4.219) if we identify the current density functions as that of two infinitesimal electrical dipoles with the same magnitude of current moment but different directions; that is,

$$\overline{J}_a(\overline{R}) = \overline{c}_i\delta(\overline{R} - \overline{R}_a) = c\hat{x}_i\delta(\overline{R} - \overline{R}_a) \tag{4.220}$$

$$\overline{J}_b(\overline{R}) = \overline{c}_j\delta(\overline{R} - \overline{R}_b) = c\hat{x}_j\delta(\overline{R} - \overline{R}_b). \tag{4.221}$$

By definition, the corresponding electric fields produced by these sources with $i\omega\mu_0 c = 1$ would be equal to the electric vector Green functions; that is,

$$\overline{E}_a(\overline{R}) = \overline{G}_{e1i}(\overline{R}, \overline{R}_a) \tag{4.222}$$

$$\overline{E}_b(\overline{R}) = \overline{G}_{e1j}(\overline{R}, \overline{R}_b). \tag{4.223}$$

The functions are of the first kind because the problem under consideration has a conducting scattering body, and we require the vector Green functions to satisfy the Dirichlet boundary condition. Substituting (4.220)–(4.221) to (4.219), we obtain

$$\hat{x}_i \cdot \overline{G}_{e1j}(\overline{R}_a, \overline{R}_b) = \hat{x}_j \cdot \overline{G}_{e1i}(\overline{R}_b, \overline{R}_a)$$

or

$$\left[\overline{\overline{G}}_{e1}(\overline{R}_a, \overline{R}_b)\right]_{ij} = \left[\overline{\overline{G}}_{e1}(\overline{R}_b, \overline{R}_a)\right]_{ji}, \tag{4.224}$$

where the quantity with subscript "ij" denotes the scalar component of the electric dyadic Green function $\overline{\overline{G}}_{e1}(\overline{R}_a, \overline{R}_b)$. According to the theory of dyadic analysis, (4.224) means

$$\overline{\overline{G}}_{e1}(\overline{R}_a, \overline{R}_b) = \left[\overline{\overline{G}}_{e1}(\overline{R}_b, \overline{R}_a)\right]^T. \tag{4.225}$$

This is the symmetrical relationship of the electric dyadic Green function which we have derived before. It is now clear that such a relationship is not merely a mathematical transformation; its physical significance is manifested by the Rayleigh-Carson reciprocity theorem.

If the volume of integration in (4.217) excludes the volumes occupied by $\overline{J}_a$ and $\overline{J}_b$, then the surface of the surface integral would consist of S_∞ and two surfaces enclosing $\overline{J}_a$ and $\overline{J}_b$, denoted by S_a and S_b. Consequently, (4.217) reduces to

$$\iint_{S_a+S_b} \hat{n} \cdot \left[\overline{E}_b(\overline{R}) \times \overline{H}_a(\overline{R}) - \overline{E}_a(\overline{R}) \times \overline{H}_b(\overline{R})\right] dS = 0. \tag{4.226}$$

The integral evaluated at S_∞ goes to zero as a result of the radiation condition. Equation (4.226) represents the Lorentz reciprocity theorem in electromagnetic theory. It can also be written in the form

$$\iiint_V \nabla \cdot \left[\overline{E}_b(\overline{R}) \times \overline{H}_a(\overline{R}) - \overline{E}_a(\overline{R}) \times \overline{H}_b(\overline{R})\right] dV = 0, \tag{4.227}$$

where V is the volume exterior to S_a and S_b.

Case 2. $\overline{P} = \overline{E}_a(\overline{R}), \quad \overline{Q} = \overline{H}_b(\overline{R})$

By using these two functions in (4.212), we obtain

$$\begin{aligned}
&\iiint_V \left\{\overline{E}_a(\overline{R}) \cdot \left[k^2 \overline{H}_b(\overline{R}) + \nabla \times \overline{J}_b(\overline{R})\right]\right. \\
&\left. -\overline{H}_b(\overline{R}) \cdot \left[k^2 \overline{E}_a(\overline{R}) + i\omega\mu_0 \overline{J}_a(\overline{R})\right]\right\} dV \\
&= -\oiint_S \left\{\left[\hat{n} \times \nabla \times \overline{E}_a(\overline{R})\right] \cdot \overline{H}_b(\overline{R})\right. \\
&\left. + \left[\hat{n} \times \overline{E}_a(\overline{R})\right] \cdot \nabla \times \overline{H}_b(\overline{R})\right\} dS
\end{aligned}$$

or

$$\iiint_V \left[\overline{E}_a(\overline{R}) \cdot \nabla \times \overline{J}_b(\overline{R}) - i\omega\mu_o \overline{J}_a(\overline{R}) \cdot \overline{H}_b(\overline{R})\right] dV$$
$$= -\oiint_S \left\{ i\omega\mu_0 \left[\hat{n} \times \overline{H}_a(\overline{R})\right] \cdot \overline{H}_b(\overline{R}) + \left[\hat{n} \times \overline{E}_a(\overline{R})\right] \cdot \left[\overline{J}_b(\overline{R}) - i\omega\epsilon_0 \overline{E}_b(\overline{R})\right] \right\} dS. \tag{4.228}$$

The first term in the volume integral of the above equation can be split into two terms; that is,

$$\iiint_V \overline{E}_a(\overline{R}) \cdot \nabla \times \overline{J}_b(\overline{R}) dV$$
$$= \iiint_V \left\{ \nabla \cdot \left[\overline{J}_b(\overline{R}) \times \overline{E}_a(\overline{R})\right] + \overline{J}_b(\overline{R}) \cdot \nabla \times \overline{E}_a(\overline{R}) \right\} dV$$
$$= \oiint_S \hat{n} \cdot \left[\overline{J}_b(\overline{R}) \times \overline{E}_a(\overline{R})\right] dS + \iiint_V \overline{J}_b(\overline{R}) \cdot \left[i\omega\mu_0 \overline{H}_a(\overline{R})\right] dV. \tag{4.229}$$

The term of surface integral in (4.228) and in (4.229), containing the *volume* density of current $\overline{J}_b$ is null. By combining (4.228) and (4.229), we obtain

$$\iiint_V \left[\overline{J}_b(\overline{R}) \cdot \overline{H}_a(\overline{R}) - \overline{J}_a(\overline{R}) \cdot \overline{H}_b(\overline{R})\right] dV$$
$$= \oiint_S \hat{n} \cdot \left[\overline{E}_a(\overline{R}) \times \overline{E}_b(\overline{R})/Z_0^2 - \overline{H}_a(\overline{R}) \times \overline{H}_b(\overline{R})\right] dS, \tag{4.230}$$

where $Z_0 = (\mu_0/\epsilon_0)^{1/2}$, the wave impedance, assuming the medium to be air. There are now two reciprocity theorems which can be derived from (4.230). If we consider the environment of the two sets of field to be the same, that is, a medium containing an electrically perfect conducting body placed in an otherwise unbounded space, then the first term in the surface integral vanishes but the second term does not, because $\hat{n} \times \overline{H}_a$ or $\hat{n} \times \overline{H}_b$ is nonvanishing on the surface of an electrically conducting body. In order to eliminate the entire surface integral we need two *complementary models* or two *complementary surfaces*, denoted by S_e and S_m, such that

$$\hat{n} \times \overline{E}_a = 0 \text{ on } S_e, \quad \hat{n} \times \overline{H}_b = 0 \text{ on } S_m. \tag{4.231}$$

The first model with S_e represents the original environment of the problem. The second model has the same geometry, but S_m now represents the surface of a magnetically perfect conducting body. The model with S_m is electromagnetically nonphysical but is quite acceptable as a means to formulate the new reciprocity theorems which we are seeking. The radiation condition at infinity is satisfied by $\left(\overline{E}_a, \overline{H}_a\right)$ and $\left(\overline{E}_b, \overline{H}_b\right)$ individually. Under these conditions the entire surface integral in (4.230) vanishes. Thus, in the common volume exterior to S_e and S_m, we have

$$\iiint_V \left[\overline{J}_b(\overline{R}) \cdot \overline{H}_a(\overline{R}) - \overline{J}_a(\overline{R}) \cdot \overline{H}_b(\overline{R})\right] dV = 0. \tag{4.232}$$

The relationship stated by (4.232) is designated as the $\bar{J} \cdot \bar{H}$ complementary reciprocity theorem or simply $\bar{J} \cdot \bar{H}$ theorem for short, in contrast to the $\bar{J} \cdot \bar{E}$ reciprocity theorem of Rayleigh-Carson. The word "complementary" is used here to emphasize the fact that there are two complementary surfaces invoked in the formulation. The $\bar{J} \cdot \bar{H}$ theorem is of course valid in free space. In that case, there is only one simple environment; no boundary condition is involved.

For a volume *excluding* the regions occupied by $\bar{J}_a$ and $\bar{J}_b$, we obtain from (4.230) the relationship

$$\iint_{S_a+S_b} \hat{n} \cdot \left[\frac{\bar{E}_a(\bar{R}) \times \bar{E}_b(\bar{R})}{Z_0^2} - \bar{H}_a(\bar{R}) \times \bar{H}_b(\bar{R}) \right] dS = 0. \qquad (4.233)$$

Equation (4.233) is designated as the complementary $(\bar{E}, \bar{H})$ reciprocity theorem or $(\bar{E}, \bar{H})_c$ theorem for short, in contrast to the $(\bar{E}, \bar{H})$ reciprocity theorem due to Lorentz.

The $\bar{J} \cdot \bar{H}$ reciprocity theorem thus derived is closely related to the symmetrical relationship of the magnetic dyadic Green functions. To show their connection, we let $\bar{J}_a$ and $\bar{J}_b$ be the same as the ones defined by (4.220) and (4.221). They are, however, located in two complementary environments. Substituting them into (4.232), we obtain

$$\hat{x}_i \cdot \bar{H}_b(\bar{R}_a) = \hat{x}_j \cdot \bar{H}_a(\bar{R}_b). \qquad (4.234)$$

The magnetic field $\bar{H}_b(\bar{R}_a)$ is produced by a current element located at $\bar{R}_b$, pointed in the $\hat{x}_j$-direction in an environment with a magnetically perfect conducting surface. By definition it is equal to $\bar{G}_{m1j}(\bar{R}_a, \bar{R}_b)$, the vector component of $\bar{\bar{G}}_{m1}(\bar{R}_a, \bar{R}_b)$ due to a current source in the $\hat{x}_j$-direction placed at $\bar{R}_b$. The magnetic field $\bar{H}_a(\bar{R}_b)$ is produced by a current element located at $\bar{R}_a$, pointed in the $\hat{x}_i$-direction in an environment with an electrically perfect conducting surface. The geometry of the surface and the medium are the same. By definition, $\bar{H}_a(\bar{R}_b)$ is equal to $\bar{G}_{m2i}(\bar{R}_b, \bar{R}_a)$, the vector component of $\bar{\bar{G}}_{m2}(\bar{R}_b, \bar{R}_a)$, due to a current element in the $\hat{x}_i$-direction placed at $\bar{R}_a$ with the current moment properly normalized. In the full dyadic notation, (4.234) leads to

$$\bar{\bar{G}}_{m1}(\bar{R}_a, \bar{R}_b) = \left[\bar{\bar{G}}_{m2}(\bar{R}_b, \bar{R}_a)\right]^T$$

or

$$\left[\bar{\bar{G}}_{m2}(\bar{R}', \bar{R})\right]^T = \bar{\bar{G}}_{m1}(\bar{R}, \bar{R}'). \qquad (4.235)$$

In terms of the electric dyadic Green functions, (4.235) can be changed to

$$\left[\nabla' \times \bar{\bar{G}}_{e1}(\bar{R}', \bar{R})\right]^T = \nabla \times \bar{\bar{G}}_{e2}(\bar{R}, \bar{R}'). \qquad (4.236)$$

Equation (4.236) is identical to (4.177), which was derived previously with the aid of the dyadic-dyadic Green theorem without introducing the concept of complementary models as required by the application of the complementary $\bar{J} \cdot \bar{H}$

reciprocity theorem. The two complementary reciprocity theorems introduced here were first presented by this author at a symposium [Tai, 1987] without much elaboration. In order to digest the $\bar{J} \cdot \bar{H}$ theorem in a more descriptive, and, perhaps, more physical manner, a transmission line version of this theorem will be presented. The application of the complementary $\bar{J} \cdot \bar{H}$ theorem to derive the symmetrical relationships of dyadic Green functions with two isotropic media in contact with a conducting body in one of the media will be discussed in Chapter 11. The problem is a three-dimensional model of the complementary transmission line theory which is discussed in the next section.

4-6 TRANSMISSION LINE MODEL OF THE COMPLEMENTARY RECIPROCITY THEOREMS

We consider two sections of transmission lines ($d \geq x \geq 0$) of the same line constants: one is short-circuited at $x = 0$ and terminated by an impedance Z_a at $x = d$; the other is open-circuited at $x = 0$ and terminated by an impedance Z_b at $x = d$. The two lines are excited by two distributed current sources as shown in Fig. 4-3. The differential equations governing the voltage and the current on these two lines are

$$\frac{di_a(x)}{dx} = i\omega C v_a(x) + K_a(x) \tag{4.237}$$

$$\frac{dv_a(x)}{dx} = i\omega L i_a(x) \tag{4.238}$$

$$\frac{di_b(x)}{dx} = i\omega C v_b(x) + K_b(x) \tag{4.239}$$

$$\frac{dv_b(x)}{dx} = i\omega L i_b(x). \tag{4.240}$$

The boundary conditions for the line voltage and current are

$$v_a(0) = 0, \quad i_b(0) = 0, \tag{4.241}$$

$$v_a(d) = Z_a i_a(d), \quad v_b(d) = Z_b i_b(d). \tag{4.242}$$

The model already spells out the complementary nature of the problem at $x = 0$ by multiplying (4.237) by i_b and (4.239) by i_b, adding the two resultant equations, and making use of (4.238) and (4.240), we obtain

$$\begin{aligned} &i_a(x)K_b(x) + i_b(x)K_a(x) \\ &\quad = \frac{d}{dx}\left[i_a(x)i_b(x) - \frac{v_a(x)v_b(x)}{Z_c^2}\right], \end{aligned} \tag{4.243}$$

where $Z_c = (L/C)^{1/2}$ denotes the characteristic impedance of the lines, which are identical except the different terminal conditions. An integration of (4.243) with respect to x from zero to d yields

$$\int_0^d [i_a(x)K_b(x) + i_b(x)K_a(x)]\,dx = \left[i_a(x)i_b(x) - \frac{v_a(x)v_b(x)}{Z_c^2}\right]_0^d. \tag{4.244}$$

In view of the boundary conditions at $x = 0$ and $x = d$, (4.244) can be written in the form

$$\int_0^d [i_a(x)K_b(x) + i_b(x)K_a(x)]\,dx = i_a(d)i_b(d)\left(1 - \frac{Z_a Z_b}{Z_c^2}\right). \tag{4.245}$$

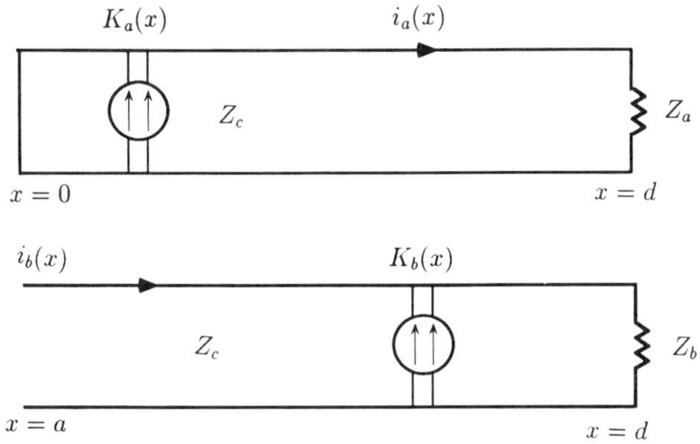

Fig. 4-3 Two sections of line with the complementary boundary condition $Z_a Z_b = Z_c^2$

Now if we impose a relationship between the terminal impedances such that

$$Z_a Z_b = Z_c^2, \tag{4.246}$$

then

$$\int_0^d [i_a(x)K_b(x) + i_b(x)K_a(x)]\,dx = 0. \tag{4.247}$$

If an integration is applied to (4.243) in the regions outside of both $K_a(x)$ and $K_b(x)$, then we obtain

$$[i_a(x)i_b(x) - v_a(x)v_b(x)]_{a_1}^{a_2} + [i_a(x)i_b(x) - v_a(x)v_b(x)]_{b_1}^{b_2} = 0, \tag{4.248}$$

where $a_2 - a_1$ denotes the interval covered by $K_a(x)$ and $b_2 - b_1$ that of $K_b(x)$. Equation (4.247) is designated as the complementary Ki reciprocity theorem and (4.248) as the complementary (v, i) reciprocity theorem for the transmission line. They are analogous to the complementary $\bar{J} \cdot \bar{H}$ and $(\bar{E}, \bar{H})_c$ theorems derived previously. In fact, we can derive the Ki and $(v, i)_c$ theorems from the $\bar{J} \cdot \bar{H}$ and $(\bar{E}, \bar{H})_c$ theorems, but a separate derivation is more tutorial. For convenience, the condition which has been imposed on the terminal impedances will be designated as the complementary impedance condition. Two special cases of this condition should be pointed out.

Case 1. $$Z_a = Z_b = Z_c.$$

In this case, the terminal impedances would correspond to that of a semi-infinite line or the same as letting d go to infinity.

Case 2. $$Z_a = 0 \text{ and } Z_b \to \infty \quad \text{or} \quad Z_a \to \infty \text{ and } Z_b = 0.$$

This condition shows very clearly the physical significance of this model. It is, therefore, quite appropriate to treat the two lines as complementary lines.

If the distributed currents $K_a(x)$ and $K_b(x)$ are localized so that

$$K_a(x) = I_a\,(x_a)\,\delta\,(x - x_a)$$
$$K_b(x) = I_b\,(x_b)\,\delta\,(x - x_b)\,,$$

then the Ki theorem yields

$$i_a\,(x_b)\,I_b\,(x_b) = -i_b\,(x_a)\,I_a\,(x_a)\,. \tag{4.249}$$

This is a network relationship similar to

$$V\,(x_b)\,I\,(x_b) = V\,(x_a)\,I\,(x_a)\,, \tag{4.250}$$

which can be obtained by applying the Rayleigh-Carson theorem to a *single* line of any terminations. In (4.250), we have used $V(x)$ to denote the line voltage. The currents $I\,(x_a)$ and $I\,(x_b)$ are the driving currents applied to the same line at two different locations. The transmission line model shows clearly the significance of the complementary reciprocal theorems. An exercise is being assigned to verify (4.249) based on the solutions for $i_a(x)$ and $i_b(x)$ for a semi-infinite line with a short circuit at $x = 0$ and its complementary line with an open circuit at $x = 0$. The solutions for $i_a(x)$ and $i_b(x)$ are available in Chapter 3.

4-7 DYADIC GREEN FUNCTIONS FOR A HALF SPACE BOUNDED BY A PLANE CONDUCTING SURFACE

The electric dyadic Green functions of the first kind for a half space bounded by a plane conducting surface can be found by superposing the electric free-space dyadic Green functions due to the original dyadic current source in the upper half space and their images in the lower half space as a result of the theory of images. The original problem and its equivalent are shown in Fig. 4-4(a) and

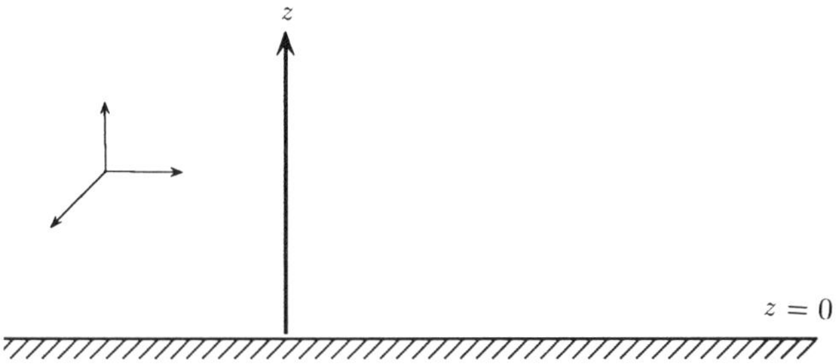

Fig. 4-4 (a) Half space with an infinite conducting plane

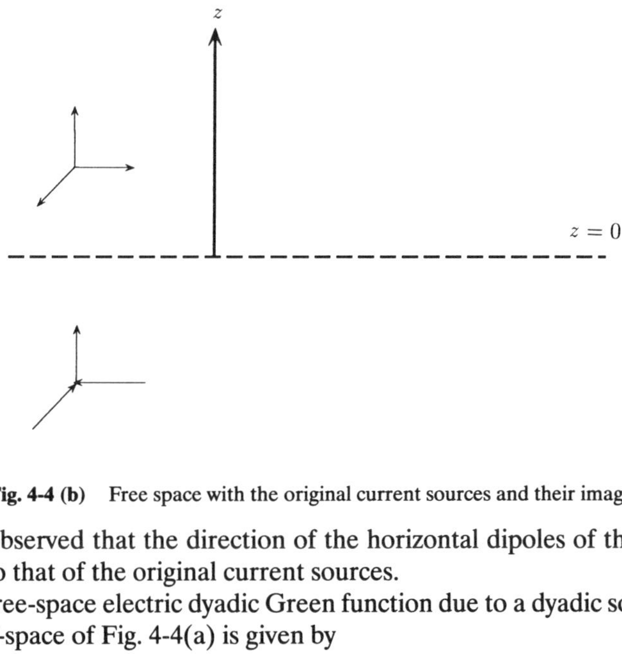

Fig. 4-4 (b) Free space with the original current sources and their images

(b). It is observed that the direction of the horizontal dipoles of the images is opposite to that of the original current sources.

The free-space electric dyadic Green function due to a dyadic source in the upper half-space of Fig. 4-4(a) is given by

$$\overline{\overline{G}}_{e0}(\overline{R}, \overline{R}') = \left(\overline{\overline{I}} + \frac{1}{k^2} \nabla\nabla \right) G_0(\overline{R}, \overline{R}'), \tag{4.251}$$

where

$$G_0(\overline{R}, \overline{R}') = \frac{e^{ik|\overline{R}-\overline{R}'|}}{4\pi \mid \overline{R} - \overline{R}' \mid},$$

with

$$\mid \overline{R} - \overline{R}' \mid = \left[(x - x')^2 + (y - y')^2 + (z - z')^2 \right]^{\frac{1}{2}}.$$

The free-space electric dyadic Green function at $\overline{R}$ due to the image dyadic source in the lower half space is given by

$$\overline{\overline{G}}_{e0i} = \left[(-\hat{x}\hat{x} - \hat{y}\hat{y} + \hat{z}\hat{z}) + \frac{1}{k^2}\nabla\left(-\hat{x}\frac{\partial}{\partial x} - \hat{y}\frac{\partial}{\partial y} + \hat{z}\frac{\partial}{\partial z}\right)\right] G_0(\overline{R}, \overline{R}_i'), \tag{4.252}$$

where

$$G_0(\overline{R}, \overline{R}_i') = \frac{e^{ik|\overline{R}-\overline{R}_i'|}}{4\pi \mid \overline{R} - \overline{R}_i' \mid},$$

with

$$\mid \overline{R} - \overline{R}_i' \mid = \left[(x - x')^2 + (y - y')^2 + (z + z')^2\right]^{\frac{1}{2}}.$$

The negative signs in (4.252) are due to the fact that the x- and y-directed dipoles of the image sources have opposite directions. Then we can write

$$\overline{\overline{G}}_{e1}(\overline{R}, \overline{R}') = \overline{\overline{G}}_{e0}(\overline{R}, \overline{R}') + \overline{\overline{G}}_{e0i}(\overline{R}, \overline{R}_i'). \tag{4.253}$$

Equation (4.252) can be changed to

$$\begin{aligned}\overline{\overline{G}}_{e0i}(\overline{R}, \overline{R}') &= \left[-\overline{\overline{I}} + 2\hat{z}\hat{z} + \frac{1}{k^2}\nabla\left(\hat{x}\frac{\partial}{\partial x'} + \hat{y}\frac{\partial}{\partial y'} + \hat{z}\frac{\partial}{\partial z'}\right)\right] G_0(\overline{R}, \overline{R}_i') \\ &= \left[-\overline{\overline{I}} + \frac{1}{k^2}\nabla\nabla'\right] G_0(\overline{R}, \overline{R}_i') + 2\hat{z}\hat{z}G_0(\overline{R}, \overline{R}_i')\end{aligned} \tag{4.254}$$

and (4.251) can be written in the form

$$\overline{\overline{G}}_{e0}(\overline{R}, \overline{R}') = \left(\overline{\overline{I}} - \frac{1}{k^2}\nabla\nabla'\right) G_0(\overline{R}, \overline{R}'); \tag{4.255}$$

hence

$$\begin{aligned}\overline{\overline{G}}_{e1}(\overline{R}, \overline{R}') = {} & \left(\overline{\overline{I}} - \frac{1}{k^2}\nabla\nabla'\right) [G_0(\overline{R}, \overline{R}') - G_0(\overline{R}, \overline{R}_i')] \\ & + 2\hat{z}\hat{z}G_0(\overline{R}, \overline{R}_i').\end{aligned} \tag{4.256}$$

It can be verified that at $z = 0$, the site of the conducting plane,

$$\hat{z} \times \overline{\overline{G}}_{e1}(\overline{R}, \overline{R}') = 0, \tag{4.257}$$

which is the boundary condition required for $\overline{\overline{G}}_{e1}$. If we introduce the complementary unit dyadic defined by

$$\overline{\overline{I}}_i = -\hat{x}\hat{x} - \hat{y}\hat{y} + \hat{z}\hat{z} = -\overline{\overline{I}} + 2\hat{z}\hat{z};$$

then

$$
\begin{aligned}
\nabla \times \bar{\bar{G}}_{e1}(\bar{R}, \bar{R}') &= \bar{\bar{G}}_{m2}(\bar{R}, \bar{R}') \\
&= \nabla \times \left\{ \bar{\bar{I}} \left[G_0(\bar{R}, \bar{R}') - G_0(\bar{R}, \bar{R}'_i) \right] + 2\hat{z}\hat{z} G_0(\bar{R}, \bar{R}'_i) \right\} \\
&= \nabla \times \left[\bar{\bar{I}} G_0(\bar{R}, \bar{R}') + \bar{\bar{I}}_i G_0(\bar{R}, \bar{R}'_i) \right] \\
&= \nabla G_0(\bar{R}, \bar{R}') \times \bar{\bar{I}} + \nabla G_0(\bar{R}, \bar{R}'_i) \times \bar{\bar{I}}_i.
\end{aligned}
\tag{4.258}
$$

This is the expression for the magnetic dyadic Green function of the second kind.

The method of images can be extended to a conducting wedge with an angle equal to π/n where n is an integer. For $n = 1$, it becomes a plane conducting surface just considered. When n is an integer greater than unity the number of images is finite; the electric dyadic Green function of the first kind can be found accordingly. This problem is assigned as an exercise for a 60° angular wedge.

So far we have only derived the expressions for the free-space dyadic Green functions and the functions for the half space. Starting in the next chapter, we will derive the eigenfunction expansion of the dyadic Green functions for many canonical problems.

Before we conclude this chapter a few words must be said of the singularity property of the electric dyadic Green function. This subject was first investigated by Van Bladel in 1961 and was later expanded to a monograph, *Singular Electromagnetic Fields and Sources* [1991]. A very penetrating and useful analysis is given by Collin [1991, pp. 99–102]. His remarks at the end of Sec. 2-12 of his book are particularly illuminating. His discussion would be better apprehended after the eigenfunction expansions of various dyadic Green functions are found in the subsequent chapters. At the end of Chapter 5, this subject will be discussed using the eigenfunction expansion of the electric dyadic Green function for a rectangular waveguide as a model.

5

Rectangular Waveguides

In this chapter the expressions for the dyadic Green functions of the first and second kinds of a rectangular waveguide will be derived. The method and the general procedure would apply equally well to all other bodies treated in the remaining chapter. The readers should therefore grasp the concept and follow the key steps in a firm manner to assure a smooth passage to the rest of the book.

5-1 RECTANGULAR VECTOR WAVE FUNCTIONS

The vector wave functions are the building blocks of the eigenfunction expansions of various kinds of dyadic Green functions. These functions were first introduced by Hansen [1935, 1936, 1937] in formulating certain electromagnetic problems. The effectiveness of these functions was recognized by Stratton [1941] who, for example, reformulated Mie's theory of the diffraction of a plane electromagnetic wave by a sphere using the spherical vector wave function. In his original work Hansen introduced three kinds of vector wave functions, denoted by $\bar{L}$, $\bar{M}$, and $\bar{N}$, which are solutions of the homogeneous vector Helmholtz equation. Such a presentation was followed by Stratton [1941] and by Morse and Feshbach [1953]. To derive the eigenfunction expansion of the magnetic dyadic Green functions that are solenoidal and satisfy with the vector wave equation, the $\bar{L}$ functions are not needed. If we try to find eigenfunction expansion of the electric dyadic Green functions then the $\bar{L}$ functions are also needed. It will be shown later that once the expressions for $\bar{\bar{G}}_m$ are found, it is relatively simple to find $\bar{\bar{G}}_e$. A detailed analysis will be given to show the direct method of finding $\bar{\bar{G}}_e$. The complexity of this approach will be evident.

A vector wave function, by definition, is an eigenfunction or a characteristic function which is a solution of the homogeneous vector wave equation

$$\nabla \times \nabla \times \overline{F} - \kappa^2 \overline{F} = 0, \tag{5.1}$$

where κ is so far arbitrary. There are two independent sets of vector wave functions which can be constructed using the characteristic function pertaining to a scalar wave equation as the generating function. One kind of vector wave function, called the Cartesian or rectilinear vector wave function, is formed if we let

$$\overline{F} = \nabla \times (\psi_1 \overline{c}), \tag{5.2}$$

where ψ_1 denotes a characteristic function which satisfies the scalar wave equation

$$\nabla^2 \psi + \kappa^2 \psi = 0 \tag{5.3}$$

and $\overline{c}$ denotes a constant vector, such as $\hat{x}, \hat{y}$, or $\hat{z}$. For convenience we shall designate $\overline{c}$ as the *piloting vector* and ψ as the *generating function*. Another kind, designated as the *spherical vector wave function*, will be introduced later, whereby the piloting vector is identified as the spherical radial vector $\overline{R}$. Except for spherical problems, we are always dealing with the Cartesian vector wave functions. When (5.2) is substituted into (5.1), we obtain

$$\nabla \times \left[\nabla \times \nabla \times (\psi_1 \overline{c}) - \kappa^2 (\psi_1 \overline{c})\right] = 0.$$

In view of identity (*A*.18) of Appendix A, this is equivalent to

$$\nabla \times \left[\overline{c} \left(\nabla^2 \psi_1 + \kappa^2 \psi_1\right)\right] = 0;$$

hence (5.2) is a solution for (5.1) if ψ_1 is a solution for (5.3). The set of functions so obtained will be denoted by the letter $\overline{M}$; that is,

$$\overline{M}_1 = \nabla \times (\psi_1 \overline{c}). \tag{5.4}$$

Another set of vector wave functions, denoted by $\overline{N}$, is obtained by letting

$$\overline{N}_2 = \frac{1}{\kappa} \nabla \times \nabla \times (\psi_2 \overline{c}), \tag{5.5}$$

where ψ_2 denotes a characteristic function which also satisfies (5.3) but may be different from the function used to define $\overline{M}_1$. By substituting (5.5) into (5.1), we obtain

$$\nabla \times \nabla \times \left[\overline{c} \left(\nabla^2 \psi_2 + \kappa^2 \psi_2\right)\right] = 0;$$

hence $\overline{N}_2$ is a solution for (5.1), if ψ_2 satisfies (5.3). In the case in which an identical generating function is used for both $\overline{M}$ and $\overline{N}$, we have the following symmetrical relations between these two types of functions:

$$\overline{N} = \frac{1}{\kappa} \nabla \times \overline{M} \tag{5.6}$$

$$\overline{M} = \frac{1}{\kappa} \nabla \times \overline{M}. \tag{5.7}$$

Equation (5.6) follows directly from (5.4) and (5.5), and (5.7) is obtained if we take the curl of (5.6), which yields

$$\nabla \times \overline{N} = \frac{1}{\kappa} \nabla \times \nabla \times \overline{M} = \kappa \overline{M}.$$

Equations (5.6) and (5.7) show why the constant κ is introduced in the definition for the N functions; otherwise the relations between the two sets will not be perfectly symmetrical.

The exact expressions for the two sets of vector wave functions depend, of course, not only on the specific expression for the scalar wave function which is being used but also on the choice of the piloting vector $\bar{c}$. For the rectangular waveguide problems to be discussed in this chapter, we will use the configuration shown in Fig. 5-1 for the orientation of the guide with respect to the rectangular coordinate system, and we will choose the unit vector $\hat{z}$ to represent the piloting vector $\bar{c}$. By doing so, the two sets of vector wave functions thus constructed would provide us the TE and TM modes described in the theory of rectangular waveguides.

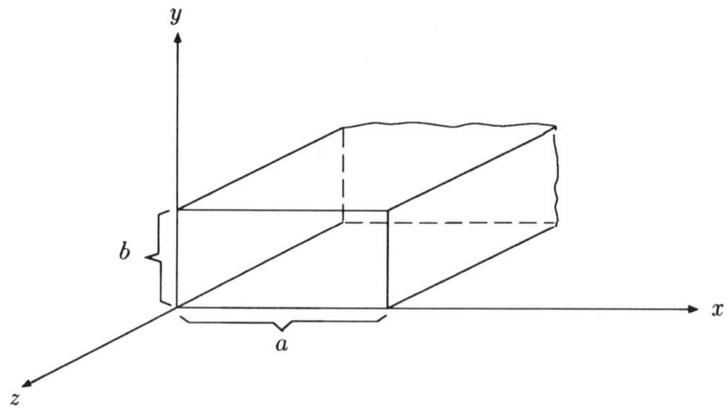

Fig. 5-1 A rectangular waveguide

The scalar wave equation (5.3), when solved by the method of separation of variables, yields

$$\psi = (A \cos k_x x + B \sin k_x x)\,(C \cos k_y y + D \sin k_y y)\, e^{ihz}, \tag{5.8}$$

where $k_x^2 + k_y^2 + h^2 = \kappa^2$.

Now if we demand that the vector wave functions, both $\overline{M}$ and $\overline{N}$, so generated would satisfy the vector Dirichlet boundary condition that on the walls of the guide

$$\hat{n} \times \overline{M} = 0 \tag{5.9}$$

and

$$\hat{n} \times \overline{N} = 0, \tag{5.10}$$

then the constants k_x and k_y must take on certain characteristic values or eigenvalues. The boundary condition specified by (5.9) or (5.10) corresponds to the one satisfied by the electric field on a perfectly conducting surface. Using (5.4), with $\bar{c}$ replaced by $\hat{z}$, it is not difficult to show that the only allowed functions in this case are the cosine functions or the even functions and the constants k_x and k_y should have the following characteristic values,

$$k_x = \frac{m\pi}{a}, \quad m = 0, 1, \dots \tag{5.11}$$

$$k_y = \frac{n\pi}{b}, \quad n = 0, 1, \dots. \tag{5.12}$$

The complete expression and the notation for the set of $\overline{M}$ functions which satisfy the vector Dirichlet condition are

$$\begin{aligned} \overline{M}_{emn}(h) &= \nabla \times [\psi_{emn}(h)\hat{z}] = \nabla \times \left(C_x C_y e^{ihz}\hat{z}\right) \\ &= \left(-k_y C_x S_y \hat{x} + k_x S_x C_y \hat{y}\right) e^{ihz}, \end{aligned} \tag{5.13}$$

where

$$\begin{aligned} k_x &= \frac{m\pi}{a}, \quad k_y = \frac{n\pi}{b} \\ S_x &= \sin k_x x, \quad C_x \cos k_x x \\ S_y &= \sin k_y y, \quad C_y = \cos k_y y \end{aligned}$$

and the constant κ in (5.1) is related to k_x, k_y, h by

$$\kappa^2 = k_x^2 + k_y^2 + h^2 = k_c^2 + h^2 \tag{5.14}$$

with

$$k_c^2 = k_x^2 + k_y^2.$$

The constant k_c denotes the cut-off wave number of a rectangular waveguide. The above abbreviated notations will be used from now on to save some writing. The subscript "e" attached to $\overline{M}_{emn}$ is an abbreviation for the word "even". In a similar manner, the set of $\overline{N}$ functions satisfying the vector Dirichlet condition are given by

$$\begin{aligned} \overline{N}_{omn}(h) &= \frac{1}{\kappa}\nabla \times \nabla \times [\psi_{omn}(h)\hat{z}] = \frac{1}{\kappa}\nabla \times \nabla \times \left(S_x S_y e^{ihz}\hat{z}\right) \\ &= \frac{1}{\kappa}\left(ihk_x C_x S_y \hat{x} + ihk_y S_x C_y \hat{y} + k_c^2 S_x S_y \hat{z}\right) e^{ihz}. \end{aligned} \tag{5.15}$$

The subscript "o" is an abbreviation for the word "odd," and we can still set $m = 0, 1, 2, \dots; n = 0, 1, 2, \dots$ with the modes $m = 0$ or $n = 0$ treated as null modes. It is obvious that $\overline{M}_{emn}(h)$ represents the electric field of the TE_{mn} mode while $\overline{N}_{omn}(h)$ represents that of the TM_{mn} mode. In view of (5.6) and (5.7) we have

$$\overline{M}_{omn}(h) = \frac{1}{\kappa}\nabla \times \overline{N}_{omn}(h) = \nabla \times [\psi_{omn}(h)\hat{z}]$$
$$= (k_y S_x C_y \hat{x} - k_x C_x S_y \hat{y})\, e^{ihz} \tag{5.16}$$

and

$$\overline{N}_{emn}(h) = \frac{1}{\kappa}\nabla \times \overline{M}_{emn}(h) = \frac{1}{\kappa}\nabla \times \nabla \times [\psi_{emn}(h)\hat{z}]$$
$$= \frac{1}{\kappa}\left(-ihk_x S_x C_y \hat{x} - ihk_y C_x S_y \hat{y} + k_c^2 C_x C_y \hat{z}\right) e^{ihz}. \tag{5.17}$$

The vector functions $\overline{M}_{omn}(h)$ and $\overline{N}_{emn}(h)$ are the proper functions to represent the H field in a rectangular waveguide, and they satisfy the vector Neumann condition on the boundary, namely,

$$\hat{n} \times \left\{ \begin{array}{c} \nabla \times \overline{M}_{omn}(h) \\ \nabla \times \overline{N}_{emn}(h) \end{array} \right\} = 0. \tag{5.18}$$

In summary, the vector wave functions which can be used to represent the electromagnetic field inside a rectangular waveguide are of the form

$$\overline{M}_{\substack{e\\o}mn}(h) = \nabla \times \left[\psi_{\substack{e\\o}mn}(h)\hat{z}\right] \tag{5.19}$$

$$\overline{N}_{\substack{e\\o}mn}(h) = \frac{1}{\kappa}\nabla \times \nabla \times \left[\psi_{\substack{e\\o}mn}(h)\hat{z}\right], \tag{5.20}$$

where

$$\psi_{\substack{e\\o}mn}(h) = \begin{pmatrix} \cos\frac{m\pi x}{a}\cos\frac{n\pi y}{b} \\ \sin\frac{m\pi x}{a}\sin\frac{n\pi y}{b} \end{pmatrix} e^{ihz} = \begin{pmatrix} C_x C_y \\ S_x S_y \end{pmatrix} e^{ihz}$$

and

$$\kappa^2 = \left(\frac{m\pi}{a}\right)^2 + \left(\frac{n\pi}{b}\right)^2 + h^2 = k_c^2 + h^2$$

with

$$m, n = 0, 1, 2, \ldots.$$

It is understood that the odd functions with m or $n = 0$ are null modes. For convenience, we will call these functions the *rectangular vector wave functions*.

It should be pointed out that the above functions have been formed as a result of our proper choice of the piloting vector $\bar{c}$ in (5.4) and (5.5). If we had chosen another piloting vector for $\bar{c}$ such as $\hat{x}$, then we would generate a hybrid rectangular vector wave function. For example, the function defined by

$$\overline{M}^{(x)}_{mn}(h) = \nabla \times \left(S_x C_y e^{ihz}\hat{x}\right)$$
$$= (ihS_x C_y \hat{y} + k_y S_x S_y \hat{z})\, e^{ihz}$$

also satisfies the vector Dirichlet condition at $x = 0$ and a, and $y = 0$ and b, but it does not represent a pure TE or TM mode with respect to the z-axis. In fact, it can be shown that

$$\overline{M}_{mn}^{(x)}(h) = \frac{1}{k_c^2}\left[\kappa k_y \overline{N}_{omn}(h) + ihk_x \overline{M}_{emn}(h)\right];$$

thus the function represents a superposition of the TE_{mn} mode and the TM_{mn} mode. In Appendix B, we list a number of these hybrid rectangular vector wave functions together with their relations to the $\overline{M}_{\substack{e\\o}mn}$ and $\overline{N}_{\substack{e\\o}mn}$ functions. The complete expressions for various vector wave functions are also tabulated there for convenient reference.

Having defined the rectangular vector wave functions, we can now discuss their orthogonal properties. It is relatively simple to show that

$$\left.\begin{aligned} &\iiint \overline{M}_{\substack{e\\o}mn}(h)\cdot \overline{N}_{\substack{e\\o}m'n'}(-h')\,dV = 0 \\ &\iiint \overline{M}_{\substack{e\\o}mn}(h)\cdot \overline{N}_{\substack{e\\o}m'n'}(-h')\,dV = 0 \end{aligned}\right\} \tag{5.21}$$

for any combination of even and odd functions and

$$\left.\begin{aligned} &\iiint \overline{M}_{emn}(h)\cdot \overline{M}_{om'n'}(-h')\,dV = 0 \\ &\iiint \overline{N}_{emn}(h)\cdot \overline{N}_{om'n'}(-h')\,dV = 0 \end{aligned}\right\} \tag{5.22}$$

where the m, n, h and m', n', h' denote two sets of eigenvalues which may be distinct or the same and the volume of integration extends from $x = 0$ to a, $y = 0$ to b, and $z = -\infty$ to $+\infty$. Actually, these orthogonal relationships require only the integration with respect to x and y. However, we include the integration with respect to z for completeness because in the eigenfunction expansion for the dyadic Green function we would encounter the volume integral instead of the surface integral. When $m \neq n'$ or $n \neq n'$, it is also quite simple to show that

$$\left.\begin{aligned} &\iiint \overline{M}_{emn}(h)\cdot \overline{M}_{em'n'}(-h')\,dV = 0 \\ &\iiint \overline{M}_{omn}(h)\cdot \overline{M}_{om'n'}(-h')\,dV = 0 \\ &\iiint \overline{N}_{emn}(h)\cdot \overline{N}_{em'n'}(-h')\,dV = 0 \\ &\iiint \overline{N}_{omn}(h)\cdot \overline{N}_{om'n'}(-h')\,dV = 0 \end{aligned}\right\}. \tag{5.23}$$

Thus all the rectangular vector wave functions are orthogonal to each other. What remains is the determination of the normalization factor for the case that $m = m'$ and $n = n'$. We will still let h and h' be different for good reason.

After performing the integration with respect to x and y, we obtain

$$\int_0^a \int_0^b \int_{-\infty}^{\infty} \overline{M}_{emn}(h) \cdot \overline{M}_{emn}(-h')\, dx\, dy\, dz$$
$$= (1+\delta_0)\frac{abk_c^2}{4}\int_{-\infty}^{\infty} e^{i(h-h')z} dz$$
$$= (1+\delta_0)\frac{\pi abk_c^2}{2}\delta(h-h'), \tag{5.24}$$

where the delta function $\delta(h-h')$ results from (1.82) and δ_0 denotes the Kronecker delta function defined as follows:

$$\delta_0 = \begin{cases} 1, & m \text{ or } n = 0 \\ 0, & m \text{ and } n \neq 0. \end{cases}$$

Similarly, we have

$$\int_0^a \int_0^b \int_{-\infty}^{\infty} \overline{M}_{omn}(h)\overline{M}_{omn}(-h')\, dx\, dy\, dz$$
$$= \frac{\pi abk_c^2}{2}\delta(h-h'), \quad m \neq 0, n \neq 0. \tag{5.25}$$

It is recalled that when m or n is equal to zero, the function $\overline{M}_{omn}$ is a null function, and its normalization factor is equal to zero; thus we can include the null modes in (5.25) so the normalization factor for the $\overline{M}_{omn}$ function can also be written in the same form as that of $\overline{M}_{emn}$; that is,

$$\int_0^a \int_0^b \int_{-\infty}^{\infty} \overline{M}_{omn}(h) \cdot \overline{M}_{omn}(-h')\, dx\, dy\, dz$$
$$= (1+\delta_0)\frac{\pi abk_c^2}{2}\delta(h-h'), \; n \neq 0 \text{ and } m \neq 0. \tag{5.26}$$

The normalization factors for the $\overline{N}_{emn}$ and $\overline{N}_{omn}$ functions can be found in the same way. By carrying out the integration, we obtain

$$\int_0^a \int_0^b \int_{-\infty}^{\infty} \overline{N}_{emn}(h) \cdot \overline{N}_{emn}(-h')\, dx\, dy\, dz$$
$$= (1+\delta_0)\frac{\pi abk_c^2}{2\kappa\kappa'}\left(k_c^2 + hh'\right)\delta(h-h') \tag{5.27}$$

$$\kappa^2 = k_c^2 + h^2$$
$$\kappa'^2 = k_c^2 + h'^2.$$

Because of the presence of the delta function $\delta(h-h')$ in (5.27), it is not necessary to distinguish h' and κ' from h and κ in the coefficient in front of the $\delta(h-h')$ function. Equation (5.27), therefore, is equivalent to

$$\int_0^a \int_0^b \int_{-\infty}^{\infty} \overline{N}_{emn}(h) \cdot \overline{N}_{emn}(-h')\, dx\, dy\, dz$$

$$= (1+\delta_0) \frac{\pi ab k_c^2}{2} \delta(h-h'). \tag{5.28}$$

Similarly,

$$\int_0^a \int_0^b \int_{-\infty}^{\infty} \overline{N}_{omn}(h) \cdot \overline{N}_{omn}(-h')\, dx\, dy\, dz$$

$$= (1+\delta_0) \frac{\pi ab k_c^2}{2} \delta(h-h'), \tag{5.29}$$

where the null modes of $\overline{N}_{omn}$ function are included in (5.29). In summary, we found that the $\overline{M}_{\substack{e\\o}mn}$ and $\overline{N}_{\substack{e\\o}mn}$ functions have the same normalization factor as expressed by (5.24), (5.26), (5.28), and (5.29).

5-2 THE METHOD OF $\overline{\overline{G}}_m$

We shall now apply the Ohm-Rayleigh method to first derive the magnetic dyadic Green function of the second kind for a rectangular waveguide. For convenience, we designate this approach as the method of $\overline{\overline{G}}_m$. The function $\overline{\overline{G}}_{m2}(\overline{R}, \overline{R}')$ satisfies the equation

$$\nabla \times \nabla \times \overline{\overline{G}}_{m2}(\overline{R}, \overline{R}') - k^2 \overline{\overline{G}}_{m2}(\overline{R}, \overline{R}') = \nabla \times \left[\overline{\overline{I}}\delta(\overline{R} - \overline{R}')\right], \tag{5.30}$$

defined in the domain $a \geq x \geq 0, b \geq y \geq 0, \infty > z > -\infty$, and the boundary condition

$$\hat{n} \times \nabla \times \overline{\overline{G}}_{m2}(\overline{R}, \overline{R}') = 0 \tag{5.31}$$

at $x = 0$ and $a, y = 0$ and b. At the open ends of the waveguide, the function satisfies a radiation condition which is different from the radiation condition in open space; it corresponds to an outgoing guided wave in both directions from a source placed inside the waveguide. Its specific form will be discussed later. The propagation constant k in (5.30) is considered to be given, and we assume the medium to be air so $k = \omega(\mu_0\epsilon_0)^{\frac{1}{2}}$. For a dielectric medium, we simply replace ϵ_0 by ϵ, which could be complex.

According to the Ohm-Rayleigh method, we first seek an eigenfunction expansion for the source function $\nabla \times \left[\overline{\overline{I}}\delta(\overline{R} - \overline{R}')\right]$ using the solenoidal vector wave functions introduced in the previous section. The proper functions to be used are $\overline{N}_{emn}(h)$ and $\overline{M}_{omn}(h)$ since they satisfy the boundary condition specified by (5.31). Thus we let

$$\nabla \times \left[\bar{\bar{I}}\delta(\bar{R} - \bar{R}')\right] = \int_{-\infty}^{\infty} dh \sum_{m=0}^{\infty} \sum_{n=0}^{\infty} \left[\bar{N}_{emn}(h)\bar{A}_{emn}(h)\right.$$
$$\left. + \bar{M}_{omn}(h)\bar{B}_{omn}(h)\right], \tag{5.32}$$

where $\bar{A}_{emn}(h)$ and $\bar{B}_{omn}(h)$ are two unknown functions or vector coefficients to be determined. These two unknown functions are found by the same method as the one described in Sec. 2-2 for the infinitely long transmission line, except that we are dealing with a three-dimensional problem with a dyadic singular function and the eigenfunctions to be used are the solenoidal vector wave functions. By taking the anterior scalar product of (5.32) with a function $\bar{N}_{em'n'}(-h')$ with certain fixed eigenvalues m', n', and h' and integrating the resultant equation through the entire volume of the waveguide, we obtain

$$\iiint_V \bar{N}_{em'n'}(-h') \cdot \nabla \times \left[\bar{\bar{I}}\delta(\bar{R} - \bar{R}')\right] dV$$
$$= \iiint_V dV \int_{-\infty}^{\infty} dh \sum_m \sum_n \bar{N}_{em'n'}(-h')$$
$$\cdot \left[\bar{N}_{emn}(h)\bar{A}_{emn}(h) + \bar{M}_{omn}(h)\bar{B}_{omn}(h)\right]. \tag{5.33}$$

The integral at the left-hand side of (5.33) can be split into two terms with the aid of (A.18) in Appendix A, which yields

$$\iiint_V \bar{N}_{em'n'}(-h') \cdot \nabla \times \left[\bar{\bar{I}}\delta(\bar{R} - \bar{R}')\right] dV$$
$$= \iiint_V \left\{\nabla \times \bar{N}_{em'n'}(-h') \cdot \bar{\bar{I}}\delta(\bar{R} - \bar{R}')\right.$$
$$\left. - \nabla \cdot \left[\bar{N}_{em'n'}(-h') \times \bar{\bar{I}}\delta(\bar{R} - \bar{R}')\right]\right\} dV$$
$$= \nabla' \times \bar{N}'_{em'n'}(-h')$$
$$- \oiint_S \hat{n} \cdot \left[\bar{N}_{em'n'}(-h') \times \bar{\bar{I}}\delta(\bar{R} - \bar{R}')\right] dS, \tag{5.34}$$

where we have already made use of the dyadic Gauss theorem to convert one volume integral into a surface integral. The surface integral in (5.34) vanishes because $\bar{R}'$ is located inside V. The functions $\nabla' \times \bar{N}'$ are defined in the primed variables x', y', z' associated with the position vector $\bar{R}$. As a result of the orthogonal relationships between the two sets of vector wave functions, (5.23) and (5.27), we obtain

$$\nabla' \times \overline{N}'_{em'n'}(-h') = (1+\delta_0)\frac{\pi abk_c'^2}{2}\bar{A}_{em'n'}(h'), \tag{5.35}$$

where

$$k_c'^2 = \left(\frac{m'\pi}{a}\right)^2 + \left(\frac{n'\pi}{b}\right)^2$$

or

$$\begin{aligned}\bar{A}_{em'n'}(h') &= \frac{(2-\delta_0)}{\pi abk_c'^2}\nabla' \times \overline{N}'_{em'n'}(-h') \\ &= \frac{2-\delta_0}{\pi abk_c'^2}\kappa'\overline{M}'_{em'n'}(-h').\end{aligned} \tag{5.36}$$

By deleting the prime for the eigenvalues, but not the prime in the $\overline{M}'$ function, we can change (5.36) to the form

$$\bar{A}_{emn}(h) = \frac{(2-\delta_0)\,\kappa}{\pi abk_c^2}\overline{M}'_{emn}(-h). \tag{5.37}$$

This is the expression for the unknown coefficient $\bar{A}_{emn}(h)$ which we are seeking. In a similar manner, by taking the anterior scalar product of (5.32) with $\overline{M}_{om'n'}(-h')$ and doing the same routine, we find

$$\bar{B}_{omn}(h) = \frac{(2-\delta_0)\,\kappa}{\pi abk_c^2}\overline{N}'_{omn}(-h). \tag{5.38}$$

The eigenfunction expansion of $\nabla \times \left[\bar{\bar{I}}\delta(\bar{R}-\bar{R}')\right]$ is, therefore, given by

$$\begin{aligned}\nabla \times \left[\bar{\bar{I}}\delta(\bar{R}-\bar{R}')\right] &= \int_{-\infty}^{\infty} dh \sum_{m,n} \frac{(2-\delta_0)\,\kappa}{\pi abk_c^2} \\ &\quad \cdot \left[\overline{N}_{emn}(h)\overline{M}'_{emn}(-h)+\overline{M}_{omn}(h)\overline{N}'_{omn}(-h)\right],\end{aligned} \tag{5.39}$$

where

$$\kappa = \left(k_c^2 + h^2\right)^{\frac{1}{2}}$$

$$k_c^2 = \left(\frac{m\pi}{a}\right)^2 + \left(\frac{n\pi}{b}\right)^2 .$$

To find $\bar{\bar{G}}_{m2}(\bar{R},\bar{R}')$ we let

$$\begin{aligned}\bar{\bar{G}}_{m2}(\bar{R},\bar{R}') = \int_{-\infty}^{\infty} dh \sum_{m,n} & \frac{(2-\delta_0)\,\kappa}{\pi abk_c^2} \\ & \cdot \left[a(h)\overline{N}_{emn}(h)\overline{M}'_{emn}(-h)\right. \\ & + \left. b(h)\overline{M}_{omn}(h)\overline{N}'_{omn}(-h)\right],\end{aligned} \tag{5.40}$$

where the coefficient $a(h)$ and $b(h)$ can be determined by substituting (5.39) and (5.40) into (5.30) that yields

$$a(h) = b(h) = \frac{1}{\kappa^2 - k^2}, \tag{5.41}$$

where we have made use of the identity

$$\nabla \times \nabla \times \begin{bmatrix} \overline{N}_{emn} \\ \overline{M}_{omn} \end{bmatrix} = \kappa^2 \begin{bmatrix} \overline{N}_{emn} \\ \overline{M}_{omn} \end{bmatrix}.$$

The eigenfunction expansion of $\overline{\overline{G}}_{m2}(\overline{R}, \overline{R}')$, therefore, is given by

$$\overline{\overline{G}}_{m2}(\overline{R}, \overline{R}') = \int_{-\infty}^{\infty} dh \sum_{m,n} \frac{(2-\delta_0)\,\kappa}{\pi ab k_c^2 \left(\kappa^2 - k^2\right)} \cdot \left[\overline{N}_{emn}(h)\overline{M}'_{emn}(-h) + \overline{M}_{omn}(h)\overline{N}'_{omn}(-h)\right]. \tag{5.42}$$

The Fourier integral in (5.42) can be evaluated in a closed form by applying the method of contour integration. The integrand has two poles at $h = \pm\left(k^2 - k_c^2\right)^{\frac{1}{2}}$ because $\kappa^2 - k^2 = k_c^2 + h^2 - k^2$, and at infinity it fulfills the requirement of the Jordan lemma in the theory of complex variables. The result gives

$$\begin{aligned} \overline{\overline{G}}{}^{\pm}_{m2}(\overline{R}, \overline{R}') =& \frac{ik}{ab} \sum_{m,n} \frac{2-\delta_0}{k_c^2 k_g} \\ & \cdot \left[\overline{N}_{emn}\left(\pm k_g\right) \overline{M}'_{emn}\left(\mp k_g\right)\right. \\ & \left. + \overline{M}_{omn}\left(\pm k_g\right) \overline{N}_{omn}\left(\mp k_g\right)\right], \quad z \gtrless z', \end{aligned} \tag{5.43}$$

where

$$k_g = \left(k^2 - k_c^2\right)^{\frac{1}{2}} \text{ with Real } k_g > 0, \text{Imag } k_g > 0.$$

In (5.43) the top line applies to $z > z'$ and the bottom line for $z < z'$. At $z = z'$, the function is discontinuous. For a discontinuous magnetic dyadic Green function, we have derived the equation (4.36),

$$\hat{n} \times \left(\overline{\overline{G}}{}^{+}_{m} - \overline{\overline{G}}{}^{-}_{m}\right) = \overline{\overline{I}}_s \delta\left(\overline{r} - \overline{r}'\right). \tag{5.44}$$

For the present problem, it becomes

$$\hat{z} \times \left(\overline{\overline{G}}{}^{+}_{m2} - \overline{\overline{G}}{}^{-}_{m2}\right) = \left(\overline{\overline{I}} - \hat{z}\hat{z}\right) \delta\left(x - x'\right) \delta\left(y - y'\right), \tag{5.45}$$

where $\overline{\overline{G}}{}^{+}_{m2}$ is for $z > z'$ and $\overline{\overline{G}}{}^{-}_{m2}$ is for $z < z'$. The point source is located at $\overline{R}'$.

To find $\overline{\overline{G}}_{e1}(\overline{R}, \overline{R}')$, we use the relationship

$$\nabla \times \overline{\overline{G}}_{m2}(\overline{R}, \overline{R}') = \overline{\overline{I}}\delta(\overline{R} - \overline{R}') + k^2\overline{\overline{G}}_{e1}(\overline{R}, \overline{R}'). \tag{5.46}$$

Since $\overline{\overline{G}}_{m2}$ is discontinuous at $z = z'$, we can write

$$\overline{\overline{G}}_{m2}(\overline{R}, \overline{R}') = \overline{\overline{G}}_{m2}^{+}(\overline{R}, \overline{R}')U(z - z') + \overline{\overline{G}}_{m2}^{-}(\overline{R}, \overline{R}')U(z' - z),$$

where the two unit step functions are defined by

$$U(z - z') = \begin{cases} 1, & z > z' \\ 0, & z < z' \end{cases}$$
$$U(z' - z) = \begin{cases} 1, & z < z' \\ 0, & z > z'. \end{cases}$$

Thus

$$\begin{aligned}\nabla \times \overline{\overline{G}}_{m2}(\overline{R}, \overline{R}') &= \left[\nabla \times \overline{\overline{G}}_{m2}^{+}(\overline{R}, \overline{R}')\right] U(z - z') \\ &\quad + \nabla U(z - z') \times \overline{\overline{G}}_{m2}^{+}(\overline{R}, \overline{R}') \\ &\quad + \left[\nabla \times \overline{\overline{G}}_{m2}^{-}(\overline{R}, \overline{R}')\right] U(z' - z) \\ &\quad + \nabla U(z' - z) \times \overline{\overline{G}}_{m2}^{-}(\overline{R}, \overline{R}'),\end{aligned}$$

where we have made use of the dyadic identity (A.24) in Appendix A. According to the theory of generalized functions,

$$\nabla U(z - z') = \hat{z}\delta(z - z')$$
$$\nabla U(z' - z) = -\hat{z}\delta(z - z');$$

hence

$$\begin{aligned}\nabla \times \overline{\overline{G}}_{m2}(\overline{R}, \overline{R}') &= \left[\nabla \times \overline{\overline{G}}_{m2}^{+}(\overline{R}, \overline{R}')\right] U(z - z') \\ &\quad + \left[\nabla \times \overline{\overline{G}}_{m2}^{-}(\overline{R}, \overline{R}')\right] U(z' - z) \\ &\quad + \hat{z}\delta(z - z') \times \left[\overline{\overline{G}}_{m2}^{+}(\overline{R}, \overline{R}') - \overline{\overline{G}}_{m2}^{-}(\overline{R}, \overline{R}')\right].\end{aligned} \tag{5.47}$$

In view of (5.45), the above equation can be written in the form

$$\begin{aligned}\nabla \times \overline{\overline{G}}_{m2}(\overline{R}, \overline{R}') &= \left[\nabla \times \overline{\overline{G}}_{m2}^{+}(\overline{R}, \overline{R}')\right] U(z - z') \\ &\quad + \left[\nabla \times \overline{\overline{G}}_{m2}^{-}(\overline{R}, \overline{R}')\right] U(z' - z) \\ &\quad + \left(\overline{\overline{I}} - \hat{z}\hat{z}\right)\delta(x - x')\delta(y - y')\delta(z - z'),\end{aligned} \tag{5.48}$$

which is applicable to all values of x, y, z. Since $\delta(x-x')\,\delta(y-y')\,\delta(z-z')$ is the same as $\delta(\overline{R}-\overline{R}')$, by substituting (5.48) into (5.46) we obtain

$$\begin{aligned}\overline{\overline{G}}_{e1}(\overline{R},\overline{R}') =& \frac{1}{k^2}\left\{-\hat{z}\hat{z}\delta(\overline{R}-\overline{R}') + \left[\nabla\times\overline{\overline{G}}^{+}_{m2}(\overline{R},\overline{R}')\right]U(z-z')\right.\\ &\left.+\left[\nabla\times\overline{\overline{G}}^{-}_{m2}(\overline{R},\overline{R}')\right]U(z'-z)\right\}\\ =& -\frac{1}{k^2}\hat{z}\hat{z}\delta(\overline{R}-\overline{R}')\\ &+\frac{i}{ab}\sum_{m,n}\frac{2-\delta_0}{k_c^2 k_g}\left[\overline{M}_{emn}(\pm k_g)\,\overline{M}'_{emn}(\mp k_g)\right.\\ &\left.+\overline{N}_{omn}(\pm k_g)\,\overline{N}'_{omn}(\mp k_g)\right],\ z \gtrless z'. \end{aligned} \tag{5.49}$$

In the double series, the top line applies to $z > z'$, the bottom line to $z < z'$. The $\overline{M}_{emn}$ and $\overline{N}_{omn}$ functions in (5.49) result from the relationships

$$\nabla\times\overline{N}_{emn}(\pm k_g) = k\overline{M}_{emn}(\pm k_g) \tag{5.50}$$

$$\nabla\times\overline{M}_{omn}(\pm k_g) = k\overline{N}_{omn}(\pm k_g). \tag{5.51}$$

Knowing the expression of $\overline{\overline{G}}_{e1}(\overline{R},\overline{R}')$, the electric field in the waveguide can be calculated by using the formula

$$\overline{E}(\overline{R}) = i\omega\mu_0\iiint\overline{\overline{G}}_{e1}(\overline{R},\overline{R}')\cdot\overline{J}(\overline{R}')\,dV'. \tag{5.52}$$

Based on the structure of $\overline{\overline{G}}_{e1}$, it is convenient to designate the posterior functions $\overline{M}'_{emn}$ and $\overline{N}'_{omn}$ as the excitation functions and the anterior functions $\overline{M}_{emn}$ and $\overline{N}_{omn}$ as the field functions. The former corresponds to a TE_{mn} mode and the latter a TM_{mn} mode because $\overline{M}_{emn}$ does not have a z-component and the magnetic field is proportional to $\nabla\times\overline{E}(\overline{R})$ so the z component of $\nabla\times\overline{N}_{omn}(\pm k_g) = k\overline{M}_{omn}(\pm k_g)$ is absent; hence the name TM_{mn} mode. When the integral of the scalar product between $\overline{J}(\overline{R}')$ and a certain excitation function is zero, it implies that the corresponding mode is not excited. For example, if $\overline{J}(\overline{R}')$ has only a longitudinal component like that of a Hertzian dipole pointed in the z-direction, then there is no coupling between $\overline{J}(\overline{R}')$ and $\overline{M}'_{emn}$ so only the TM modes are excited.

When a waveguide is excited by an aperture or slot field along the wall, we can apply the formula derived in Chapter 4, (4.183), to calculate the field inside the waveguide; namely,

$$\overline{E}(\overline{R}) = -\oiint_{S_A} \left[\nabla \times \overline{\overline{G}}_{e2}(\overline{R}, \overline{R}')\right] \cdot \left[\hat{n}' \times \overline{E}(\overline{R}')\right] dS'. \tag{5.53}$$

By definition $\nabla \times \overline{\overline{G}}_{e2} = \overline{\overline{G}}_{m1}$, thus, we can apply the same method used in deriving $\overline{\overline{G}}_{m2}$ to find

$$\begin{aligned}\overline{\overline{G}}_{m1}(\overline{R}, \overline{R}') =& \frac{ik}{ab} \sum_{m,n} \frac{2-\delta_0}{k_c^2 k_g} \\ &\cdot \left[\overline{M}_{emn}(\pm k_g)\, \overline{N}'_{emn}(\pm k_g)\right. \\ &\left. + \overline{N}_{omn}(\pm k_g)\, \overline{M}_{omn}(\mp k_g)\right], \quad z \gtrless z'.\end{aligned} \tag{5.54}$$

Finally, we shall discuss the radiation condition and demonstrate the vanishing of the surface integral evaluated at the infinite ends of a waveguide for $z \gtrless z'$. The surface integral in (4.71), with $\overline{\overline{G}}_e$ replaced by $\overline{\overline{G}}_{e1}$ and $\hat{n} = \hat{z}$ for that portion of the surface, has the form

$$\begin{aligned}-\oiint_{S_c} \Big\{&\left[\hat{z} \times \nabla \times \overline{E}(\overline{R})\right] \cdot \overline{\overline{G}}_{e1}(\overline{R}, \overline{R}') \\ &- \overline{E}(\overline{R}) \cdot \left[\hat{z} \times \nabla \times \overline{\overline{G}}_{e1}(\overline{R}, \overline{R}')\right]\Big\} dS,\end{aligned} \tag{5.55}$$

where S_c denotes the cross-sectional area of the waveguide. The electric field of a typical TE_{mn} mode of the total field $\overline{E}(\overline{R})$ will be denoted by $a_{mn}\overline{M}_{emn}(k_g)$, and the term in $\overline{\overline{G}}_{e1}(\overline{R}, \overline{R}')$ which is responsible for the excitation of this mode will be denoted by $\overline{M}_{emn}(k_g)\, \overline{A}'_{emn}(k_g)$; then, we find

$$\begin{aligned}\hat{z} \times \nabla \times \left[a_{mn}\overline{M}_{emn}(k_g)\right] =& \hat{z} \times \left[k a_{mn}\overline{N}_{emn}(k_g)\right] \\ =& -ik_g a_{mn}\overline{M}_{emn}(k_g)\end{aligned} \tag{5.56}$$

$$\begin{aligned}\hat{z} \times \nabla \times \left[\overline{M}_{emn}(k_g)\, \overline{A}'_{emn}(k_g)\right] =& \hat{z} \times \left[k\overline{N}_{emn}(k_g)\overline{A}'_{emn}(k_g)\right] \\ =& -ikg\overline{M}_{emn}(k_g)\overline{A}'_{emn}(k_g).\end{aligned} \tag{5.57}$$

Substituting these two terms into (5.55), the two scalar products cancel each other. The terms in $\overline{\overline{G}}_{e1}(\overline{R}, \overline{R}')$ with different eigenvalues and those belonging to the TM modes do not interact with $\overline{M}_{emn}$ because of the orthogonal properties of these vector wave functions. For the TM_{mn} modes, we start with $\overline{N}_{omn}(k_g)$. The procedure and the result are the same. We have now verified the so-called radiation condition for the field at one end of the waveguide. For the other end, $z < z'$, the same conclusion can be obtained by using the functions $\overline{M}_{emn}(-k_g)$ and $\overline{N}_{omn}(-k_g)$.

5-3 THE METHOD OF $\overline{\overline{G}}_e$

The expression for $\overline{\overline{G}}_{e1}$ which was obtained by the method of $\overline{\overline{G}}_m$ can be obtained directly by applying the Ohm-Rayleigh method to the differential equation for $\overline{\overline{G}}_{e1}$, namely,

$$\nabla \times \nabla \times \overline{\overline{G}}_{e1}(\overline{R}, \overline{R}') - k^2 \overline{\overline{G}}_{e1}(\overline{R}, \overline{R}') = \overline{\overline{I}}\delta(\overline{R} - \overline{R}'), \tag{5.58}$$

at the expense of a much more complicated formulation. The complication is partly due to the fact that $\overline{\overline{G}}_{e1}$, unlike $\overline{\overline{G}}_{m2}$, is not a solenoidal dyadic function because

$$\begin{aligned} \nabla \cdot \overline{\overline{G}}_{e1}(\overline{R}, \overline{R}') &= -\frac{1}{k^2} \nabla \cdot \left[\overline{\overline{I}}\delta(\overline{R} - \overline{R}')\right] \\ &= -\frac{1}{k^2} \nabla \delta(\overline{R} - \overline{R}'), \end{aligned} \tag{5.59}$$

which is not equal to zero except for $\overline{R} \neq \overline{R}'$. For this reason, the solenoidal vector wave functions $\overline{M}_{emn}$ and $\overline{N}_{omn}$ are not sufficient; we need another non-solenoidal set. The additional set of vector wave functions will be denoted by $\overline{L}_{omn}$. As had been mentioned before, $\overline{L}$, $\overline{M}$, and $\overline{N}$ are the three sets originally introduced by Hansen [1935] in formulating electromagnetic problems.

The function $\overline{L}_{omn}(h)$ is defined by

$$\begin{aligned} \overline{L}_{omn}(h) &= \nabla \left[\psi_{omn}(h)\right] = \nabla \left(S_x S_y e^{ihz}\right) \\ &= \left(k_x C_x S_y \hat{x} + k_y S_x C_y \hat{y} + ih S_x S_y \hat{z}\right) e^{ihz} \end{aligned} \tag{5.60}$$

for $m, n = 1, 2, \ldots$. They are solutions of the homogeneous vector Helmholtz equation

$$\nabla^2 \overline{F} + \kappa^2 \overline{F} = 0, \tag{5.61}$$

where

$$\kappa^2 = k_x^2 + k_y^2 + h^2 = k_c^2 + h^2.$$

The constants k_x and k_y and the functions S_x, C_x, S_y, and C_y are the same as the ones appearing in Sec. 5-1. The orthogonal relationships of this set of functions themselves and with the two other sets are listed below:

$$\iiint_V \overline{L}_{omn}(h) \cdot \overline{L}_{om'n'}(-h')\, dV = \begin{cases} 0, m \neq n', n \neq n' \\ \dfrac{\pi ab \kappa^2}{2} \delta(h - h'), m = m', n = n' \end{cases} \tag{5.62}$$

$$\iiint_V \overline{L}_{omn}(h) \cdot \overline{M}_{em'n'}(-h')\, dV = 0 \tag{5.63}$$

$$\iiint_V \overline{L}_{omn}(h) \cdot \overline{N}_{om'n'}(-h)\, dV$$

$$= \begin{cases} 0, m \neq m', n \neq n' \\ \dfrac{i\pi ab k_c^2}{2\kappa}(h-h')\,\delta(h-h') \text{ for } m=m', n=n'. \end{cases} \tag{5.64}$$

It is observed that $\overline{L}_{omn}(h)$ and $\overline{N}_{omn}(-h')$ are formally not orthogonal in the spatial domain, but when the h domain is included,

$$\int_{-\infty}^{\infty} dh \iiint_V \overline{L}_{omn}(h) \cdot \overline{N}_{omn}(-h)\, dV = 0. \tag{5.65}$$

According to the Ohm-Rayleigh method we let

$$\overline{\overline{I}}\delta(\overline{R}-\overline{R}') = \int_{-\infty}^{\infty} dh \sum_{m,n} \Big[\overline{L}_{omn}(h)\overline{A}_{omn}(h) + \overline{M}_{emn}(h)\overline{B}_{emn}(h) + \overline{N}_{omn}(h)\overline{C}_{omn}(h) \Big].$$

As a result of the orthogonal property of the three sets of vector wave functions in the spatial and h domain we can readily determine the coefficients $\overline{A}, \overline{B}, \overline{C}$ in (5.65). They are

$$\overline{A}_{omn}(h) = \frac{2-\delta_0}{\pi ab\kappa^2}\overline{L}'_{omn}(-h)$$

$$\overline{B}_{emn}(h) = \frac{2-\delta_0}{\pi ab k_c^2}\overline{M}'_{emn}(-h)$$

$$\overline{C}_{omn}(h) = \frac{2-\delta_0}{\pi ab k_c^2}\overline{N}'_{omn}(-h).$$

The factor $2-\delta_0$ in $\overline{A}_0$ and $\overline{C}_0$ is always equal to 2 because for $m=0$ and/or $n=0$, the functions $\overline{L}_0$ and $\overline{N}_0$ are null. The eigenfunction expansion for $\overline{\overline{I}}\delta(\overline{R}-\overline{R}')$, therefore, is given by

$$\overline{\overline{I}}\delta(\overline{R}-\overline{R}') = \int_{-\infty}^{\infty} dh \sum_{m,n} C_{mn} \Big[\frac{k_c^2}{\kappa^2}\overline{L}_{omn}(h)\overline{L}'_{omn}(-h) + \overline{M}_{emn}(h)\overline{M}'_{emn}(-h) + \overline{N}_{omn}(h)\overline{N}'_{omn}(-h) \Big] \tag{5.66}$$

where

$$C_{mn} = \frac{2-\delta_0}{\pi ab k_c^2}, \quad \delta_0 = \begin{cases} 1, & n \text{ or } m = 0 \\ 0, & n \text{ and } m \neq 0 \end{cases}$$

$$k_c^2 = k_x^2 + k_y^2.$$

We now let

$$\overline{\overline{G}}_{e1}(\overline{R}, \overline{R}') = \int_{-\infty}^{\infty} dh \sum_{m,n} C_{mn} \left(a\frac{k_c^2}{\kappa^2}\overline{L}_0\overline{L}'_0 + b\overline{M}_e\overline{M}'_e + c\overline{N}_0\overline{N}'_0 \right). \tag{5.67}$$

The subscript "mn" attached to the functions has been deleted to simplify the writing. Substituting (5.66) and (5.67) into (5.58), we find

$$a = -\frac{1}{k^2}, \quad b = c = \frac{1}{\kappa^2 - k^2} = \frac{1}{h^2 + k_c^2 - k^2}.$$

Thus the complete expression for $\overline{\overline{G}}_{e1}(\overline{R}, \overline{R}')$ is given by

$$\overline{\overline{G}}_{e1}(\overline{R}, \overline{R}') = \int_{-\infty}^{\infty} dh \sum_{m,n} C_{mn} \left[-\frac{k_c^2}{k^2\kappa^2} \overline{L}_o \overline{L}_o' + \frac{1}{\kappa^2 - k^2} \left(\overline{M}_e \overline{M}_e' + \overline{N}_o \overline{N}_o' \right) \right]. \tag{5.68}$$

In order to apply the residue theorem to (5.68) we must first extract the part in (5.68) which does not satisfy the Jordan lemma. To do so, we write

$$\overline{L}_o = \overline{L}_{ot} + \overline{L}_{oz}$$
$$\overline{N}_o = \overline{N}_{ot} + \overline{N}_{oz},$$

where $\overline{L}_{ot}$ and $\overline{N}_{ot}$ denote the transversal vector components of these two functions and $\overline{L}_{oz}$ and $\overline{N}_{oz}$ their z-components. By definition

$$\overline{L}_{ot} = (k_x C_x S_y \hat{x} + k_y S_x C_y \hat{y})\, e^{ihz} \tag{5.69}$$
$$\overline{L}_{oz} = ih S_x S_y e^{ihz} \hat{z} \tag{5.70}$$
$$\overline{N}_{ot} = \frac{1}{\kappa} ih\, (k_x C_x S_y \hat{x} + k_y S_x C_y \hat{y})\, e^{ihz} \tag{5.71}$$
$$\overline{N}_{oz} = \frac{1}{\kappa} k_c^2 S_x S_y e^{ihz} \hat{z}. \tag{5.72}$$

In terms of these functions, (5.68) can be written in the form

$$\overline{\overline{G}}_{e1}(\overline{R}, \overline{R}') = \int_{-\infty}^{\infty} dh \sum_{m,n} C_{mn} \left\{ \frac{1}{\kappa^2 - k^2} \overline{M}_e \overline{M}_e' + \frac{\kappa^2}{k^2(\kappa^2 - k^2)} \left[\frac{k^2 - k_c^2}{h^2} \overline{N}_{ot} \overline{N}_{ot}' + \overline{N}_{ot} \overline{N}_{oz}' + \overline{N}_{oz} \overline{N}_{ot}' + \frac{k^2 - h^2}{k_c^2} \overline{N}_{oz} \overline{N}_{oz}' \right] \right\}, \tag{5.73}$$

where we have expressed $\overline{L}_{ot}$ and $\overline{L}_{oz}$ in terms of $\overline{N}_{ot}$ and $\overline{N}_{oz}$, and similarly, for the primed functions; namely,

$$\overline{L}_{ot} = \frac{-i\kappa}{h} \overline{N}_{ot}, \quad \overline{L}_{ot}' = \frac{i\kappa}{h} \overline{N}_{ot}'$$

and

$$\overline{L}_{oz} = \frac{ih\kappa}{k_c^2} \overline{N}_{oz}, \quad \overline{L}_{oz}' = \frac{-ih\kappa}{k_c^2} \overline{N}_{oz}'.$$

The singular term in (5.73) is contained in the component $\overline{N}_{oz}\overline{N}'_{oz}$. From (5.66), we have

$$\hat{z}\hat{z}\delta(\overline{R}-\overline{R}') = \int_{-\infty}^{\infty} dh \sum_{m,n} C_{mn} \left[\frac{k_c^2}{\kappa^2} \overline{L}_{oz}\overline{L}'_{oz} + \overline{N}_{oz}\overline{N}'_{oz} \right]$$
$$= \int_{-\infty}^{\infty} dh \sum_{m,n} C_{mn} \frac{\kappa^2}{k_c^2} \overline{N}_{oz}\overline{N}'_{oz}. \tag{5.74}$$

Thus (5.73) can be split into

$$\overline{\overline{G}}_{e1}(\overline{R},\overline{R}') = -\int_{-\infty}^{\infty} dh \sum_{m,n} C_{mn} \frac{\kappa^2}{k^2 k_c^2} \overline{N}_{oz}\overline{N}'_{oz}$$
$$+ \int_{-\infty}^{\infty} dh \sum_{m,n} C_{mn} \left\{ \frac{1}{\kappa^2 - k^2} \overline{M}_e \overline{M}'_e \right.$$
$$+ \frac{\kappa^2}{k^2(\kappa^2-k^2)} \left[\frac{k^2-k_c^2}{h^2} \overline{N}_{ot}\overline{N}'_{ot} + \overline{N}_{ot}\overline{N}'_{oz} \right.$$
$$\left.\left. + \overline{N}_{oz}\overline{N}'_{ot} + \overline{N}_{oz}\overline{N}'_{oz} \right] \right\}. \tag{5.75}$$

The splitting can be verified because we have the algebraic relations

$$\frac{\kappa^2}{\kappa^2-k^2} = 1 + \frac{k^2}{\kappa^2-k^2}$$
$$= \frac{\kappa^2}{h^2} + \left(\frac{\kappa^2}{\kappa^2-k^2} \right) \left(\frac{k^2-k_c^2}{h^2} \right). \tag{5.76}$$

In view of (5.74), the first integral in (5.75) is equal to

$$-\frac{1}{k^2} \hat{z}\hat{z}\delta(\overline{R}-\overline{R}')$$

and the second integral, denoted by $\overline{\overline{S}}_{e1}(\overline{R},\overline{R}')$, can be evaluated in a closed form by the method of contour integration as the integrand decays to zero at infinity in the upper and the lower h-plane. The final result is given by

$$\overline{\overline{S}}_{e1}(\overline{R},\overline{R}') = \frac{i}{ab} \sum_{m,n} \frac{2-\delta_0}{k_c^2 k_g} \left[\overline{M}_{emn}(\pm k_g) \overline{M}'_{emn}(\mp k_g) \right.$$
$$\left. + \overline{N}_{omn}(\pm k_g) \overline{N}'_{omn}(\mp k_g) \right], \quad z \gtrless z'. \tag{5.77}$$

Thus

$$\overline{\overline{G}}_{e1}(\overline{R},\overline{R}') = -\frac{1}{k^2} \hat{z}\hat{z}\delta(\overline{R}-\overline{R}') + \overline{\overline{S}}_{e1}(\overline{R},\overline{R}'), \tag{5.78}$$

which is the same as (5.54).

It should be pointed out that in the first edition of this book [Tai, 1971], the method of $\overline{\overline{G}}_e$ was improperly formulated. Consequently, the singular term $-\hat{z}\hat{z}\delta(\overline{R} - \overline{R}')/k^2$ was missing in that work. This mistake was later corrected [Tai, 1973]. The correction is essentially based on the method of $\overline{\overline{G}}_m$ but is not carried out in the same manner as described here.

5-4 THE METHOD OF $\overline{\overline{G}}_A$

In addition to the two methods discussed so far in finding the eigenfunction expansion of $\overline{\overline{G}}_{e1}$, it is of interest to mention briefly the method of $\overline{\overline{G}}_A$ which is based on the dyadic version of the method of potentials. The system of equations for $\overline{\overline{G}}_e$ and $\overline{\overline{G}}_m$ according to (4.23)–(4.26) are

$$\nabla \times \overline{\overline{G}}_e = \overline{\overline{G}}_m \tag{5.79}$$

$$\nabla \times \overline{\overline{G}}_m = \overline{\overline{I}}\delta(\overline{R} - \overline{R}') + k^2\overline{\overline{G}}_e \tag{5.80}$$

$$\nabla \cdot \overline{\overline{G}}_e = -\frac{1}{k^2}\nabla\delta(\overline{R} - \overline{R}') \tag{5.81}$$

$$\nabla \cdot \overline{\overline{G}}_m = 0. \tag{5.82}$$

In view of (5.82) we can define a dyadic Green function of the potential type, denoted by $\overline{\overline{G}}_A$, such that

$$\overline{\overline{G}}_m = \nabla \times \overline{\overline{G}}_A. \tag{5.83}$$

Substituting it into (5.79), we find that the difference between $\overline{\overline{G}}_e$ and $\overline{\overline{G}}_A$ is a solenoidal dyadic that enables us to define a vector function $\overline{\Phi}$ such that

$$\overline{\overline{G}}_e = \overline{\overline{G}}_A - \nabla\overline{\Phi}. \tag{5.84}$$

$\overline{\Phi}$ is then the vector form of the dynamic scalar potential function. Putting (5.81) and (5.84) into (5.80) and introducing the gauge condition

$$\nabla \cdot \overline{\overline{G}}_A = -k^2\overline{\Phi} \tag{5.85}$$

yields

$$\nabla^2\overline{\overline{G}}_A + k^2\overline{\overline{G}}_A = \overline{\overline{I}}\delta(\overline{R} - \overline{R}'), \tag{5.86}$$

which is the wave equation for $\overline{\overline{G}}_A$. For the waveguide problem under consideration the function should be of the first kind, denoted by $\overline{\overline{G}}_{A1}$. Its eigenfunction expansion can be found by a similar procedure as had been done for $\overline{\overline{G}}_{e1}$. The result gives

$$\overline{\overline{G}}_{A1}(\overline{R}, \overline{R}') = \int_{-\infty}^{\infty} dh \sum_{m,n} C_{mn} \frac{1}{\kappa^2 - k^2} \left[\frac{k_c^2}{\kappa^2} \overline{L}_0 \overline{L}_0' \right.$$

$$\left. + \overline{M}_e \overline{M}_e' + \overline{N}_0 \overline{N}_0' \right]. \tag{5.87}$$

Now

$$\overline{\overline{G}}_{e1}(\overline{R}, \overline{R}') = \overline{\overline{G}}_{A1} + \frac{1}{k^2} \nabla\nabla \cdot \overline{\overline{G}}_{A1}. \tag{5.88}$$

Substituting (5.87) into (5.88), we obtain

$$\overline{\overline{G}}_{e1}(\overline{R}, \overline{R}') = \int_{-\infty}^{\infty} dh \sum_{m,n} C_{mn} \left[-\frac{k_c^2}{k^2 \kappa^2} \overline{L}_0 \overline{L}_0' \right.$$

$$\left. + \frac{1}{\kappa^2 - k^2} \left(\overline{M}_e \overline{M}_e' + \overline{N}_0 \overline{N}_0' \right) \right], \tag{5.89}$$

which is the same as (5.68). The rest is identical to the treatment following that equation.

In conclusion, it is quite clear that the method of $\overline{\overline{G}}_m$ is the simplest; its formulation does not involve the nonsolenoidal vector wave function $\overline{L}_{omn}$ while the methods of $\overline{\overline{G}}_e$ and $\overline{\overline{G}}_A$ are much more complicated; their formulations require the use of $\overline{L}_{omn}$, although the final result does not contain that set of functions explicitly. From now on the method of $\overline{\overline{G}}_m$ will be used exclusively to derive the eigenfunction expansions of other canonical problems.

5-5 PARALLEL PLATE WAVEGUIDE

To prepare for the formulation of a waveguide filled with two dielectrics, it is convenient to start with the dyadic Green functions for a parallel plate waveguide and then apply the method of scattering superposition to construct the functions for the composite waveguide [Tai, 1988]. The functions of the parallel plate waveguide itself have their own applications in practical problems.

For a parallel plate waveguide bounded at $y = 0$ and $y = b$ by two conducting plates, the magnetic dyadic Green function of the second kind satisfies the following equation:

$$\nabla \times \nabla \times \overline{\overline{G}}_{m2} - k_1^2 \overline{\overline{G}}_{m2} = \nabla \times \left[\overline{\overline{I}} \delta(\overline{R} - \overline{R}') \right]. \tag{5.90}$$

The wave number in the region under consideration is denoted by k_1, which could be an empty space. The function $\overline{\overline{G}}_{m2}$ satisfies the boundary condition

$$\hat{y} \times \nabla \times \overline{\overline{G}}_{m2} = 0$$

at $y = 0$ and $y = b$. To find $\overline{\overline{G}}_{m2}$ we use the vector wave functions

$$\overline{M}_{om}(h_1,h) = \nabla \times \left[\sin\left(\frac{m\pi y}{b}\right) e^{i(h_1 x + hz)}\hat{x}\right] \tag{5.91}$$

$$\overline{N}_{em}(h_1,h) = \frac{1}{\kappa}\nabla \times \nabla \times \left[\cos\left(\frac{m\pi y}{b}\right) e^{(ih_1 x + hz)}\hat{x}\right], \tag{5.92}$$

where

$$\kappa = \left[h_1^2 + h_2^2 + h^2\right]^{\frac{1}{2}} \qquad h_2 = \frac{m\pi}{b}$$

h_1 and h are two continuous eigenvalues, and m = integer, including $m = 0$ for $\overline{N}_{em}$. Two additional vector wave functions to be used later to construct the electric dyadic Green functions are defined by

$$\overline{M}_{em}(h_1,h) = \nabla \times \left[\cos\left(\frac{m\pi y}{b}\right) e^{i(h_1 x + hz)}\hat{x}\right] \tag{5.93}$$

$$\overline{N}_{om}(h_1,h) = \frac{1}{\kappa}\nabla \times \nabla \times \left[\sin\left(\frac{m\pi y}{b}\right) e^{i(h_1 x + hz)}\hat{x}\right]. \tag{5.94}$$

The relations between these functions are

$$\nabla \times \overline{M}_{\substack{e\\o}m} = \kappa \overline{N}_{\substack{e\\o}m} \tag{5.95}$$

$$\nabla \times \overline{N}_{\substack{e\\o}m} = \kappa \overline{M}_{\substack{e\\o}m}. \tag{5.96}$$

The four vector wave functions defined by (5.91)–(5.94) are solutions of the homogeneous vector wave equation

$$\nabla \times \nabla \times \overline{F} - \kappa^2 \overline{F} = 0.$$

The orthogonal property of these functions are

$$\iiint_V \overline{M}_{\substack{e\\o}m}(h_1,h) \cdot \overline{N}_{\substack{e\\o}m'}(h_1',h')\, dV = 0$$

for any combination of even and odd functions and for any two sets of eigenvalues (m, h_1, h) and (m', h_1', h'). The volume of integration corresponds to the entire space inside the parallel plate waveguide. The normalization constants of these functions are stated by the following relations:

$$\begin{aligned}
&\iiint_V \overline{M}_{em}(h_1,h) \cdot \overline{M}_{em'}(-h_1',-h')\, dV \\
&= \iiint_V \overline{N}_{em}(h_1,h) \cdot \overline{N}_{em'}(-h_1',-h')\, dV \\
&= \begin{cases} 0, \quad m \neq m' \\ (1+\delta_0)\, 2\pi^2 \left(h_2^2 + h^2\right) \delta(h_1 - h_1')\, \delta(h - h'), \\ \qquad m = m' = 0, 1, 2, \ldots, \end{cases}
\end{aligned} \tag{5.97}$$

where

$$\delta = \begin{cases} 1, & m = 0 \\ 0, & m \neq 0 \end{cases}$$

$$\iiint_V \overline{M}_{om}(h_1, h) \cdot \overline{M}_{om'}(-h_1', -h')\, dV$$
$$= \iiint_V \overline{N}_{om}(h_1, h) \cdot \overline{N}_{om'}(-h_1', -h')\, dV$$
$$= \begin{cases} 0, & m \neq m' \\ 2\pi^2 \left(h_2^2 + h^2\right) \delta(h_1 - h_1')\, \delta(h - h'), & m = m' = 1, 2, \ldots. \end{cases} \tag{5.98}$$

To find $\overline{\overline{G}}_{m2}$ we first let

$$\nabla \times \left[\overline{\overline{I}}\delta(\overline{R} - \overline{R}')\right] = \int_{-\infty}^{\infty} dh_1 \int_{-\infty}^{\infty} dh \sum_{m=0}^{\infty} \cdot \left[\overline{M}_{om}(h_1, h)\, \overline{A}_{om}(h_1, h) + \overline{N}_{em}(h_1, h)\, \overline{B}_{em}(h_1, h)\right]. \tag{5.99}$$

By taking the anterior scalar product of (5.99) with $\overline{M}_{om}(-h_1', -h')$ and $\overline{N}_{em}(-h_1', -h')$, respectively, and integrating through V we can determine the vector coefficients $\overline{A}_{om}$ and $\overline{B}_{em}$ as a result of the orthogonal property of these vector wave functions. The results are

$$\overline{A}_{om}(h_1, h) = \frac{(2 - \delta_0)\, \kappa \overline{N}'_{om}(-h_1, -h)}{4\pi^2 b \left(h_2^2 + h^2\right)} \tag{5.100}$$

$$\overline{B}_{em}(h_1, h) = \frac{(2 - \delta_0)\, \kappa \overline{M}'_{em}(-h_1, -h)}{4\pi^2 b \left(h_2^2 + h^2\right)}. \tag{5.101}$$

In (5.100) and (5.101) the primed functions are defined with respect to (x', y', z'), the site of $\overline{R}'$. For $m = 0$ the function $\overline{N}'_{o0}$ vanishes; hence $\overline{A}_{o0}$. The reason that we include $m = 0$, as implied by the factor $(2 - \delta_0)$ in (5.98), is to put it in similar form to (5.101). Substituting (5.100) and (5.101) into (5.99), we obtain

$$\nabla \times \overline{\overline{I}}\delta(\overline{R} - \overline{R}') = \int_{-\infty}^{\infty} dh_1 \int_{\infty}^{\infty} dh \sum_{m=0}^{\infty} \frac{(2 - \delta_0)\, \kappa}{4\pi^2 b \left(h_2^2 + h^2\right)} \cdot \left[\overline{M}_{om}(h_1, h)\, \overline{N}'_{om}(-h_1, -h) + \overline{N}_{em}(h_1, h)\, \overline{M}'_{em}(-h_1, -h)\right]. \tag{5.102}$$

Now we let

$$\overline{G}_{m2} = \int_{-\infty}^{\infty} dh_1 \int_{-\infty}^{\infty} dh \sum_{m=0}^{\infty} \frac{(2 - \delta_0)\, \kappa}{4\pi^2 b \left(h_2^2 + h^2\right)} \cdot \left[a_{om} \overline{M}_{om}(h_1, h)\, \overline{N}'_{om}(-h_1, -h) + b_{em} \overline{N}_{em}(h_1, h)\, \overline{M}'_{em}(-h_1, -h)\right]. \tag{5.103}$$

Substituting (5.102) and (5.103) into (5.90), and making use of relation (5.98), one finds

$$a_{om} = b_{em} = \frac{1}{\kappa^2 - k_1^2}.$$

The integration with respect to h_1 can be carried out in a closed form by applying the residue theorem that yields

$$\begin{aligned}\overline{\overline{G}}_{m2} = &\int_{-\infty}^{\infty} dh \sum_{m=0}^{\infty} \frac{i\,(2-\delta_0)\,k_1}{4\pi b\beta_1\,(h_2^2+h^2)} \left[\overline{M}_{om}\,(\pm\beta_1, h)\,\overline{N}'_{om}\,(\mp\beta_1, -h)\right.\\ &\left.+\,\overline{N}_{em}\,(\pm\beta_1, h)\,\overline{M}'_{em}\,(\mp\beta_1, -h)\right], \quad x \gtrless x', \qquad (5.104)\end{aligned}$$

where $\beta_1 = \left(k_1^2 - h_2^2 - h^2\right)^{\frac{1}{2}}$. The top line applies to $x > x'$ and the bottom line to $x < x'$. Following the method described in Sec. 5-2 relating the electric dyadic Green function and the magnetic dyadic Green function, one finds

$$\begin{aligned}\overline{\overline{G}}_{e1} =& \frac{1}{k_1^2}\hat{x}\hat{x}\delta(\overline{R}, \overline{R}') + \int_{-\infty}^{\infty} dh \sum_{m=0}^{\infty} \frac{i\,(2-\delta_0)}{4\pi b\beta_1\,(h_2^2+h^2)}\\ &\cdot \left[\overline{M}_{em}\,(\pm\beta_1, h)\,\overline{M}'_{em}\,(\mp\beta_1, -h)\right.\\ &\left.+\,\overline{N}_{om}\,(\pm\beta_1, h)\,\overline{N}'_{om}\,(\mp\beta_1, -h)\right], \quad x \gtrless x'. \qquad (5.105)\end{aligned}$$

This function will be used to build the dyadic Green functions for a waveguide filled with two dielectrics in the following section.

5-6 RECTANGULAR WAVEGUIDE FILLED WITH TWO DIELECTRICS

The waveguide under consideration is shown in Fig. 5-2.

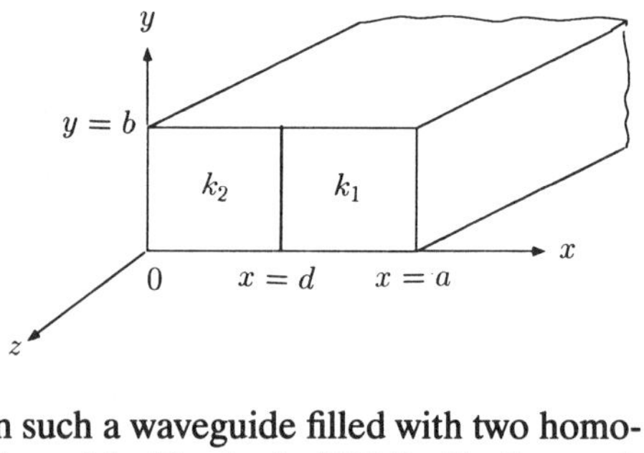

Fig. 5-2 A rectangular waveguide filled with two dielectrics with wave number denoted, respectively, by k_1 and k_2

The guided waves or the modes in such a waveguide filled with two homogeneous media were previously investigated by Pincherle [1944]. Similar work also appeared in the book by Marcuvitz [1951, p. 385]. Our task is to derive the dyadic Green function of this structure by using two sets of solenoidal vector wave functions with a piloting vector pointed in the direction normal to the interface, in contrast to the method of using vector wave functions for an empty waveguide with the piloting vector pointed in the longitudinal direction. One residue series resulting from the Fourier integral representation of the dyadic Green functions yields the guided wave previously studied by Pincherle. The present formulation furnishes the excitation coefficients for these waves for any current source, including aperture source, placed inside the waveguide.

In Fig. 5-2, region 1 ($a \geq x \geq d$) is filled with a dielectric with wave number k_1 which could be air; region 2 ($0 \leq x \leq d$) is filled with another dielectric with wave number k_2.

If a current source is placed in region 1 the functions which we want to find are the electric dyadic Green functions of the third kind as well as the first kind $\overline{\overline{G}}_{e1}^{(11)}$ and $\overline{\overline{G}}_{e1}^{(21)}$. Knowing these functions the electric field in the two regions can be determined by using the formulas:

$$\overline{E}_1(\overline{R}) = i\omega\mu_0 \iiint_{V_1} \overline{\overline{G}}_{e1}^{(11)}(\overline{R}, \overline{R}') \cdot \overline{J}_1(\overline{R}')\, dV' \tag{5.106}$$

$$\overline{E}_2(\overline{R}) = i\omega\mu_0 \iiint_{V_1} \overline{\overline{G}}_{e1}^{(21)}(\overline{R}, \overline{R}') \cdot \overline{J}_2(\overline{R}')\, dV', \tag{5.107}$$

where V_1 denotes the volume occupied by $\overline{J}_1$. It is understood that the current source $\overline{J}_1$ is located in region 1 and both media are nonmagnetic. To construct $\overline{\overline{G}}_{e1}^{(11)}$ and $\overline{\overline{G}}_{e1}^{(21)}$ we need some new vector wave functions defined as

$$\overline{M}_{oem}(\beta_2, h) = \nabla \times \left(\sin \beta_2 x \sin h_2 y e^{ihz} \hat{x}\right) \tag{5.108}$$

$$\overline{N}_{oem}(\beta_2, h) = \frac{1}{k_2} \nabla \times \overline{M}_{oem}(\beta_2, h) \tag{5.109}$$

$$\overline{M}_{eom}(\beta_2, h) = \nabla \times \left(\cos \beta_2 x \cos h_2 y e^{ihz} \hat{x}\right) \tag{5.110}$$

$$\overline{N}_{eom}(\beta_2, h) = \frac{1}{k_2} \nabla \times \overline{M}_{eom}(\beta_2, h), \tag{5.111}$$

where $\beta_2 = \left(k_2^2 - h_2^2 - h^2\right)^{\frac{1}{2}}$ and $h_2 = m\pi/b$. These functions are solutions of the homogeneous vector wave equation

$$\nabla \times \nabla \times \overline{F} - k_2^2 \overline{F} = 0$$

defined in region 2. Furthermore, we have the relations

$$\nabla \times \overline{N}_{oem} = k_2 \overline{M}_{oem} \tag{5.112}$$

$$\nabla \times \overline{N}_{eom} = k_2 \overline{M}_{eom}. \tag{5.113}$$

The boundary conditions satisfied by these functions are, at $x = 0, y = 0$, and $y = b$,

$$\begin{aligned}
\hat{n} \times \overline{M}_{oem} &= 0 \\
\hat{n} \times \overline{N}_{eom} &= 0 \\
\hat{n} \times \nabla \times \overline{N}_{oem} &= 0 \\
\hat{n} \times \nabla \times \overline{M}_{eom} &= 0,
\end{aligned}$$

where $\hat{n}$ denotes the normal to the walls in region 2.

By means of the method of scattering superposition we let

$$\overline{\overline{G}}_{e1}^{(11)} = \overline{\overline{G}}_{e1} + \overline{\overline{G}}_{es}^{(11)}, \tag{5.114}$$

where $\overline{\overline{G}}_{e1}$ is given by (5.105). The scattered part, $\overline{\overline{G}}_{es}^{(11)}$, can be written in the form

$$\begin{aligned}
\overline{\overline{G}}_{es}^{(11)} = \int_{-\infty}^{\infty} dh \sum_{m=0}^{\infty} \frac{i\,(2-\delta_0)}{4\pi b \beta_1 \left(h_2^2 + h^2\right)} \big[& \overline{M}_{em}\left(\beta_1, h\right) \overline{A}_1^{+} \\
& + \overline{M}_{em}\left(-\beta_1, h\right) \overline{A}_1^{-} + \overline{N}_{om}\left(\beta_1, h\right) \overline{B}_1^{+} \\
& + \overline{N}_{om}\left(-\beta_1, h\right) \overline{B}_1^{-} \big].
\end{aligned} \tag{5.115}$$

The functions $\overline{M}_{em}\left(\pm\beta_1, h\right)$ and $\overline{N}_{om}\left(\pm\beta_1, h\right)$ were defined in the expression for $\overline{\overline{G}}_{e1}$ of the parallel plate waveguide. The function $\overline{\overline{G}}_{e1}^{(21)}$ must be of the form

$$\begin{aligned}
\overline{\overline{G}}_{e1}^{(21)} = \int_{-\infty}^{\infty} dh \sum_{m=0}^{\infty} \frac{i\,(2-\delta_0)}{4\pi b \beta_1 \left(h_2^2 + h^2\right)} \big[& \overline{M}_{oem}\left(\beta_2, h\right) \overline{A}_2 \\
& + \overline{N}_{eom}\left(\beta_2, h\right) \overline{B}_2 \big]
\end{aligned} \tag{5.116}$$

in order to satisfy the boundary condition

$$\hat{n} \times \overline{\overline{G}}_{e1}^{(11)} = 0$$

at the walls of the waveguide in region 2. The physical meaning of (5.115) and (5.116) is illustrated graphically in Fig. 5-3 for the *TE* modes with respect to the x-axis. A similar one applies to the *TM* modes. The six unknown vector coefficients $\overline{A}_1^{\pm}, \overline{B}_1^{\pm}$, and $\overline{A}_2, \overline{B}_2$ are determined by invoking the remaining boundary conditions

$$\hat{x} \times \overline{\overline{G}}_{e1}^{(11)} = 0 \quad \text{at } x = 0 \tag{5.117}$$

$$\hat{x} \times \left[\overline{\overline{G}}_{e1}^{(11)} - \overline{\overline{G}}_{e1}^{(21)} \right] = 0 \tag{5.118}$$

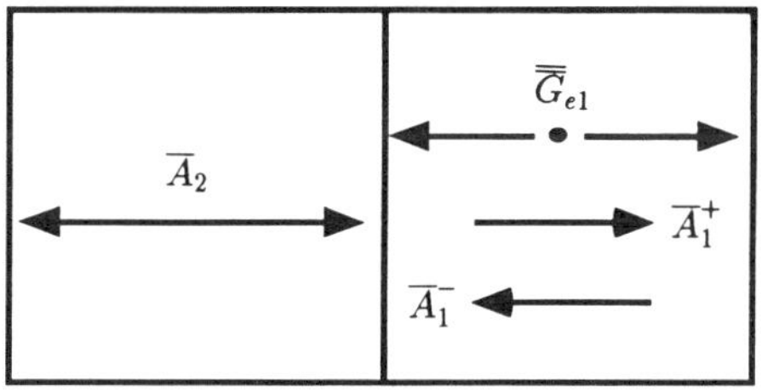

Fig. 5-3 Scattered waves for the TE modes

$$\hat{x} \times \left[\nabla \times \overline{\overline{G}}_{e1}^{(11)} - \nabla \times \overline{\overline{G}}_{e1}^{(21)}\right) = 0 \tag{5.119}$$

at the interface, corresponding to $x = d$.

The formulation is straightforward, although quite tedious. The six linear equations resulting from applying these boundary conditions are

$$e^{i\beta_1 a}\overline{M}'_{em}(-\beta_1, -h) + e^{i\beta_1 a}\overline{A}_1 + e^{-i\beta_1 a}\overline{A}_1^- = 0$$

$$e^{-i\beta_1 d}\overline{M}'_{em}(\beta_1, -h) + e^{i\beta_1 d}\overline{A}_1^+ + e^{-i\beta_1 d}\overline{A}_1^- - \sin\beta_2 d\overline{A}_2 = 0$$

$$-\beta_1 e^{-i\beta_1 d}\overline{M}'_{em}(\beta_1, -h) + \beta_1 e^{i\beta_1 d}\overline{A}_1^+ - \beta_1 e^{-i\beta_1 d}\overline{A}_1^- + i\beta_2\cos\beta_2 d\overline{A}_2 = 0$$

$$e^{i\beta_1 a}\overline{N}'_{om}(-\beta_1, -h) + e^{i\beta_1 a}\overline{B}_1^+ - e^{-i\beta_1 a}\overline{B}_1^- = 0$$

$$-e^{i\beta_1 d}\overline{N}'_{om}(\beta_1, -h) + e^{i\beta_1 d}\overline{B}_1^+ - e^{-i\beta_1 d}\overline{B}_1^- - i\frac{\beta_2 k_1}{\beta_1 k_2}\sin\beta_2 d\overline{B}_2 = 0$$

$$e^{i\beta_1 d}\overline{N}'_{om}(\beta_1, -h) + e^{i\beta_1 d}\overline{B}_1^+ + e^{-i\beta_1 d}\overline{B}_1^- - \left(\frac{k_2}{k_1}\right)\cos\beta_2 d\overline{B}_2 = 0.$$

The solutions for the six vector coefficients from these equations are

$$\overline{A}_1^+ = \frac{1}{\Gamma_1}e^{-iD}\left[e^{iA}\overline{M}'_{em}(\beta_1, -h) - e^{iA}\overline{M}'_{em}(-\beta_1, -h)\right]$$

$$\overline{A}_1^+ = \frac{1}{\Gamma_1}e^{-iA}\left[e^{iD}\overline{M}'_{em}(\beta_1, -h) + T_1 e^{iD}\overline{M}'_{em}(-\beta_1, -h)\right]$$

$$\overline{A}_2 = \frac{2e^{iD}}{\Delta_1}\overline{A}_1^+$$

$$\overline{B}_1^+ = -\frac{1}{\Gamma_2}e^{-iD}\left[e^{-iA}\overline{N}'_{om}(\beta_1, -h) + e^{iA}\overline{N}'_{om}(-\beta, -h)\right]$$

$$\overline{B}_1^- = -\frac{1}{\Gamma_2}e^{-iA}\left[e^{-iD}\overline{N}'_{om}(\beta_1, -h) - T_2 e^{iD}\overline{N}'_{om}(-\beta, -h)\right]$$

$$\overline{B}_2 = \frac{2ie^{iD}}{\Delta_2}\overline{B}_1^+,$$

where

$$A = \beta_1 a, \quad D = \beta_1 d$$

$$\beta_1 = \left(k_1^2 - h_2^2 - h^2\right)^{\frac{1}{2}}, \quad h_2 = \frac{m\pi}{b}$$

$$\beta_2 = \left(k_2^2 - h_2^2 - h^2\right)^{\frac{1}{2}}$$

$$\Gamma_1 = e^{i(A-D)} + T_1 e^{-i(A-D)}$$

$$\Gamma_2 = e^{i(A-D)} + T_2 e^{-i(A-D)}$$

$$T_1 = \frac{1 + i\,(\beta_2/\beta_1)\cot D_2}{1 - i\,(\beta_2/\beta_1)\cot D_2}, \quad D_2 = \beta_2 d$$

$$T_2 = \frac{1 + i\,(k_2/k_1)^2\,(\beta_1/\beta_2)\cot D_2}{1 - i\,(k_2/k_1)^2\,(\beta_1/\beta_2)\cot D_2}$$

$$\Delta_1 = \sin D_2 - i\left(\frac{\beta_2}{\beta_1}\right)\cos D_2$$

$$\Delta_2 = \frac{k_1\beta_2}{k_2\beta_1}\left[\sin D_2 - i\left(\frac{k_2}{k_1}\right)^2\left(\frac{\beta_1}{\beta_2}\right)\cos D_2\right].$$

The expressions for $\overline{\overline{G}}_{e1}^{(11)}$ given by (5.105), (5.114), and (5.115) with the coefficients $\overline{A}_1^{\pm}$ and $\overline{B}_1^{\pm}$ thus determined can be written in a more compact form.

For $x \gtrless x'$, one finds,

$$\overline{M}_{em}(\pm\beta_1, -h)\,\overline{M}'_{em}(\mp\beta_1, -h) + \overline{M}_{em}(\beta_1, h)\,\overline{A}_1^{+} + \overline{M}_{em}(-\beta_1, h)\,\overline{A}_1^{-}$$

$$= \begin{Bmatrix} -\frac{2i}{\Gamma_1}\overline{M}_{oem}(a-x)\left[e^{-iD}\overline{M}'_{em}(\beta_1, -h) + T_1 e^{iD}\overline{M}'_{em}(-\beta_1, -h)\right] \\ -\frac{2i}{\gamma_1}\left[e^{-iD}\overline{M}_{em}(\beta_1, h) + T_1 e^{iD}\overline{M}_{em}(-\beta_1 - h)\right]\overline{M}'_{oem}(a-x') \end{Bmatrix} \tag{5.120}$$

$$\overline{N}_{om}(\pm\beta_1, -h)\,\overline{N}'_{om}(\mp\beta_1, h) + \overline{N}_{om}(\beta_1, h)\,\overline{B}_1^{+} + \overline{N}_{om}(-\beta_1, h)\,\overline{B}_1^{-}$$

$$= \begin{Bmatrix} -\frac{2}{\Gamma 1}\overline{N}_{eom}(a-x)\left[e^{-iD}\overline{N}'_{om}(\beta_1, -h) - T_2 e^{iD}\overline{N}'_{om}(-\beta_1, -h)\right] \\ -\frac{2}{\Gamma_1}\left[e^{-iD}\overline{N}_{om}(\beta_1, h) + T_2 e^{iD}\overline{N}_{om}(-\beta_1, -h)\right]\overline{N}'_{eom}(a-x') \end{Bmatrix}, \tag{5.121}$$

where Γ_1 and Γ_2 have been defined previously and

$$\overline{M}_{oem}(a-x) = \nabla \times \left[\sin\beta_1(a-x)\cos h_2 y e^{ihz}\hat{x}\right] \tag{5.122}$$

$$\overline{N}_{eom}(a-x) = \frac{1}{k_1}\nabla \times \nabla \times \left[\cos\beta_1(a-x)\sin h_2 h e^{ihz}\hat{x}\right]. \tag{5.123}$$

The two vector wave functions defined by (5.122) and (5.123) satisfy the Dirichlet boundary condition at $x = a$; that is,

$$\hat{x} \times \overline{M}_{oem}(0) = \hat{x} \times \overline{N}_{eom}(0) = 0. \tag{5.124}$$

The symmetrical property of $\overline{\overline{G}}_{e1}^{(11)}$ is also evident from (5.120) and (5.121).

In view of the composition of the two expressions given by (5.120) and (5.121), it is quite obvious that we could use $\overline{M}_{oem}(a-x)$ and $\overline{N}_{eom}(a-x)$ at the very beginning to derive $\overline{\overline{G}}_{e1}^{(11)}$ instead of using $\overline{M}_{em}\,(\pm\beta_1, h)$ and $\overline{N}_{om}\,(\pm\beta_1, h)$, but the work would be much more tedious because the discontinuity condition at $\overline{R} = \overline{R}'$ must be invoked to determine the unknown excitation coefficients.

In view of (5.120) and (5.121) the expressions for $\overline{\overline{G}}_{e1}^{(11)}$ can be written in the form

$$\begin{aligned}
\overline{\overline{G}}_{e1}^{(11)} =& -\frac{1}{k^2}\hat{x}\hat{x}\delta(\overline{R}, \overline{R}') + \int_{-\infty}^{+\infty} dh \sum_{m=0}^{\infty} \frac{(2-\delta_0)}{2\pi b\beta_1\,(h_2^2 + h^2)} \\
&\cdot \left\{ \frac{1}{\Gamma_1} \begin{bmatrix} \overline{M}_{oem}(a-x)\left[e^{-iD}\overline{M}'_{em}\,(\beta_1, -h) + T_1 e^{iD}\overline{M}'_{em}\,(-\beta_1, -h)\right] \\ \left[e^{-iD}\overline{M}_{em}\,(\beta_1, h) + T_1 e^{iD}\overline{M}_{em}\,(-\beta_1, h)\right]\overline{M}'_{oem}\,(a-x') \end{bmatrix} \right. \\
&\left. + \frac{i}{\Gamma_2} \begin{bmatrix} \overline{N}_{eom}(a-x)\left[e^{-iD}\overline{N}'_{om}\,(\beta_1, -h) + T_2 e^{iD}\overline{N}'_{om}\,(-\beta_1, -h)\right] \\ \left[e^{-iD}\overline{N}_{om}\,(\beta_1, h) + T_2 e^{iD}\overline{N}_{om}\,(-\beta_1, h)\right]\overline{N}'_{eom}\,(a-x') \end{bmatrix} \right\},
\end{aligned}$$

$$\text{for } x \gtrless x'. \tag{5.125}$$

The expression for $\overline{\overline{G}}_{e1}^{(21)}$ can be obtained by substituting $\overline{A}_2$ and $\overline{B}_2$ into (5.116). The remaining Fourier integration with respect to h in $\overline{\overline{G}}_{e2}^{(11)}$ and $\overline{\overline{G}}_{e1}^{(21)}$ can be evaluated in a closed form by means of a contour integration. The residue terms related to the poles of the integrand can be found. Two sets of poles of the integrand are governed by the transcendental equations

$$\Gamma_1 = e^{i(A-D)} + T_1 e^{-i(A-D)} = 0 \tag{5.126}$$

for the *TE* modes $(\overline{M}'s)$ and

$$\Gamma_2 = e^{i(A-D)} + T_2 e^{-i(A-D)} = 0 \tag{5.127}$$

for the *TM* modes $(\overline{N}'s)$. They are equivalent to

$$\tan\beta_2 d = -\left(\frac{\beta_1}{\beta_2}\right)\tan\beta_1\,(a-d) \tag{5.128}$$

for the *TE* modes and

$$\tan\beta_2 d = -\left(\frac{k^2}{k^1}\right)^2\left(\frac{\beta_1}{\beta_2}\right)\tan\beta_1\,(a-d) \tag{5.129}$$

for the *TM* modes. By substituting

$$\beta_1 = \left[k_1^2 - h_2^2 - h^2\right]^{\frac{1}{2}}$$

and

$$\beta_2 = \left[k_2^2 - h_2^2 - h^2\right]^{\frac{1}{2}}$$

into these equations and solving for h, one can determine the guided wave number for these normal modes. The cut-off frequencies calculated by Pincherle correspond to solutions for ω_c in our formulation by putting $h = 0, k_1 = \omega_c/v_1$, and $k_2 = \omega_c/v_2$ into (5.128) and (5.129), where v_1 and v_2 denote, respectively, the velocity of light in the two media, assuming to be purely dielectric without loss.

5-7 RECTANGULAR CAVITY

The simplest approach to derive the dyadic Green functions for a rectangular cavity is to start with the functions available for a rectangular waveguide with the same cross-sectional dimension and apply the method of scattering superposition to find the desired function. The procedure to accomplish it can be carried out in two steps. We consider first the functions for a semi-infinite waveguide defined in a region $\infty > z \geq 0$. At $z = 0$, the waveguide, the function $\overline{\overline{G}}_{e1}$ is given by (5.49). Its expression is

$$\begin{aligned}\overline{\overline{G}}_{e1}(\overline{R}, \overline{R}') = & -\frac{1}{k^2}\hat{z}\hat{z}\delta(\overline{R} - \overline{R}') \\ & + \frac{i}{ab}\sum_{m,n}\frac{2-\delta_0}{k_c^2 k_g}\left[\overline{M}_e\left(\pm k_g\right)\overline{M}'_e\left(\mp k_g\right)\right. \\ & \left. + \overline{N}_0\left(\pm k_g\right)\overline{N}'_0\left(\mp k_g\right)\right].\end{aligned} \tag{5.130}$$

The upper line in the series is for $z > z'$, and the bottom line is for $z < z'$. To simplify the writing, the subscript "mn" attached to the vector wave functions has been deleted. To find the electric dyadic Green function of the first kind for the semi-infinite waveguide, denoted by $\overline{\overline{G}}_{E1}$, we let

$$\overline{\overline{G}}_{E1}(\overline{R}, \overline{R}') = \overline{\overline{G}}_{e1}(\overline{R}, \overline{R}') + \overline{\overline{G}}_{es}(\overline{R}, \overline{R}'), \tag{5.131}$$

where the scattered term $\overline{\overline{G}}_{es}$ can be written in the form

$$\begin{aligned}\overline{\overline{G}}_{es}(\overline{R}, \overline{R}') = & \frac{i}{ab}\sum_{m,n}\frac{2-\delta_0}{k_c^2 k_g}\left[A_e\overline{M}_e(k_g)\overline{M}'_e(k_g)\right. \\ & \left. + B_0\overline{N}_0(k_g)\overline{N}'_0(k_g)\right],\end{aligned} \tag{5.132}$$

The field functions $\overline{M}_e(k_g)$ and $\overline{N}_o(k_g)$ represent waves propagating in the positive z-direction. The excitation functions $\overline{M}'_e(k_g)$ and $\overline{N}'_o(k_g)$ are chosen because the scattered waves are originated by the primary waves propagating to the negative direction; their excitation functions are of these two kinds. Now let us consider the couplets representing the *TE* modes. In order to satisfy the Dirichlet boundary condition at $z = 0$,

$$\hat{z} \times \left[\overline{M}_e\left(-k_g\right)\overline{M}'_e(k_g) + A_e \overline{M}_e(k_g)\overline{M}'_e(k_g)\right]_{z=0}$$

or

$$\hat{z} \times \left[\overline{M}_e(-k_g) + A_e \overline{M}_e(k_g)\right]_{z=0} = 0. \tag{5.133}$$

Since

$$\overline{M}_e(-k_g) = \nabla \times \left(C_x C_y e^{-ik_g z}\hat{z}\right)$$

$$\overline{M}_e(k_g) = \nabla \times \left(C_x C_y e^{ik_g z}\hat{z}\right),$$

Equation (5.133) can be satisfied if $A_e = -1$. Similarly, for the *TM* modes one finds that the boundary condition at $z = 0$ can be satisfied if $B_0 = 1$. The expression for $\overline{\overline{G}}_{E1}$ is, therefore, given by

$$\overline{\overline{G}}_{E1}(\overline{R}, \overline{R}') = -\frac{1}{k^2}\hat{z}\hat{z}\delta(\overline{R} - \overline{R}') + \frac{i}{ab}\sum_{m,n}\frac{2-\delta_0}{k_c^2 k_g}\left\{\begin{array}{l} \overline{M}_e(k_g)\left[\overline{M}'_e(-k_g) - \overline{M}'_e(k_g)\right] \\ +\overline{N}_0(k_g)\left[\overline{N}_0(-k_g) + \overline{N}'_0(k_g)\right], \; z > z \\ \left[\overline{M}_e(-k_g) - \overline{M}_e(k_g)\right]\overline{M}'_e(k_g) \\ +\left[\overline{N}_0(-k_g) + \overline{N}_0(k_g)\right]\overline{N}'_0(k_g), \; z < z' \end{array}\right\}. \tag{5.134}$$

We now define two standing wave vector wave functions, denoted by $\overline{M}_{eo}(z)$ and $\overline{N}_{oe}(z)$, in the form

$$\overline{M}_{eo}(z) = \nabla \times (C_x C_y \sin k_g z\hat{z}) \tag{5.135}$$

$$\overline{N}_{oe}(z) = \frac{1}{k}\nabla \times \nabla \times (S_x S_y \cos k_g z\hat{z}). \tag{5.136}$$

In terms of these functions

$$\overline{M}_e(-k_g) - \overline{M}_e(k_g) = -2i\overline{M}_{eo}(z) \tag{5.137}$$

$$\overline{N}_o(-k_g) + \overline{N}_0(k_g) = 2\overline{N}_{oe}(z), \tag{5.138}$$

(5.134) thus can be written in the form

$$\overline{\overline{G}}_{E1}(\overline{R}, \overline{R}') = -\frac{1}{k^2}\hat{z}\hat{z}\delta(\overline{R}-\overline{R}') + \frac{2}{ab}\sum_{m,n}\frac{2-\delta_0}{k_c^2 k_g}\left\{\begin{array}{l}\overline{M}_e(k_g)\overline{M}'_{eo}(z') + i\overline{N}_o(k_g)\overline{N}'_{oe}(z'),\ z > z' \\ \overline{M}_{eo}(z)\overline{M}'_e(k_g) + i\overline{N}_{oe}(z)\overline{N}'_o(k_g),\ z < z'\end{array}\right\}. \tag{5.139}$$

This is the expression for the electric dyadic Green function of the first kind for the semi-infinite waveguide. The symmetrical property of this function is evident.

For the cavity shown in Fig. 5-4(b), its electric dyadic Green function of the first kind will be denoted by $\overline{\overline{G}}_{E'1}$, which can be written in the form

$$\overline{\overline{G}}_{E'1}(\overline{R}, \overline{R}') = \overline{\overline{G}}_{E1}(\overline{R}, \overline{R}') + \overline{\overline{G}}_{Es}(\overline{R}, \overline{R}'). \tag{5.140}$$

The scattered term $\overline{\overline{G}}_{Es}$ can be cast in the form

$$\overline{\overline{G}}_{Es}(\overline{R}, \overline{R}') = \frac{2}{ab}\sum_{m,n}\frac{2-\delta_0}{k_c^2 kg}\left[A_s\overline{M}_{eo}(z)\overline{M}'_{eo}(z') + B_s\overline{N}_{oe}(z)\overline{N}'_{oe}(z')\right]. \tag{5.141}$$

The field functions in $\overline{\overline{G}}_{Es}$ are so chosen that they have already satisfied the Dirichlet boundary condition at $z = 0$, and the excitation functions must be the same as those of the $\overline{\overline{G}}_{E1}$ for $z > z'$ because they are responsible for the excitation of the scattered waves. At $z = c$, the boundary conditions are

$$\hat{z}\times\left[\overline{M}_e(k_g) + A_s\overline{M}_{e0}(z)\right]_{z=c} = 0 \tag{5.142}$$

and

$$\hat{z}\times\left[i\overline{N}_o(k_g) + B_s\overline{N}_{e0}(z)\right]_{z=c} = 0. \tag{5.143}$$

Equation (5.142) yields

$$A_s = -\frac{e^{ikgc}}{\sin kgc}, \tag{5.144}$$

and (5.143) yields

$$B_s = -\frac{ei^{kgc}}{\sin kgc}. \tag{5.145}$$

By substituting (5.144) and (5.145) into (5.141), then combining $\overline{\overline{G}}_{Es}$ with $\overline{\overline{G}}_{E1}$, we find that for $z > z'$

$$\overline{M}_e(k_g) + A_s\overline{M}_{eo}(z) = \frac{1}{\sin kgc}\overline{M}_{eo}(c-z), \tag{5.146}$$

where the function $\overline{M}_{eo}(c-z)$ is defined by

$$\overline{M}_{eo}(c-z) = \nabla\times\left[C_x C_y \sin k_g(c-z)\hat{z}\right], \tag{5.147}$$

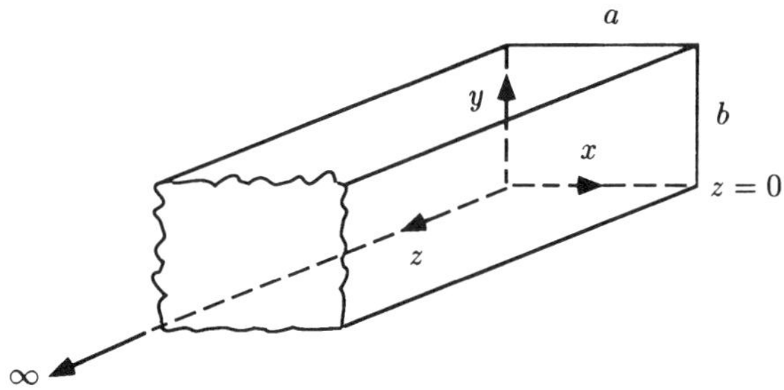

Fig. 5-4 (a) A semi-infinite waveguide terminated at $z = 0$ by a conducting wall

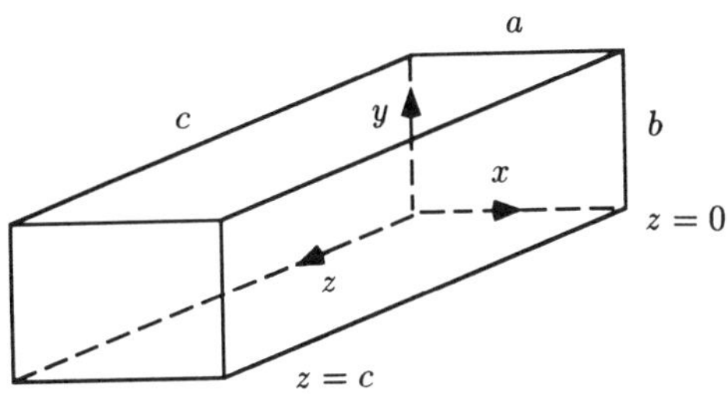

Fig. 5-4 (b) A rectangular cavity

and

$$i\overline{N}_0(k_g) + B_s\overline{N}_{oe}(z) = -\frac{1}{\sin k_g c}\overline{N}_{oe}(c-z), \tag{5.148}$$

where the function $\overline{N}_{oe}(c-z)$ is defined by

$$\overline{N}_{oe}(c-z) = \frac{1}{k}\nabla \times \nabla \times [S_x S_y \cos k_g(c-z)\hat{z}]\,. \tag{5.149}$$

We can do the same reduction for $z < z'$. The final result of $\overline{\overline{G}}_{E'1}$ is given by

$$\begin{aligned}\overline{\overline{G}}_{E'1}(\overline{R},\overline{R}') = &-\frac{1}{k^2}\hat{z}\hat{z}\delta(\overline{R}-\overline{R}')\\ &+\frac{2}{ab}\sum_{m,n}\frac{(2-\delta_0)}{k_c^2 k_g \sin k_g c}\left\{\begin{array}{l}\overline{M}_{eo}(c-z)\overline{M}'_{eo}(z')\\ \quad -\overline{N}_{oe}(c-z)\overline{N}'_{oe}(z')\\ \overline{M}_{eo}(z)\overline{M}'_{eo}(c-z')\\ \quad -\overline{N}_{oe}(z)\overline{N}'_{oe}(c-z')\end{array}\right\} z \gtrless z'.\end{aligned} \tag{5.150}$$

This is the expression for the electric dyadic Green function of the first kind for a rectangular cavity. In a previous work [Tai and Rozenfeld, 1976], the same function was derived by the methods of $\overline{\overline{G}}_A$ and $\overline{\overline{G}}_e$ which were very laborious mainly because the nonsolenoidal vector wave functions $\overline{L}'_s$ are needed. The method of scattering superposition was also used in that work without breaking into two steps as have been done now. The function for the semi-infinite waveguide derived here is, by itself, a useful formula for formulating problems dealing with a terminated waveguide.

When the frequency of excitation corresponds to the resonant frequency of the cavity such that

$$k_g c = n\pi, \quad n = 0, 1, 2, \ldots$$

or

$$\left[\left(\frac{2\pi}{\lambda}\right)^2 - \left(\frac{m\pi}{a}\right)^2 - \left(\frac{n\pi}{b}\right)^2\right]^{\frac{1}{2}} c = n\pi,$$

we encounter the phenomenon of resonant catastrophe as described by Sommerfeld [1949, p. 187]. This is the consequence of a lossless system similar to the resonance of a lossless series $L - C$ circuit where the current goes to infinity. In practice, such a phenomenon does not occur when the system has loss.

5-8 THE ORIGIN OF THE ISOLATED SINGULAR TERM IN $\overline{\overline{G}}_e$

The eigenfunction expansion of the electric dyadic Green function as given by (5.78) contains an isolated singular term in the form of $-\hat{z}\hat{z}\delta(\overline{R} - \overline{R}')/k^2$. From the point of view of the method of $\overline{\overline{G}}_m$, this term results from the discontinuous condition of $\overline{\overline{G}}_m$ across a point source and the Ampère-Maxwell equation relating $\overline{\overline{G}}_e$ and $\overline{\overline{G}}_m$ in the dyadic form; namely,

$$\nabla \times \overline{\overline{G}}_m = \overline{\overline{I}}\delta(\overline{R} - \overline{R}') + k^2\overline{\overline{G}}_e. \tag{5.151}$$

In general $\overline{\overline{G}}_m$ is discontinuous across a surface containing a point source with unit normal vector $\hat{n}$ defined in a Dupin coordinate system [Tai, 1992]. The discontinuity is stated by

$$\hat{n} \times \left(\overline{\overline{G}}_m^+ - \overline{\overline{G}}_m^-\right) = \overline{\overline{I}}_t\delta(\overline{r} - \overline{r}'), \tag{5.152}$$

where $\delta(\overline{r} - \overline{r}')$ denotes a two-dimensional delta function defined at the surface of discontinuity and $\overline{\overline{I}}_t$ the two-dimensional idem factor defined by

$$\overline{\overline{I}}_t = \overline{\overline{I}} - \hat{n}\hat{n}. \tag{5.153}$$

Thus, if we write

$$\overline{\overline{G}}_m = \overline{\overline{G}}_m^{+} U(n-n') + \overline{\overline{G}}_m^{-} U(n'-n), \tag{5.154}$$

where U denotes a unit step function, then according to the theory of generalized functions,

$$\begin{aligned}
\nabla \times \overline{\overline{G}}_m =& (\nabla \times \overline{\overline{G}}_m^{+}) U(n-n') + (\nabla \times \overline{\overline{G}}_m^{-}) U(n'-n) \\
&+ \nabla U(n-n') \times \overline{\overline{G}}_m^{+} + \nabla U(n'-n) \times \overline{\overline{G}}_m^{-} \\
=& (\nabla \times \overline{\overline{G}}_m^{+}) U(n-n') + \left(\nabla \times \overline{\overline{G}}_m^{-}\right) U(n'-n) \\
&+ \delta(n-n') \hat{n} \times \left(\overline{\overline{G}}_m^{+} - \overline{\overline{G}}_m^{-}\right) \\
=& \left(\nabla \times \overline{\overline{G}}_m^{+}\right) U(n-n') + (\nabla \times \overline{\overline{G}}_m^{-}) U(n'-n) \\
&+ \overline{\overline{I}}_t \delta(n-n') \delta(\bar{r}-\bar{r}') \\
=& \nabla \times \overline{\overline{G}}_m^{\pm} + \overline{\overline{I}}_t \delta(\overline{R}-\overline{R}'), \; n \gtrless n'.
\end{aligned} \tag{5.155}$$

Substituting it into (5.151), we obtain

$$\overline{\overline{G}}_e^{\pm} = \frac{1}{k^2} \left[\nabla \times \overline{\overline{G}}_m^{\pm} - \hat{n}\hat{n}\delta(\overline{R}-\overline{R}') \right]. \tag{5.156}$$

The singular term $\hat{n}\hat{n}\delta(\overline{R}-\overline{R}')/k^2$ is therefore always present in such a representation though the explicit expression for $\overline{\overline{G}}_m$ may not be available.

In the case of a rectangular waveguide, the method of $\overline{\overline{G}}_m$ yields first the solution for $\overline{\overline{G}}_{m2}$ in the form

$$\begin{aligned}
\overline{\overline{G}}_{m2}^{\pm}(\overline{R}, \overline{R}') =& \frac{ik}{ab} \sum_{m,n} \frac{2-\delta_0}{k_c^2 k_g} \\
& \cdot \left[\overline{N}_e(\pm k_g) \overline{M}_e'(\mp k_g) + \overline{M}_0(\pm k_g) \overline{N}_0'(\mp k_g) \right], \quad z \gtrless z'.
\end{aligned} \tag{5.157}$$

Hence

$$\begin{aligned}
\overline{\overline{G}}_{e1}(\overline{R}, \overline{R}') =& -\frac{1}{k^2} \hat{z}\hat{z}\delta(\overline{R}-\overline{R}') \\
=& \sum_{m,n} C_{mn} \left[\overline{M}_e(\pm k_g) \overline{M}_e'(\mp k_g) + \overline{N}_0(\pm k_g) \overline{N}_0'(\mp k_g) \right], \quad z \gtrless z'.
\end{aligned} \tag{5.158}$$

For completeness an exercise is assigned in this chapter to verify

$$\nabla \times \overline{\overline{G}}_{e1} = \overline{\overline{G}}_{m2}.$$

with $\overline{\overline{G}}_{m2}$ and $\overline{\overline{G}}_{e1}$ given by (5.157) and (5.158).

From the point of view of the method of $\overline{\overline{G}}_e$, the singular term $-\hat{z}\hat{z}\delta(\overline{R} - \overline{R}')/k^2$ as shown by (5.74) and (5.75) results from the longitudinal terms of $\overline{L}_0\overline{L}_0'$ and $\overline{N}_0\overline{N}_0'$; that is,

$$\begin{aligned}
&-\frac{1}{k^2}\hat{z}\hat{z}\delta(\overline{R} - \overline{R}') \\
&= -\frac{1}{k^2}\int_{-\infty}^{\infty} dh \sum_{m,n} C_{mn}\left(\frac{k_c^2}{\kappa^2}\overline{L}_{0z}\overline{L}_{0z}' + \overline{N}_{0z}\overline{N}_{0z}'\right) \\
&= -\frac{1}{k^2}\int_{-\infty}^{\infty} dh \sum_{m,n} C_{mn}\frac{\kappa^2}{k_c^2}\overline{N}_{0z}\overline{N}_{0z}' \\
&= -\frac{1}{k^2}\int_{-\infty}^{\infty} dh \sum_{m,n} C_{mn}\frac{k_c^2}{h^2}\overline{L}_{0z}\overline{L}_{0z}'.
\end{aligned} \tag{5.159}$$

The last two lines in (5.159) are due to the linear relations between $\overline{L}_{0z}$ and $\overline{N}_{0z}$ shown after (5.73).

The role played by $-\hat{z}\hat{z}\delta(\overline{R}-\overline{R}')/k^2$ can also be viewed from another point of view as demonstrated by Johnson, Howard, and Dudley [1979]. According to these authors one can treat $\overline{\overline{G}}_{e1}(\overline{R}, \overline{R})$ as consisting of two parts, one solenoidal and another irrotational, so (5.68) can be written as

$$\overline{\overline{G}}_{e1}(\overline{R}, \overline{R}') = \overline{\overline{G}}_{eI}(\overline{R}, \overline{R}') + \overline{\overline{G}}_{es}(\overline{R}, \overline{R}'), \tag{5.160}$$

where

$$\overline{\overline{G}}_{eI} = -\int_{-\infty}^{\infty} dh \sum_{m,n} C_{mn}\left[\frac{k_c^2}{k^2\kappa^2}\overline{L}_0\overline{L}_0'\right] \tag{5.161}$$

$$\overline{\overline{G}}_{es} = \int_{-\infty}^{\infty} dh \sum_{m,n} C_{mn}\left[\frac{1}{\kappa^2 - k^2}\left(\overline{M}_e\overline{M}_e' + \overline{N}_0\overline{N}_0'\right)\right]. \tag{5.162}$$

Now we decompose $\overline{L}_0$ into two parts, that is,

$$\overline{L}_0 = \overline{L}_{0t} + \overline{L}_{0z}; \tag{5.163}$$

then (5.161) can be written in the form

$$\begin{aligned}
\overline{\overline{G}}_{eI} = -\int_{-\infty}^{\infty} dh \sum_{m,n} C_{mn}\left(\frac{k_c^2}{k^2\kappa^2}\right) \\
\cdot\left[\overline{L}_{0t}\overline{L}_{0t}' + \overline{L}_{0t}\overline{L}_{0z}' + \overline{L}_{0z}\overline{L}_{0t}' + \overline{L}_{0z}\overline{L}_{0z}'\right].
\end{aligned} \tag{5.164}$$

As expected, there is a singular term associated with $\bar{L}_{0z}\bar{L}'_{0z}$ which can be sorted out; the result is

$$\bar{\bar{G}}_{eI} = -\frac{1}{k^2}\int_{-\infty}^{\infty} dh \sum_{m,n} C_{mn}\frac{k_c^2}{h^2}\bar{L}_{0z}\bar{L}'_{0z} - \frac{1}{k^2}\int_{-\infty}^{\infty} dh \sum_{m,n} C_{mn}\frac{k_c^2}{\kappa^2}\Big[\bar{L}_{0t}\bar{L}'_{0t} + \bar{L}_{0t}\bar{L}'_{0z} + \bar{L}_{0z}\bar{L}'_{0t} - \frac{k_c^2}{h^2}\bar{L}_{0z}\bar{L}'_{0z}\Big]. \tag{5.165}$$

The first integral represents $-\hat{z}\hat{z}\delta(\bar{R}-\bar{R}')/k^2$ as shown in (5.159). The second integral has poles at $h = \pm ik_c$, corresponding to $\kappa = 0$. Denoting the residue series resulting from these poles by $\bar{\bar{S}}_{eI}$ we can convert (5.165) to

$$\bar{\bar{G}}_{eI} = -\hat{z}\hat{z}\delta(\bar{R}-\bar{R}')/k^2 + \bar{\bar{S}}_{eI}(\bar{R},\bar{R}'), \tag{5.166}$$

where

$$\bar{\bar{S}}_{eI}(\bar{R},\bar{R}') = -\sum_{m,n}\frac{2-\delta_0}{abk^2k_c^2}\left(\nabla_t\nabla'_t + \nabla_t\nabla'_z + \nabla_z\nabla'_t + \nabla_z\nabla'_z\right)\left(S_xS_yS'_xS'_ye^{-k_c|z-z'|}\right). \tag{5.167}$$

The solenoidal part, $\bar{\bar{G}}_{es}$, represented by (5.162) can be evaluated in a similar manner. There is however no isolated singular term. The result is

$$\bar{\bar{G}}_{es} = \bar{\bar{S}}_{e1}(\bar{R},\bar{R}') - \bar{\bar{S}}_{eI}(\bar{R},\bar{R}'), \tag{5.168}$$

where $\bar{\bar{S}}_{e1}$ is the residue series given by (5.77) or (5.158). It is observed that the series $\bar{\bar{S}}_{eI}$ contained in (5.166) now also appears in (5.168) but with a negative sign. The sum of (5.166) and (5.168) yields

$$\bar{\bar{G}}_{e1}(\bar{R},\bar{R}') = -\frac{1}{k^2}\hat{z}\hat{z}\delta(\bar{R}-\bar{R}') + \bar{\bar{S}}_{e1}(\bar{R},\bar{R}'), \tag{5.169}$$

which is the same as (5.49) or (5.78). It should be pointed out that the "static" residue series $\bar{\bar{S}}_{eI}(\bar{R},\bar{R}')$ resulting from the static poles, $h = \pm ik_c$, appeared as a result of splitting the function $\bar{\bar{G}}_{e1}$ into two parts. As far as the end result is concerned, this series is not involved in $\bar{\bar{G}}_{e1}$. In the applications of the electric dyadic Green function to calculate the electromagnetic field, the entire function $\bar{\bar{G}}_{e1}$ is used. If there is a need to use $\bar{\bar{G}}_{eI}$ and $\bar{\bar{G}}_{es}$ separately, the two series $\pm\bar{\bar{S}}_{eI}$ cancel each other anyway. The fact that such a series occurs as a result of splitting $\overline{\overline{G}}_{e1}$ is an interesting mathematical phenomenon. Physically, there is no known problem involving these "static" modes.

For problems using explicitly the free-space electric dyadic function in the formulation, such as in solving the integral equation, in scattering theory the splitting of $\overline{\overline{G}}_{eo}$ into the irrotational and the rotational parts is very useful for numerical calculations. The structure of $\overline{\overline{G}}_{eo}$ shows that the irrotational term $\nabla\nabla G_0/k^2$ is highly singular as compared to the rotational part $G_0\overline{\overline{I}}$. The nature of the integrals resulting from such a splitting has been discussed very thoroughly by Johnson, Howard, and Dudley [1979] and by Lee, Boersma, Law, and Deschamps [1980].

6

Cylindrical Waveguides

The derivation of the dyadic Green function for a cylindrical waveguide follows the same procedures as that for a rectangular waveguide. The only difference is that cylindrical vector wave functions are used in the eigenfunction expansion. Once the orthogonal relations between these vector wave functions are known, the remaining steps are exactly the same as in the rectangular case.

6-1 CYLINDRICAL VECTOR WAVE FUNCTIONS WITH DISCRETE EIGENVALUES

The cylindrical vector wave functions with discrete eigenvalues can be used to describe the electromagnetic field inside a cylindrical waveguide with circular cross section, the geometry of which is shown in Fig. 6-1. Before we define these functions, we shall review the roots of the Bessel functions with integer order, $J_n(x)$, and the roots of their derivatives. We will designate the roots of the equation

$$J_n(x) = 0 \tag{6.1}$$

by p_{nm}. Thus p_{23} represents the third root of the Bessel function of second order. The roots of the equation

$$\frac{dJ_n(x)}{dx} = 0 \tag{6.2}$$

will be designated by q_{nm}. Tables 6-1 and 6-2 list a number of these roots in ascending order.

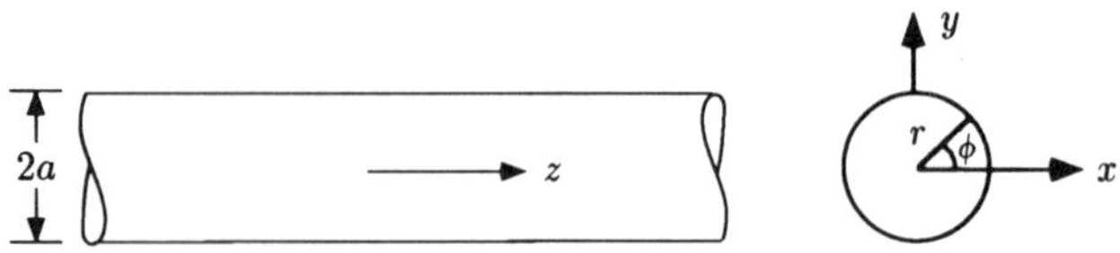

Fig. 6-1 A cylindrical waveguide with circular cross section

TABLE 6-1 Roots of $J_n(x) = 0 : p_{nm}$ (TM_{nm} modes)

$n\,/\,m$	1	2	3
0	2.405	5.520	8.654
1	3.832	7.106	10.173
2	5.136	8.417	11.620
3	6.380	9.761	13.015

TABLE 6-2 Roots of $\frac{dJ_n(x)}{dx} = 0 : q_{nm}$ (TE_{nm} modes)

$n\,/\,m$	1	2	3
0	0	3.832	7.016
1	1.841	5.331	8.536
2	3.054	6.706	9.969
3	4.201	8.015	11.346

With such a designation, the values of p_{nm} and q_{nm} increase with either n or m in an orderly manner as shown more vividly in Fig. 6-2. This figure can also be used to interpolate the roots of the Bessel function of fractional order or the roots of their derivatives. We now define two types of cylindrical vector wave functions, both of which satisfy the vector Dirichlet boundary condition at $r = a$, corresponding to the site of the wall of the cylindrical waveguide shown in Fig. 6-1. They are

$$\overline{M}_{\substack{e\\o}m\mu}(h) = \nabla \times \left[J_n(\mu r) \begin{matrix}\cos\\ \sin\end{matrix}\, n\phi e^{ihz} \hat{z} \right], \tag{6.3}$$

with $\mu = q_{nm}/a$, and

$$\overline{N}_{\substack{e\\o}n\lambda}(h) = \frac{1}{\kappa_\lambda} \nabla \times \nabla \times \left[J_n(\lambda r) \begin{matrix}\cos\\ \sin\end{matrix}\, n\phi e^{ihz} \hat{z} \right], \tag{6.4}$$

with $\lambda = p_{nm}/a, \quad \kappa_\lambda^2 = \lambda^2 + h^2$.

It is obvious that these two sets of functions are generated in the same manner as the rectangular vector wave functions except that the scalar wave equation

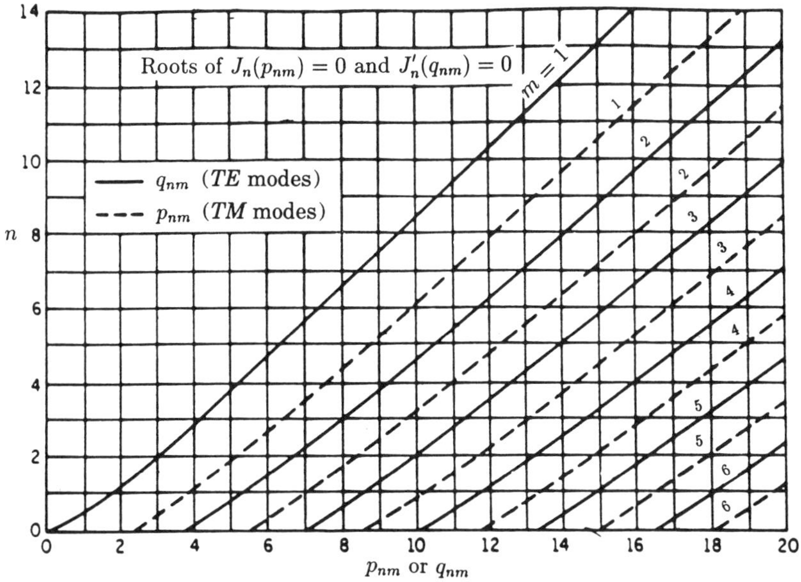

Fig. 6-2 Roots of Bessel functions and their derivatives

is now solved, by the method of separation of variables, in cylindrical coordinate system, and the eigenvalues λ and μ are so chosen that the functions would satisfy the vector Dirichlet boundary condition at $r = a$. The complete expressions of these two sets are

$$\overline{M}_{\substack{e\\o}m\mu}(h) = \left[\mp \frac{nJ_n(\mu r)}{r} \begin{matrix}\sin\\ \cos\end{matrix} n\phi\hat{r} - \frac{\partial J_n(\mu r)}{\partial r} \begin{matrix}\cos\\ \sin\end{matrix} n\phi\hat{\phi}\right] e^{ihz} \tag{6.5}$$

$$\overline{N}_{\substack{e\\o}n\lambda}(h) = \frac{1}{\kappa_\lambda}\left[ih\frac{\partial J_n(\lambda r)}{\partial r} \begin{matrix}\cos\\ \sin\end{matrix} n\phi\hat{r} \right.$$

$$\left. \mp \frac{ihn}{r} J_n(\lambda r) \begin{matrix}\sin\\ \cos\end{matrix} n\phi\hat{\phi} + \lambda^2 J_n(\lambda r) \begin{matrix}\cos\\ \sin\end{matrix} n\phi\hat{z}\right] e^{ihz} j, \tag{6.6}$$

where $\kappa_\lambda^2 = \lambda^2 + h^2$.

It should be noted that both λ and μ are denumerable numbers depending on n and m as defined previously. The functions $\overline{M}_{\substack{e\\o}m\mu}(h)$ can be used to describe the electric field of the TE_{nm} mode in a cylindrical waveguide with radius equal to a and the function $\overline{N}_{\substack{e\\o}n\lambda}(h)$ for the TM_{nm} mode. According to our nomenclature, the mode TE_{01} is a null mode because $\mu_{01} = 0$ and the dominant mode is TE_{11}. Such a nomenclature appears to be more logical than the one used in the current literature. This view is shared by Ramo, Whinnery, and Van Duzer [1965, p. 432]. To describe the magnetic field in the guide, the proper vector wave functions to be used are

$$\overline{M}_{\substack{e\\o}n\lambda}(h) = \nabla \times \left[J_n(\lambda r) \begin{matrix}\cos\\\sin\end{matrix} n\phi e^{ihz}\hat{z}\right]$$

$$= \left[\mp \frac{nJ_n(\lambda r)}{r} \begin{matrix}\sin\\\cos\end{matrix} n\phi\hat{r} - \frac{\partial J_n(\lambda r)}{\partial r} \begin{matrix}\cos\\\sin\end{matrix} n\phi\hat{\phi}\right] e^{ihz} \tag{6.7}$$

$$\overline{N}_{\substack{e\\o}m\mu}(h) = \frac{1}{\kappa_\mu} \nabla \times \nabla \times \left[J_n(\mu r) \begin{matrix}\cos\\\sin\end{matrix} n\phi e^{ihz}\hat{z}\right]$$

$$= \frac{1}{\kappa_\mu}\left[ih \frac{\partial J_n(\mu r)}{\partial r} \begin{matrix}\cos\\\sin\end{matrix} n\phi\hat{r} \mp \frac{ihn}{r} J_n(\mu r) \begin{matrix}\sin\\\cos\end{matrix} n\phi\hat{\phi}\right.$$

$$\left. + \mu^2 J_n(\mu r) \begin{matrix}\cos\\\sin\end{matrix} n\phi\hat{z}\right] e^{ihz}, \tag{6.8}$$

where

$$\kappa_\mu^2 = \mu^2 + h^2, \quad \kappa_\lambda^2 = \lambda^2 + h^2$$

$$\overline{N}_{\substack{e\\o}n\mu}(h) = \frac{1}{\kappa_\mu} \nabla \times \overline{M}_{\substack{e\\o}n\mu}(h) \tag{6.9}$$

and

$$\overline{N}_{\substack{e\\o}n\lambda}(h) = \frac{1}{\kappa_\lambda} \nabla \times \overline{M}_{\substack{e\\o}n\lambda}(h), \tag{6.10}$$

and vice versa. The $\overline{M}_{\substack{e\\o}n\lambda}(h)$ and $\overline{N}_{\substack{e\\o}m\mu}(h)$ functions obviously satisfy the vector Neumann boundary condition at $r = a$.

Before we prove the orthogonality between the various vector wave functions, the orthogonal properties of $J_n(\lambda r)$ and $J_n(\mu r)$ should be discussed first. We consider two Bessel functions $J_n(\alpha r)$ and $J_n(\beta r)$ which satisfy, respectively, the differential equations

$$\frac{1}{r}\frac{d}{dr}\left[r \frac{dJ_n(\alpha r)}{dr}\right] + \left(\alpha^2 - \frac{n^2}{r^2}\right) J_n(\alpha r) = 0 \tag{6.11}$$

and

$$\frac{1}{r}\frac{d}{dr}\left[r \frac{dJ_n(\beta r)}{dr}\right] + \left(\beta^2 - \frac{n^2}{r^2}\right) J_n(\beta r) = 0. \tag{6.12}$$

Multiplying (6.11) by $J_n(\beta r)$ and (6.12) by $J_n(\alpha r)$ and taking the differences, we obtain

$$\left(\alpha^2 - \beta^2\right) J_n(\alpha r) J_n(\beta r) = \frac{1}{r}\left[J_n(\alpha r) \frac{d}{dr}\left(r \frac{dJ_n(\beta r)}{dr}\right)\right.$$

$$\left. - J_n(\beta r) \frac{d}{dr}\left(r \frac{dJ_n(\alpha r)}{dr}\right)\right].$$

An integration of the above equation with respect to $r\,dr$ from $r = 0$ to a yields

$$\left(\alpha^2 - \beta^2\right) \int_0^a J_n(\alpha r)\, J_n(\beta r)\, r dr$$
$$= \left[r J_n(\alpha r) \frac{dJ_n(\beta r)}{dr} - r J_n(\beta r) \frac{dJ_n(\alpha r)}{dr} \right]_0^a . \tag{6.13}$$

If we let $\alpha = \lambda = p_{nm}/a$ and $\beta = \lambda' = p_{nm'}/a$, then, for $m \neq m'$,

$$\int_0^a J_n(\lambda r)\, J_n(\lambda' r)\, r dr = 0 \tag{6.14}$$

because $J_n(p_{nm}) = 0$ and $J_n(p_{nm'}) = 0$. Similarly, if we let $\alpha = \mu = q_{nm}/a$ and $\beta = \mu' = q_{nm'}/a$, we obtain

$$\int_0^a J_n(\mu r)\, J_n(\mu' r)\, r dr = 0. \tag{6.15}$$

To determine the normalization factor when $\alpha = \beta = \lambda$ or $\alpha = \beta = \mu$, we let $\beta = \alpha + \Delta$. Then

$$I_\alpha = \int_0^a J_n^2(\alpha r) r dr = \lim_{\Delta \to 0} \int_0^a J_n(\alpha r) J_n\left[(\alpha + \Delta) r\right] r dr$$
$$= \frac{-1}{2\alpha} \lim_{\Delta \to 0} \frac{1}{\Delta} \left\{ r J_n(\alpha r) \frac{\partial J_n\left[(\alpha+\Delta) r\right]}{\partial r} - r J_n\left[(\alpha+\Delta) r\right] \frac{\partial J_n(\alpha r)}{\partial r} \right\}_0^a$$
$$= \frac{-1}{2\alpha} \left[r J_n(\alpha r) \frac{\partial^2 J_n(\alpha r)}{\partial \alpha \partial r} - r \frac{\partial J_n(\alpha r)}{\partial \alpha} \frac{\partial J_n(\alpha r)}{\partial r} \right]_0^a$$
$$= \frac{-a}{2\alpha} \left[J_n(\alpha r) \frac{\partial^2 J_n(\alpha r)}{\partial \alpha \partial r} - \frac{\partial J_n(\alpha r)}{\partial \alpha} \frac{\partial J_n(\alpha r)}{\partial r} \right]_{r=a}$$

Now

$$\frac{\partial^2 J_n(\alpha r)}{\partial \alpha \partial r} = \frac{\partial}{\partial \alpha} \left[\alpha \frac{dJ_n(\alpha r)}{d(\alpha r)} \right] = \frac{dJ_n(\alpha r)}{d(\alpha r)} + \alpha r \frac{d^2 J_n(\alpha r)}{d^2(\alpha r)}$$
$$= -\frac{r}{\alpha} \left(\alpha^2 - \frac{n^2}{r^2} \right) J_n(\alpha r)$$

and

$$\frac{\partial J_n(\alpha r)}{\partial \alpha} = \frac{r}{\alpha} \frac{\partial J_n(\alpha r)}{\partial r};$$

hence

$$I_\alpha = \frac{a^2}{2\alpha^2} \left\{ \left(\alpha^2 - \frac{n^2}{r^2} \right) J_n^2(\alpha r) + \left[\frac{\partial J_n(\alpha r)}{\partial r} \right]^2 \right\}_{r=a} .$$

Thus, when $\alpha = \lambda$, we obtain

$$I_\lambda = \int_0^a J_n^2(\lambda r) r dr = \frac{a^2}{2\lambda^2} \left[\frac{\partial J_n(\lambda r)}{\partial r} \right]^2_{r=a} \tag{6.16}$$

because

$$J_n(\lambda a) = 0, \quad \lambda a = p_{nm}.$$

When $\alpha = \mu$, we have

$$I_\mu = \int_0^a J_n^2(\mu r) r dr = \frac{a^2}{2\mu^2} \left(\mu^2 - \frac{n^2}{a^2} \right) J_n^2(\mu a) \tag{6.17}$$

because

$$\left. \frac{dJ_n(x)}{dx} \right|_{x=\mu a} = 0, \quad \mu a = q_{nm}.$$

The orthogonal relations (6.14) and (6.15) and the normalization factors, (6.16) and (6.17), are needed in the discussion of the orthogonal property of the vector wave functions.

The orthogonal relations between the various cylindrical vector wave functions can be stated as follows:

$$\iiint \overline{M}_{\substack{e\\o} n\mu}(h) \cdot \overline{N}'_{\substack{e\\o} n'\lambda}(-h')\, dV = 0 \tag{6.18}$$

$$\iiint \overline{M}_{\substack{e\\o} n\mu}(h) \cdot \overline{M}_{\substack{e\\o} n'\mu'}(-h')\, dV$$

$$= \iiint \overline{N}_{\substack{e\\o} n\mu}(h) \cdot \overline{N}_{\substack{e\\o} n'\mu'}(-h')\, dV$$

$$= \begin{cases} 0, n \neq n', \mu \neq \mu' \\ (1+\delta_0)\, 2\pi^2 \mu^2 I_\mu \delta(h-h'), \; n = n', \mu = \mu' \end{cases} \tag{6.19}$$

$$\iiint \overline{M}_{\substack{e\\o} n\lambda}(h) \cdot \overline{M}_{\substack{e\\o} n'\lambda'}(-h')\, dV = \iiint \overline{N}_{\substack{e\\o} n\lambda}(h) \cdot \overline{N}_{\substack{e\\o} n'\lambda'}(-h')\, dV$$

$$= \begin{cases} 0, n \neq n', \lambda \neq \lambda' \\ (1+\delta_0)\, 2\pi^2 \lambda^2 I_\lambda \delta(h-h'), \; n = n', \lambda = \lambda', \end{cases} \tag{6.20}$$

where the volume integral is extended through the entire volume of an infinitely long guide. Because of the trigonometric functions, all the cylindrical vector wave functions are orthogonal in the ϕ-domain when $n \neq n'$. It is, therefore, sufficient to discuss those cases where $n = n'$. We shall demonstrate only two of them, as the proofs of the remaining ones are similar. We consider, for example, the integral

$$I = \iiint \overline{M}_{\substack{e\\o}n\mu}(h) \cdot \overline{N}_{\substack{e\\o}n\lambda}(-h')\, dV, \;\; n \neq 0.$$

In view of (6.5) and (6.6) we obtain

$$\begin{aligned}
I = &\int_0^a rdr \int_0^{2\pi} d\phi \int_{-\infty}^{\infty} dz \frac{1}{\kappa_\lambda} \left[ih'n \frac{J_n(\mu r)}{r} \frac{\partial J_n(\lambda r)}{\partial r} \sin^2 n\phi \right. \\
&\left. + ih'n \frac{J_n(\lambda r)}{r} \frac{\partial J_n(\mu r)}{\partial r} \cos^2 n\phi \right] e^{i(h-h')z} \\
= &\int_0^a rdr \int_{-\infty}^{\infty} dz \frac{ih'n\pi}{\kappa_\lambda} \left[\frac{J_n(\mu r)}{r} \frac{\partial J_n(\lambda r)}{\partial r} + \frac{J_n(\lambda r)}{r} \frac{\partial J_n(\mu r)}{\partial r} \right] e^{i(h-h')z} \\
= &\int_0^a rdr \int_{-\infty}^{\infty} dz \frac{ih'n\pi}{\kappa_\lambda r} \frac{\partial}{\partial r} \left[J_n(\lambda r) J_n(\mu r) \right] e^{i(h-h')z} \\
= &\frac{ih'n\pi}{\kappa_\lambda} J_n(\lambda a) J_n(\mu a) 2\pi\delta(h-h').
\end{aligned}$$

Since $J_n(\lambda a) = 0$, I vanishes identically. The normalization factor for functions of the same species can be found as follows. Using the function $\overline{M}_{en\mu}$ as an example, we have

$$\begin{aligned}
&\iiint \overline{M}_{en\mu}(h) \cdot \overline{M}_{en\mu}(-h')\, dV \\
&= \iiint \left\{ \frac{n^2 J_n^2(\mu r)}{r^2} \sin^2 n\phi + \left[\frac{\partial J_n(\mu r)}{\partial r} \right]^2 \cos^2 n\phi \right\} \cdot e^{i(h-h')z}\, dV \\
&= (1+\delta_0)\, 2\pi^2\, \delta(h-h') \int_0^a \left\{ \frac{n^2 J_n^2(\mu r)}{r^2} + \left[\frac{\partial J_n(\mu r)}{\partial r} \right]^2 \right\} r\, dr,
\end{aligned}$$

where δ_0 denotes the Kronecker delta function defined with respect to n. The integral involving the square of the derivative can be simplified by integration by parts that cancels the term involving J_n^2/r^2 and yields

$$2\,(1+\delta_0)\,\pi^2\,\delta(h-h') \int_0^a \mu^2 J_n^2(\mu r) r\, dr.$$

In view of (6.17) the final results for the normalization factor is given by

$$2\,(1+\delta_0)\,\pi^2 \mu^2 I_\mu \delta(h-h').$$

The readers can verify other relations listed in (6.18) through (6.20) as exercises. Knowing these orthogonal relations we can find the eigenfunction expansion of the dyadic Green functions by the Ohm-Rayleigh method.

6-2 CYLINDRICAL WAVEGUIDE

Based on the method of $\overline{\overline{G}}_m$ we let

$$\nabla \times \left[\overline{\overline{I}}\delta(\overline{R}-\overline{R}')\right] = \int_{-\infty}^{\infty} dh \sum_{n=0}^{\infty}\sum_{m=1}^{\infty} \left[\overline{N}_{\substack{e\\o}n\mu}(h)\overline{A}_{\substack{e\\o}n\mu}(h) + \overline{M}_{\substack{e\\o}n\lambda}(h)\overline{B}_{\substack{e\\o}n\lambda}(h)\right]. \tag{6.21}$$

The index m is used here to designate the ordinal number associated with p_{nm} $(=\lambda a)$ and q_{nm} $(=\mu a)$. We adopt the simplified notation that

$$\overline{N}_{\substack{e\\o}n\mu}(h)\overline{A}_{\substack{e\\o}n\mu}(h) = \overline{N}_{en\mu}(h)\overline{A}_{en\mu}(h) + \overline{N}_{on\mu}(h)\overline{A}_{on\mu}(h)$$

and similarly for $\overline{M}_{\substack{e\\o}n\lambda}(h)\overline{B}_{\substack{e\\o}n\lambda}(h)$. Taking the anterior scalar product of (6.21) with $\overline{N}_{en\mu}(-h)$ and $\overline{N}_{on\mu}(-h')$, $\overline{M}_{en\lambda}(-h')$ and $\overline{M}_{on\lambda}(-h')$ in turn and integrating the resultant equations through the entire volume of cylindrical waveguide, we can determine, as a consequence of the orthogonal relationships of the cylindrical vector wave functions, the coefficients $\overline{A}_{\substack{e\\o}n\mu}(h)$ and $\overline{B}_{\substack{e\\o}n\lambda}(h)$. They are

$$\begin{aligned}\overline{A}_{\substack{e\\o}n\mu}(h) &= \frac{2-\delta_0}{4\pi^2\mu^2 I_\mu}\nabla' \times \overline{N}'_{\substack{e\\o}n\mu}(-h)\\ &= \frac{(2-\delta_0)\,\kappa_\mu}{4\pi^2\mu^2 I_\mu}\overline{M}'_{\substack{e\\o}n\mu}(-h)\end{aligned} \tag{6.22}$$

and

$$\begin{aligned}\overline{B}_{\substack{e\\o}n\lambda}(h) &= \frac{2-\delta_0}{4\pi^2\lambda^2 I_\lambda}\nabla' \times \overline{M}_{\substack{e\\o}n\lambda}(-h)\\ &= \frac{(2-\delta_0)\,\kappa_\lambda}{4\pi^2\lambda^2 I_\lambda}\overline{N}'_{\substack{e\\o}n\lambda}(-h).\end{aligned} \tag{6.23}$$

The primed functions in (6.22) and (6.23) are defined with respect to the primed variables (r', ϕ', z'), corresponding to the location of $\overline{R}'$. Using the wave equation for $\overline{\overline{G}}_{m2}$, we obtain

$$\begin{aligned}\overline{\overline{G}}_{m2}(\overline{R},\overline{R}') = \int_{-\infty}^{\infty} dh \sum_{n,m} \frac{2-\delta_0}{4\pi^2}&\left[\frac{\kappa_\mu}{\mu^2 I_\mu\left(\kappa_\mu^2-k^2\right)}\overline{N}_{\substack{e\\o}n\mu}(h)\overline{M}'_{\substack{e\\o}n\mu}(-h)\right.\\ &\left.+\frac{\kappa_\lambda}{\lambda^2 I_\lambda\left(\kappa_\lambda^2-k^2\right)}\overline{M}_{\substack{e\\o}n\lambda}(h)\overline{N}'_{\substack{e\\o}n\lambda}(-h)\right].\end{aligned} \tag{6.24}$$

The Fourier integration in (6.24) can be evaluated by the method of contour integration as in the case of the rectangular waveguide. The poles of the integrand, however, are different for the TE modes and TM modes; they are

$$h = \pm\left(k^2 - \mu^2\right)^{\frac{1}{2}} = \pm\, k_\mu, \quad \text{for the } TE \text{ modes}$$

and

$$h = \pm\left(k^2 - \lambda^2\right)^{\frac{1}{2}} = \pm\, k_\mu, \quad \text{for the } TM \text{ modes,}$$

where k_μ and k_λ represent the guided wave numbers for these two sets. The final expression for $\overline{\overline{G}}_{m2}$ is given by

$$\overline{\overline{G}}{}^{\pm}_{m2}(\overline{R}, \overline{R}') = \sum_{n,m} k \left[c_\mu \overline{N}_{\substack{e\\o}n\mu}(\pm k_\mu)\, \overline{M}'_{\substack{e\\o}n\mu}(\mp k_\mu) + c_\lambda \overline{M}_{\substack{e\\o}n\lambda}(\pm k_\lambda)\, \overline{N}'_{\substack{e\\o}n\lambda}(\mp k_\lambda)\right], \quad z \gtrless z', \tag{6.25}$$

where

$$\begin{aligned} c_\mu &= i\,(2 - \delta_0)\,/4\pi\mu^2 I_\mu k_\mu \\ c_\lambda &= i\,(2 - \delta_0)\,/4\pi\lambda^2 I_\lambda k_\lambda. \end{aligned} \tag{6.26}$$

The top line in (6.25) applies to $z > z'$ and the bottom line is for $z < z'$. Knowing $\overline{\overline{G}}_{m2}$, we can find $\overline{\overline{G}}_{e1}$ based on the relation

$$\overline{\overline{G}}_{e1}(\overline{R}, \overline{R}') = \frac{1}{k^2}\left[-\hat{z}\hat{z}\delta(\overline{R} - \overline{R}') + \left(\nabla \times \overline{\overline{G}}{}^{+}_{m2}\right) U(z - z') + \left(\nabla \times \overline{\overline{G}}{}^{-}_{m2}\right) U(z' - z)\right], \tag{6.27}$$

which yields

$$\overline{\overline{G}}_{e1}(\overline{R}, \overline{R}') = -\frac{1}{k^2}\hat{z}\hat{z}\delta(\overline{R} - \overline{R}') + \sum_{n,m}\left[c_\mu \overline{M}_{\substack{e\\o}n\mu}(\pm k_\mu)\, \overline{M}'_{\substack{e\\o}n\mu}(\mp k_\mu) + c_\lambda \overline{N}_{\substack{e\\o}n\lambda}(\pm k_\lambda)\, \overline{N}'_{\substack{e\\o}n\lambda}(\mp k_\lambda)\right], \quad z \gtrless z'. \tag{6.28}$$

The function $\overline{\overline{G}}_{e2}(\overline{R}, \overline{R}')$ for the cylindrical waveguide can be found by making use of the symmetrical relations between $\overline{\overline{G}}_{e1}$ and $\overline{\overline{G}}_{e2}$ without going through a lengthy derivation. The result is given by

$$\overline{\overline{G}}_{e2}(\overline{R}, \overline{R}') = -\frac{1}{k^2}\hat{z}\hat{z}\delta(\overline{R} - \overline{R}') + \sum_{n,m}\left[c_\mu \overline{N}_{\substack{e\\o}n\mu}(\pm k_\mu)\, \overline{N}'_{\substack{e\\o}n\mu}(\mp k_\mu) + c_\lambda \overline{M}_{\substack{e\\o}n\lambda}(\pm k_\lambda)\, \overline{M}'_{\substack{e\\o}n\lambda}(\mp k_\lambda)\right], \quad z \gtrless z'. \tag{6.29}$$

This function is needed to study the field inside a cylindrical waveguide excited by an aperture field distribution on the wall of the waveguide.

6-3 CYLINDRICAL CAVITY

The dyadic Green functions for a cylindrical cavity can be obtained by the method of scattering superposition as has been done for the rectangular cavity. We again break the formulation into two steps. A semi-infinite cylindrical waveguide ($\infty > z \geq 0$) terminated at $z = 0$ with a conducting wall will be considered first. Denoting the electric dyadic Green function of the first kind for this structure by $\overline{\overline{G}}_{E1}$, we find

$$\begin{aligned} \overline{\overline{G}}_{E1}(\overline{R}, \overline{R}') = & -\frac{1}{k^2}\hat{z}\hat{z}\delta(\overline{R} - \overline{R}') \\ & + \sum_{n,m}(-2i) \begin{Bmatrix} c_\mu \overline{M}_\mu\,(k_\mu)\,\overline{M}'_{\mu 0}(z') + ic_\lambda \overline{N}_\lambda\,(k_\lambda)\,\overline{N}'_{\lambda e}(z') \\ c_\mu \overline{M}_{\mu 0}\,(z)\,\overline{M}'_\mu(k_\mu) + ic_\lambda \overline{N}_{\lambda e}\,(z)\,\overline{N}'_\lambda\,(k_\lambda) \end{Bmatrix}, \ z \gtrless z', \end{aligned} \tag{6.30}$$

where we have used a condensed notation for the functions in (6.30); they are

$$\begin{aligned} \overline{M}_\mu\,(k_\mu) &= \overline{M}_{\substack{e\\o}n\mu}\,(k_\mu) \\ \overline{N}_\lambda\,(k_\lambda) &= \overline{N}_{\substack{e\\o}n\lambda}\,(k_\lambda) \\ \overline{M}_{\mu 0}(z) &= \overline{M}_{\substack{e\\o}n\mu o}(z) \\ &= \nabla \times \left[J_n\,(\mu r) \begin{matrix}\cos\\ \sin\end{matrix} n\phi \sin k_\mu z\hat{z} \right] \\ \overline{N}_{\lambda e}(z) &= \overline{N}_{\substack{e\\o}n\lambda e}(z) \\ &= \frac{1}{k}\nabla \times \nabla \times \left[J_n\,(\lambda r) \begin{matrix}\cos\\ \sin\end{matrix} n\phi \cos k_\lambda z\hat{z} \right]. \end{aligned}$$

We repeat here the coefficients involved in (6.30); they are

$$\begin{aligned} c_\mu &= \frac{i\,(2 - \delta_0)}{4\pi\mu^2 I_\mu k_\mu} \\ c_\lambda &= \frac{i\,(2 - \delta_0)}{4\pi\lambda^2 I_\lambda k_\lambda} \\ k_\mu &= \left(k^2 - \mu^2\right)^{\frac{1}{2}} \\ k_\lambda &= \left(k^2 - \lambda^2\right)^{\frac{1}{2}} \\ \delta_0 &= \begin{cases} 1, & n = 0 \\ 0, & n \neq 0. \end{cases} \end{aligned}$$

The functions $\overline{M}_{\mu 0}(z)$ and $\overline{N}_{\lambda e}(z)$ are similar to the functions defined by (5.129) and (5.130) for the rectangular waveguide except that two different wave numbers, k_μ and k_λ, are involved in the present case in contrast to a single wave number k_g for the rectangular case.

Applying once more the method of scattering superposition for the cylindrical cavity of length c we obtain the electrical dyadic Green function of the first kind for the cavity given by

$$\begin{aligned}\overline{\overline{G}}_{E'1}(\overline{R},\overline{R}') = &-\frac{1}{k^2}\hat{z}\hat{z}\delta(\overline{R}-\overline{R}')\\ &+\sum_{n,m}\frac{2-\delta_0}{2\pi}\left\{\frac{1}{\mu^2 I_\mu k_\mu \sin k_\mu c}\begin{bmatrix}\overline{M}_{\mu 0}(c-z)\overline{M}'_{\mu 0}(z')\\ \overline{M}_{\mu 0}(z)\overline{M}'_{\mu 0}(c-z')\end{bmatrix}\right.\\ &\left.-\frac{1}{\lambda^2 I_\lambda k_\lambda \sin k_\lambda c}\begin{bmatrix}\overline{N}_{\lambda e}(c-z)\overline{N}'_{\lambda 0}(z')\\ \overline{N}_{\lambda 0}(z)\overline{N}'_{\lambda 0}(c-z')\end{bmatrix}\right\},\quad z \gtrless z', \end{aligned}\tag{6.31}$$

where

$$\overline{M}_{\mu 0}(c-z) = \nabla\times\left[J_n(\mu r)\,{\cos \atop \sin}\, n\phi \sin k_\mu(c-z)\hat{z}\right]$$

$$\overline{N}_{\lambda e}(c-z) = \frac{1}{k}\nabla\times\nabla\times\left[J_n(\lambda r)\,{\cos \atop \sin}\, n\phi\cos k_\lambda(c-z)\hat{z}\right].$$

These two functions are analogous to the functions $\overline{M}_{eo}(c-z)$ and $\overline{N}_{oe}(c-z)$ defined by (5.141) and (5.143) for the rectangular cavity.

6-4 COAXIAL LINE

The vector wave functions needed to construct the dyadic Green functions for a coaxial line are

$$\overline{M}_{e00}(h) = \nabla\times\left(1nre^{ihz}\hat{z}\right) = -\frac{e^{ihz}}{r}\hat{\phi} \tag{6.32}$$

$$\overline{N}_{e00}(h) = \frac{1}{|h|}\nabla\times\overline{M}_{e00}(h) = i\frac{he^{ihz}}{|h|\,r}\hat{r} \tag{6.33}$$

$$\begin{aligned}\overline{M}_{{e \atop o}n\lambda}(h) &= \nabla\times\left[S_n(\lambda r)\,{\cos \atop \sin}\, n\phi e^{ihz}\hat{z}\right]\\ &= \left[\mp\frac{nS_n(\lambda r)}{r}\,{\sin \atop \cos}\, n\phi\hat{r} - \frac{\partial S_n(\lambda r)}{\partial r}\,{\cos \atop \sin}\, n\phi\hat{\phi}\right]e^{ihz}\end{aligned} \tag{6.34}$$

$$\begin{aligned}\overline{N}_{{e \atop o}n\lambda}(h) &= \frac{1}{\kappa_\lambda}\nabla\times\overline{M}_{{e \atop o}n\lambda}(h)\\ &= \frac{1}{\kappa_\lambda}\left[ih\frac{\partial S_n(\lambda r)}{\partial r}\,{\cos \atop \sin}\, n\phi\hat{r} \mp \frac{ihn}{r}S_n(\lambda r)\,{\sin \atop \cos}\, n\phi\hat{\phi}\right.\\ &\left.\quad +\lambda^2 S_n(\lambda r)\,{\cos \atop \sin}\, n\phi\hat{z}\right]e^{ihz}\end{aligned} \tag{6.35}$$

$$\begin{aligned}
\overline{M}_{\substack{e\\o}m\mu}(h) &= \nabla \times \left[T_n(\mu r)\begin{matrix}\cos\\ \sin\end{matrix}n\phi e^{ihz}\hat{z}\right] \\
&= \left[\mp\frac{nT_n(\mu r)}{r}\begin{matrix}\sin\\ \cos\end{matrix}n\phi\hat{r} - \frac{\partial T_n(\mu r)}{\partial r}\begin{matrix}\cos\\ \sin\end{matrix}n\phi\hat{\phi}\right]e^{ihz} \qquad (6.36)\\
\overline{N}_{\substack{e\\o}m\mu}(h) &= \frac{1}{\kappa_\mu}\nabla \times \overline{M}_{\substack{e\\o}m\mu}(h) \\
&= \frac{1}{\kappa_\mu}\left[ih\frac{\partial T_n(\mu r)}{\partial r}\begin{matrix}\cos\\ \sin\end{matrix}n\phi\hat{r} \mp \frac{ihn}{r}T_n(\mu r)\begin{matrix}\sin\\ \cos\end{matrix}n\phi\hat{\phi}\right. \\
&\quad \left. + \mu^2 T_n(\mu r)\begin{matrix}\cos\\ \sin\end{matrix}n\phi\hat{z}\right]e^{ihz}, \qquad (6.37)
\end{aligned}$$

where

$$\begin{aligned}
\kappa_\lambda &= \left(\lambda^2 + h^2\right)^{\frac{1}{2}} \\
\kappa_\mu &= \left(\mu^2 + h^2\right)^{\frac{1}{2}} \\
S_n(\lambda r) &= Y_n(\lambda a)J_n(\lambda r) - J_n(\lambda a)Y_n(\lambda r) \\
T_n(\mu r) &= Y_n'(\mu a)J_n(\mu r) - J_n'(\mu a)Y_n(\mu r)
\end{aligned}$$

and J_n and Y_n denote, respectively, Bessel function and Neumann function of integral order.

The eigenvalues λ and μ are solutions of the characteristic equations

$$S_n(\lambda b) = Y_n(\lambda a)J_n(\lambda b) - J_n(\lambda a)Y_n(\lambda b) = 0 \qquad (6.38)$$

and

$$T_n'(\mu b) = Y_n'(\mu a)J_n'(\mu b) - J_n'(\mu a)Y_n'(\mu b) = 0, \qquad (6.39)$$

where the primed functions denote the derivative of these functions with respect to their arguments, μa or μb. The calculation of the value of λ and μ is discussed in Dwight [1948] and Abramowitz and Stegun [1964]. A complete tabulation of these values, however, is not yet available. The normalization factors of the vector wave functions defined by (6.32)–(6.37) are

$$\begin{aligned}
&\iiint \overline{M}_{e00}(h)\cdot\overline{M}_{e00}(-h')dV \\
&= \iiint \overline{N}_{e00}(h)\cdot\overline{N}_{e00}(-h')dV = 4\pi^2 \ln\left(\frac{b}{a}\right)\delta\left(h-h'\right) \\
&\iiint \overline{M}_{\substack{e\\o}n\lambda}(h)\cdot\overline{M}_{\substack{e\\o}n'\lambda'}(-h')dV \\
&= \iiint \overline{N}_{\substack{e\\o}n\lambda}(h)\cdot\overline{N}_{\substack{e\\o}n'\lambda'}(-h')dV \\
&= \begin{cases} 0, & n \neq n' \text{ and/or } \lambda \neq \lambda' \\ 2\pi^2\left(1+\delta_0\right)\lambda^2 I_\lambda \delta(h-h'), & n = n', \lambda = \lambda' \end{cases} \\
&\iiint \overline{M}_{\substack{e\\o}n\lambda}(h)\cdot\overline{M}_{\substack{e\\o}n'\lambda'}(-h')dV
\end{aligned}$$

$$= \iiint \overline{N}_{\substack{e\\o}n\lambda}(h) \cdot \overline{N}_{\substack{e\\o}n'\lambda'}(-h')dV$$

$$= \begin{cases} 0, & n \neq n', \text{ and/or } \mu \neq \mu' \\ 2\pi^2 (1+\delta_0)\,\mu^2 I_\mu \delta(h-h'), & n = n', \mu = \mu', \end{cases}$$

where δ_0 denotes the Kronecker delta function with respect to n and

$$I_\lambda = \int_a^b S_n^2(\lambda r)dr \tag{6.40}$$

$$I_\mu = \int_a^b T_n^2(\mu r)dr. \tag{6.41}$$

Functions not of the same type are orthogonal; for example,

$$\iiint \overline{M}_{e00}(h) \cdot \overline{M}_{\substack{e\\o}n\lambda}(-h')\,dV = 0$$

$$\iiint \overline{N}_{e00}(h) \cdot \overline{N}_{\substack{e\\o}n'\mu'}(-h')\,dV = 0$$

$$\iiint \overline{M}_{\substack{e\\o}n\lambda}(h) \cdot \overline{N}_{\substack{e\\o}n'\lambda'}(-h')\,dV = 0$$

$$\iiint \overline{M}_{\substack{e\\o}n\mu}(h) \cdot \overline{N}_{\substack{e\\o}n'\mu'}(-h')\,dV = 0.$$

To construct the eigenfunction expansion of $\overline{\overline{G}}_{m2}$, we use $\overline{M}_{e00}$, $\overline{M}_{\substack{e\\o}n\lambda}$, $\overline{N}_{\substack{e\\o}n\mu}$. These functions satisfy the boundary condition required for $\overline{\overline{G}}_{m2}$, and they are solutions for the homogeneous wave equation

$$\nabla \times \nabla \times \overline{F} - \kappa^2 \overline{F} = 0$$

with κ^2 equal to h^2, or κ_λ^2, or κ_μ^2. They are solenoidal functions as $\nabla \cdot \overline{\overline{G}}_{m2} = 0$ and hence are sufficient to represent $\overline{\overline{G}}_{m2}$. Following the method of $\overline{\overline{G}}_m$, we let

$$\nabla \times \left[\overline{\overline{I}}\delta(\overline{R} - \overline{R}')\right]$$

$$= \int_{-\infty}^{\infty} dh \left[\overline{M}_{e00}(h)\overline{A}_{e00}(h) + \sum_{n,\lambda} \overline{M}_{\substack{e\\o}n\lambda}(h)\overline{B}_{\substack{e\\o}h\lambda}(h)\right.$$

$$\left. + \sum_{n,\mu} \overline{N}_{\substack{e\\o}n\mu}(h)\overline{C}_{\substack{e\\o}n\mu}(h)\right]. \tag{6.42}$$

As a result of the orthogonal property of the functions $\overline{M}_{e00}$, $\overline{M}_{\substack{e\\o}n\lambda}$, and $\overline{N}_{\substack{e\\o}n\mu}$, we find

$$\overline{A}_{e00}(h) = \frac{|\,h\,|}{4\pi^2 I_0} \overline{N}'_{e00}(-h) \tag{6.43}$$

$$\overline{B}_{\substack{e\\o}n\lambda}(h) = \frac{(2-\delta_0)\,K_\lambda}{4\pi^2\lambda^2 I_\lambda}\overline{N}'_{\substack{e\\o}n\lambda}(-h) \tag{6.44}$$

$$\overline{C}_{\substack{e\\o}n\mu}(h) = \frac{(2-\delta_0)\,K_\mu}{4\pi^2\mu^2 I_\mu}\overline{M}'_{\substack{e\\o}n\mu}(-h), \tag{6.45}$$

where $I_0 = 1n(b/a)$ and I_λ and I_μ are defined in (6.40) and (6.41). The primed functions in (6.43) and (6.45) are defined with respect to the primed variables (r', ϕ', z') pertaining to $\overline{R}'$. Following the Ohm-Rayleigh method we obtain

$$\begin{aligned}\overline{\overline{G}}_{m2} = \int_{-\infty}^{\infty} dh \Bigg[& \frac{C_{00}\mid h\mid}{h^2-k^2}\overline{M}_{e00}(h)\overline{N}'_{e00}(-h) \\ & + \sum_{n,\lambda}\frac{C_{n\lambda}\kappa_\lambda}{\kappa_\lambda^2-k^2}\overline{M}_{\substack{e\\o}n\lambda}(h)\overline{N}'_{\substack{e\\o}n\lambda}(-h) \\ & + \sum_{n,\lambda}\frac{C_{n\mu}\kappa_\mu}{\kappa_\mu^2-k^2}\overline{N}_{\substack{e\\o}n\mu}(h)\overline{M}'_{\substack{e\\o}n\mu}(-h)\Bigg],\end{aligned} \tag{6.46}$$

where

$$\begin{aligned} C_{00} &= \frac{1}{4\pi^2 I_0} \\ C_{n\lambda} &= \frac{2-\delta_0}{4\pi^2\lambda^2 I_\lambda} \\ C_{n\mu} &= \frac{2-\delta_0}{4\pi^2\mu^2 I_\mu}. \end{aligned}$$

By means of a contour integration, (6.46) can be reduced to

$$\overline{\overline{G}}_{m2} = \overline{\overline{G}}^{+}_{m2}U^+ + \overline{\overline{G}}^{-}_{m2}U^-, \tag{6.47}$$

where

$$\begin{aligned}\overline{\overline{G}}^{\pm}_{m2} = i\pi \Bigg[& c_{00}\overline{M}_{e00}(\pm k)\overline{N}'_{e00}(\mp k) \\ & + \sum_{n,\lambda}\frac{k}{k_\lambda}c_{n\lambda}\overline{M}_{\substack{e\\o}n\lambda}(\pm k_\lambda)\,\overline{N}'_{\substack{e\\o}n\lambda}(\mp k_\lambda) \\ & + \sum_{n,\mu}\frac{k}{k_\mu}c_{n\mu}\overline{N}_{\substack{e\\o}n\mu}(\pm k_\mu)\,\overline{M}'_{\substack{e\\o}n\mu}(\mp k_\mu)\Bigg], \quad z \gtrless z'\end{aligned} \tag{6.48}$$

$$\begin{aligned} U^+ &= U(z-z') \\ U^- &= U(z'-z) \\ k_\lambda &= (k^2-\lambda^2)^{1/2} \\ k_\mu&(k^2-\mu^2)^{1/2}. \end{aligned}$$

According to the method of $\overline{\overline{G}}_m$,

$$\overline{\overline{G}}_{e1} = \frac{1}{k^2}\left[\left(\nabla \times \overline{\overline{G}}^+_{m2}\right) U^+ + \left(\nabla \times \overline{\overline{G}}^-_{m2}\right) U^- - \hat{z}\hat{z}\delta(\overline{R} - \overline{R}')\right], \tag{6.49}$$

where

$$\begin{aligned}\nabla \times \overline{\overline{G}}^{\pm}_{m2} =& i\pi k^2 \Bigg[c_{00}\overline{N}_{e00}(\pm k)\,\overline{N}'_{e00}(\mp k) \\ &+ \sum_{n,\lambda} \frac{1}{k_\lambda} c_{n\lambda} \overline{N}_{\substack{e\\o}n\lambda}(\pm k_\lambda)\,\overline{N}'_{\substack{e\\o}n\lambda}(\mp k_\lambda) \\ &+ \sum_{n,\mu} \frac{1}{k_\mu} c_{n\mu} \overline{M}_{\substack{e\\o}m\mu}(\pm k_\mu)\,\overline{M}'_{\substack{e\\o}n\mu}(\mp k_\mu)\Bigg], \quad z \gtrless z'.\end{aligned} \tag{6.50}$$

Following the same procedure, it can be shown that

$$\overline{\overline{G}}_{m1} = \overline{\overline{G}}^+_{m1} U^+ + \overline{\overline{G}}^-_{m1} U^-, \tag{6.51}$$

with

$$\begin{aligned}\overline{\overline{G}}^{\pm}_{m1} =& i\pi \Bigg[C_{00}\overline{N}_{e00}(\pm k)\,\overline{M}'_{e00}(\mp k) \\ &+ \sum_{n,\lambda} \frac{k}{k_\lambda} \overline{N}_{\substack{e\\o}n\lambda}(\pm k_\lambda)\,\overline{M}'_{\substack{e\\o}n\lambda}(\mp k_\lambda) \\ &+ \sum_{n,\mu} \frac{k}{k_\mu} \overline{M}_{\substack{e\\o}n\mu}(\pm k_\mu)\,\overline{N}'_{\substack{e\\o}n\mu}(\mp k_\mu)\Bigg],\end{aligned} \tag{6.52}$$

and

$$\overline{\overline{G}}^{\pm}_{e2} = \frac{1}{k^2}\left[\left(\nabla \times \overline{\overline{G}}^+_{m1}\right) U^+ + \left(\nabla \times \overline{\overline{G}}^-_{m1}\right) U^- - \hat{z}\hat{z}\delta(\overline{R} - \overline{R}')\right], \tag{6.53}$$

with

$$\begin{aligned}\nabla \times \overline{\overline{G}}^{\pm}_{m1} =& i\pi k^2 \Bigg[C_{00}\overline{M}_{e00}(\pm k)\,\overline{M}'_{e00}(\mp k) \\ &+ \sum_{n,\lambda} \frac{1}{k_\lambda} C_{n\lambda} \overline{M}_{\substack{e\\o}n\lambda}(\pm k_\lambda)\,\overline{M}'_{\substack{e\\o}n\lambda}(\mp k_\lambda) \\ &+ \sum_{n,\mu} \frac{1}{k_\mu} C_{n,\mu} \overline{N}_{\substack{e\\o}n\mu}(\pm k_\mu)\,\overline{N}'_{\substack{e\\o}n\mu}(\mp k_\mu)\Bigg], \quad z \gtrless z'.\end{aligned} \tag{6.54}$$

This completes our derivation for the various dyadic Green functions for a coaxial line. The symmetrical relations between these functions are quite obvious.

For example,

$$\left[\overline{\overline{G}}_{e1}(\overline{R}', \overline{R})\right]^T = \overline{\overline{G}}_{e1}(\overline{R}, \overline{R}')$$

and

$$\left[\overline{\overline{G}}_{m1}(\overline{R}', \overline{R})\right]^T = \overline{\overline{G}}_{m2}(\overline{R}, \overline{R}').$$

Most of the material presented here was originally contained in a paper [Tai, 1983] where the method of $\overline{\overline{G}}_m$ was enunciated. Several typographical errors in that paper have now been corrected.

7

Circular Cylinder in Free Space

When we are dealing with radiation problems in the presence of an infinite cylinder as a diffracting body, the vector wave functions with continuous eigenvalues both in the h and λ domains are needed. The orthogonal properties of these functions are not quite the same as the ones with discrete eigenvalues. We will discuss these properties first, and then find the eigenfunction expansion of the free-space dyadic Green function in terms of these functions. Once the latter is known, the functions of the other kinds can be found by the method of scattering superposition.

7-1 CYLINDRICAL VECTOR WAVE FUNCTIONS WITH CONTINUOUS EIGENVALUES

In dealing with an integral representation of fields in free space, the cylindrical vector wave functions to be used have continuous eigenvalues in the λ-domain as well as in the h-domain. The formulas for $\overline{M}$ and $\overline{N}$ as defined by (6.3) and (6.4), with $\mu = \lambda$, are still valid, but no constraint is placed on λ or μ. Since λ and μ merge into one parameter, only two sets of functions,

$$\overline{M}_{\substack{e\\o}n\lambda} \text{ and } \overline{N}_{\substack{e\\o}n\lambda}$$

are left to represent both the electric and magnetic field. These functions are now defined in the entire space, corresponding to $\infty > r \geq 0, 2\pi \geq \phi \geq 0$, and $-\infty < z < \infty$. The orthogonal properties of these functions can be stated as follows:

$$\iiint \overline{M}_{\substack{e\\o}n\lambda}(h)\cdot\overline{N}_{\substack{e\\o}n'\lambda'}(-h')dV=0 \tag{7.1}$$

$$\iiint \overline{M}_{\substack{e\\o}n\lambda}(h)\cdot\overline{M}_{\substack{e\\o}n'\lambda'}(-h')dV = \begin{cases} 0, n\neq n' \\ 2\left(1+\delta_0\right)\pi^2\lambda\delta\left(\lambda-\lambda'\right)\delta\left(h-h'\right), \ n=n' \end{cases} \tag{7.2}$$

$$\iiint \overline{N}_{\substack{e\\o}n\lambda}(h)\cdot\overline{N}_{\substack{e\\o}n'\lambda'}(-h')dV = \begin{cases} 0, n\neq n' \\ 2\left(1+\delta_0\right)\pi^2\lambda\delta\left(\lambda-\lambda'\right)\delta\left(h-h'\right), \ n=n', \end{cases} \tag{7.3}$$

where the domain of integration encloses the entire space. The proofs of these formulas are slightly different from the ones for (6.18) to (6.20). Because of the angular function, all these functions, either of the same species or of different species, are orthogonal when $n \neq n'$. It is therefore sufficient to discuss those cases when $n = n'$. To prove (7.1), let us consider, for example, the integral

$$I=\iiint \overline{M}_{en\lambda}(h)\cdot\overline{N}_{on\lambda'}(-h')dV,$$

where (λ, h) and $(\lambda', -h')$ denote two distinct pairs of eigenvalues. Using (6.3) and (6.4), with $\mu = \lambda$, we obtain, after an integration with respect to ϕ,

$$\begin{aligned} I=&\frac{(1+\delta_0)\pi}{\sqrt{\lambda'^2+h'^2}}\int_0^\infty r\,dr\int_{-\infty}^\infty dz \\ &\cdot\frac{inh'}{r}\left[J_n(\lambda r)\frac{\partial J_n(\lambda' r)}{\partial r}+J_n(\lambda' r)\frac{\partial J_n(\lambda r)}{\partial r}\right]e^{i(h-h')z} \\ =&\frac{inh'(1+\delta_0)\pi}{\sqrt{\lambda'^2+h'^2}}\cdot 2\pi\delta(h-h')\left[J_n(\lambda r)J_n(\lambda' r)\right]_0^\infty=0. \end{aligned}$$

Actually, $\overline{M}_{\substack{e\\o}n\lambda}$ and $\overline{N}_{on\lambda'}$ are orthogonal in the $r-\phi$-plane, the integration with respect to z is not necessary. We include it here for completeness because, in general, we always deal with the volume integral of these functions. To prove (7.2) or (7.3), let us consider the integral

$$\begin{aligned} I=&\iiint \overline{N}_{en\lambda}(h)\cdot\overline{N}_{en\lambda'}(-h')dV \\ =&\frac{1}{\kappa\kappa'}\iiint\left[hh'\frac{\partial J_n(\lambda r)}{\partial r}\frac{\partial J_n(\lambda' r)}{\partial r}\cos^2 n\phi\right. \\ &+\frac{hh'n^2}{r^2}J_n(\lambda r)J_n(\lambda' r)\sin^2 n\phi \\ &\left.+\lambda^2\lambda'^2 J_n(\lambda r)J_n(\lambda' r)\cos^2 n\phi\right]e^{i(h-h')z}dV, \end{aligned}$$

where

$$\kappa=\left(\lambda^2+h^2\right)^{\frac{1}{2}},\ \kappa'=\left(\lambda'^2+h'^2\right)^{\frac{1}{2}}.$$

The integrations with respect to ϕ and z yield

$$I = \frac{(1+\delta_0)\, 2\pi^2 \delta\,(h-h')}{\kappa\kappa'} \cdot \int_0^\infty \left[hh' \frac{\partial J_n(\lambda r)}{\partial r} \frac{\partial J_n\,(\lambda' r)}{\partial r} + \left(\frac{hh'n^2}{r^2} + \lambda^2\lambda'^2 \right) J_n\,(\lambda r)\, J_n\,(\lambda' r) \right] r\, dr.$$

Using the recurrence relations of the Bessel functions,

$$J_n(x) = \frac{x}{2n}\left[J_{n-1}(x) + J_{n+1}(x)\right] \tag{7.4}$$

$$\frac{dJ_n(x)}{dx} = \frac{1}{2}\left[J_{n-1}(x) - J_{n+1}(x)\right], \tag{7.5}$$

we can change the integral into the form

$$I = \frac{(1+\delta_0)\, 2\pi^2 \delta\,(h-h')}{\kappa\kappa'} \cdot \int_0^\infty \left\{ \frac{1}{2}\lambda\lambda' hh' \left[J_{n-1}(\lambda r) J_{n-1}(\lambda' r) + J_{n+1}(\lambda r) J_{n+1}(\lambda' r)\right] + \lambda^2\lambda'^2 J_n(\lambda r) J_n(\lambda' r) \right\} r\, dr.$$

As a result of the integral representation of the weighted delta function given by (1.83), we obtain

$$\begin{aligned} I &= \frac{1}{\kappa\kappa'}\,(1+\delta_0)\, 2\pi^2 \left(\lambda' hh' + \lambda\lambda'^2\right) \delta\,(h-h')\, \delta\,(\lambda-\lambda') \\ &= 2\,(1+\delta_0)\, \pi^2 \lambda \delta\,(\lambda-\lambda')\, \delta\,(h-h'), \end{aligned}$$

which is the normalization factor for two cylindrical vector wave functions of the same species. The proofs for the remaining combinations are practically the same. It should be mentioned that the orthogonal properties of these functions have previously been investigated by Stratton [1941, pp. 397–399]. However, he used a mixed domain consisting of (r, ϕ, λ), which is partly spatial and partly in the eigenvalue domain. As a result, the fine structure of these properties is not spelled out. Our presentation shows that the normalization factor in the spatial domain contains two delta functions in the eigenvalue domain. It is this important feature that facilitates greatly the eigenfunction expansion of a dyadic Green function or any other vector function. In regard to (7.1), it is clear that $\overline{M}_e$ and $\overline{N}_o$ or $\overline{M}_o$ and $\overline{N}_e$ are orthogonal in the $r-\phi$-plane but not in the ϕ-plane alone as asserted by Stratton [p. 398]. This is presumably just an oversight.

7-2 EIGENFUNCTION EXPANSION OF THE FREE-SPACE DYADIC GREEN FUNCTIONS

The free-space magnetic dyadic Green function introduced previously in Sec. 4-2, by definition, satisfies the equation

$$\nabla \times \nabla \times \overline{\overline{G}}_{m0}(\overline{R}, \overline{R}') - k^2 \overline{\overline{G}}_{m0}(\overline{R}, \overline{R}') = \nabla \times \left[\overline{\overline{I}}\delta(\overline{R} - \overline{R}')\right] \tag{7.6}$$

and the radiation condition at infinity. Its explicit expression was shown to be

$$\begin{aligned} \overline{\overline{G}}_{m0}(\overline{R}, \overline{R}') &= \nabla \times \left[\overline{\overline{I}} G_0(\overline{R}, \overline{R}')\right] \\ &= \nabla G_0(\overline{R}, \overline{R}') \times \overline{\overline{I}}, \end{aligned} \tag{7.7}$$

where

$$G_0(\overline{R}, \overline{R}') = \frac{e^{ik|\overline{R} - \overline{R}'|}}{4\pi \mid \overline{R} - \overline{R}' \mid}. \tag{7.8}$$

For cylindrical problems in space, we need the eigenfunction expansion of this function in order to construct the functions of the first kind and the third kind by the method of scattering superposition. According to the Ohm-Rayleigh method we let

$$\begin{aligned} \nabla \times \left[\overline{\overline{I}}\delta(\overline{R} - \overline{R}')\right] = \int_0^\infty d\lambda \int_{-\infty}^{\infty} dh \sum_{n=0}^{\infty} \\ \left[\overline{N}_{\substack{e\\o}n\lambda}(h)\overline{A}_{\substack{e\\o}n\lambda}(h) + \overline{M}_{\substack{e\\o}n\lambda}(h)\overline{B}_{\substack{e\\o}n\lambda}(h)\right], \end{aligned} \tag{7.9}$$

where λ and h are two continuous eigenvalues. Only positive values of λ are included because $J_n(\lambda r)$ and $J_n(-\lambda r)$ are not independent functions. Equation (7.9) is treated as the Fourier transform and the Fourier-Bessel transform or the Hankel transform of $\nabla \times \left[\overline{\overline{I}}\delta(\overline{R} - \overline{R}')\right]$. By taking the anterior scalar product of (7.9) with $\overline{N}_{\substack{e\\o}n'\lambda'}(-h')$ and integrating the resultant equation through the entire space, we obtain, as a result of the orthogonal relationships described by (7.1) through (7.3),

$$\overline{A}_{en\lambda}(h) = \frac{(2 - \delta_0)}{4\pi^2 \lambda} \nabla' \times \overline{N}'_{en\lambda}(-h) = \frac{(2 - \delta_0)\,\kappa}{4\pi^2 \lambda} \overline{M}'_{en\lambda}(-h), \tag{7.10}$$

where the primed functions $\overline{M}'$ and $\overline{N}'$ are defined with respect to (r', ϕ', z') of the position vector $\overline{R}'$. Repeating the same procedure with the odd $\overline{M}$ functions and both the even and the odd $\overline{N}$ functions, we obtain

$$\overline{A}_{\substack{e\\o}n\lambda}(h) = \frac{(2 - \delta_0)\,\kappa}{4\pi^2 \lambda} \overline{M}'_{\substack{e\\o}n\lambda}(-h) \tag{7.11}$$

$$\overline{B}_{\substack{e\\o}n\lambda}(h) = \frac{(2 - \delta_0)\,\kappa}{4\pi^2 \lambda} \overline{N}'_{\substack{e\\o}n\lambda}(-h). \tag{7.12}$$

Hence the continuous eigenfunction expansion of $\nabla \times \left[\bar{\bar{I}}\delta(\bar{R}-\bar{R}')\right]$ is given by

$$\nabla \times \left[\bar{\bar{I}}\delta(\bar{R}-\bar{R}')\right] = \int_0^\infty d\lambda \int_{-\infty}^\infty dh \sum_n \frac{(2-\delta_0)\,\kappa}{4\pi^2\lambda} \cdot \left[\bar{N}_{\substack{e\\o}n\lambda}(h)\bar{M}'_{\substack{e\\o}n\lambda}(-h) + \bar{M}_{\substack{e\\o}n\lambda}(h)\bar{N}'_{\substack{e\\o}n\lambda}(-h)\right]. \tag{7.13}$$

In view of (7.6), the expansion of $\bar{\bar{G}}_{m0}$ can be written in the form

$$\bar{\bar{G}}_{m0}(\bar{R},\bar{R}') = \int_0^\infty d\lambda \int_{-\infty}^\infty dh \sum_n \frac{(2-\delta_0)\,\kappa}{4\pi^2\lambda\,(\kappa^2-k^2)} \cdot \left[\bar{N}_\lambda(h)\bar{M}'_\lambda(-h) + \bar{M}_\lambda(h)\bar{N}'_\lambda(-h)\right], \tag{7.14}$$

where we have now used the condensed notation $\bar{M}_\lambda(h)\bar{N}_\lambda(h)$, $\bar{M}'_\lambda(-h)$, and $\bar{N}'_\lambda(-h)$ for the four dyadic pairs contained in the brackets of (7.13), for example, $\bar{M}_\lambda = \bar{M}_{\substack{e\\o}n\lambda}(h)$. Although (7.14) contains a double integral, one of them can be eliminated. The elimination depends on the nature of the problem. For the construction of dyadic Green functions associated with an infinite cylinder, we want to eliminate the λ-integration. The reason for doing so will be clear later. In dealing with a flat ground, or layered media, the integration with respect to h will be eliminated.

To perform the integration with respect to λ in (7.14) we can apply the integral relation (1.103) in an operational form. Thus we write

$$\bar{N}_\lambda(h)\bar{M}'_\lambda(-h) = \bar{\bar{T}}_\lambda\left[J_n(\lambda r)J_n(\lambda r')\right], \tag{7.15}$$

where $\bar{\bar{T}}$ represents a dyadic spatial operator. Its precise form can be written out if necessary, but this is not required. Then

$$\int_0^\infty \frac{\kappa\bar{N}_\lambda(h)\bar{M}'_\lambda(-h)}{\lambda\,(\kappa^2-k^2)}d\lambda = \int_0^\infty \frac{\kappa\bar{\bar{T}}_\lambda\left[J_n(\lambda r)J_n(\lambda r')\right]}{\lambda\,(\kappa^2-k^2)}d\lambda, \tag{7.16}$$

where $\kappa = \left(\lambda^2+h^2\right)^{\frac{1}{2}}$. In view of (1.103), (7.16) is equal to

$$\begin{aligned} &\frac{ik\pi}{2\eta^2}\begin{cases} \bar{\bar{T}}_\eta\left[H_n^{(1)}(\eta r)\,J_n(\eta r')\right], & r>r' \\ \bar{\bar{T}}_\eta\left[J_n(\eta r)\,H_n^{(1)}(\eta r')\right], & r<r' \end{cases} \\ &= \frac{ik\pi}{2\eta^2}\begin{cases} \bar{N}_\eta^{(1)}(h)\bar{M}'_\eta(-h), & r>r' \\ \bar{N}_\eta(h)\bar{M}'^{(1)}_\eta(-h), & r<r', \end{cases} \end{aligned} \tag{7.17}$$

where $\eta = \left(k^2-h^2\right)^{\frac{1}{2}}$ and $H_n^{(1)}$ denotes the Hankel function of the first kind. The superscript (1) attached to $\bar{N}_\eta^{(1)}$ and $\bar{M}_\eta^{(1)}$ means that these functions are now defined with respect to the Hankel function of the first kind; that is,

$$\overline{N}_\eta^{(1)}(h) = \frac{1}{k}\nabla \times \nabla \times \left[H_n^{(1)}(\eta r) \begin{matrix}\cos\\ \sin\end{matrix} n\phi e^{ihz}\hat{z}\right] \tag{7.18}$$

$$\overline{M}_\eta^{\prime(1)}(-h) = \nabla' \times \left[H_n^{(1)}(\eta r') \begin{matrix}\cos\\ \sin\end{matrix} n\phi' e^{-ihz'}\hat{z}\right]. \tag{7.19}$$

Repeating the same procedure for $\overline{M}_\lambda(h)\overline{N}'_\lambda(-h)$, we obtain finally an expression for $\overline{\overline{G}}_{m0}$, involving only a Fourier integration with respect to h, which has the form

$$\overline{\overline{G}}_{m0}^{\pm}(\overline{R}, \overline{R}') = \frac{ik}{8\pi}\int_{-\infty}^{\infty} dh \sum_n \frac{2-\delta_0}{\eta^2}\cdot$$
$$\cdot\begin{cases} \overline{N}_\eta^{(1)}(h)\overline{M}'_\eta(-h) + \overline{M}_\eta^{(1)}(h)\overline{N}'_\eta(-h), & r > r' \\ \overline{N}_\eta(h)\overline{M}_\eta^{\prime(1)}(-h) + \overline{M}_\eta(h)\overline{N}_\eta^{\prime(1)}(-h), & r > r' \end{cases} \tag{7.20}$$

The function $\overline{\overline{G}}_{m0}$ is discontinuous at $r = r'$. According to the method of $\overline{\overline{G}}_m$, the expression for $\overline{\overline{G}}_{e0}(\overline{R}, \overline{R}')$ in this case would be given by

$$\overline{\overline{G}}_{eo}(\overline{R}, \overline{R}') = \frac{1}{k^2}\left[-\hat{r}\hat{r}\delta(\overline{R} - \overline{R}') + \left(\nabla \times \overline{\overline{G}}_{m0}^{+}\right) U(r - r')\right.$$
$$\left. + \left(\nabla \times \overline{\overline{G}}_{m0}^{-}\right) U(r' - r)\right]$$
$$= -\frac{1}{k^2}\hat{r}\hat{r}\delta(\overline{R} - \overline{R}')$$
$$+ \frac{i}{8\pi}\int_{-\infty}^{\infty} dh \sum_n \frac{(2-\delta_0)}{\eta^2}$$
$$\begin{cases} \overline{M}_\eta^{(1)}(h)\overline{M}'_\eta(-h) + \overline{N}_\eta^{(1)}(h)\overline{N}'_\eta(-h), & r < r' \\ \overline{M}_\eta(h)\overline{M}_\eta^{\prime(1)}(-h) + \overline{N}_\eta(h)\overline{N}_\eta^{\prime(1)}(-h), & r < r'. \end{cases} \tag{7.21}$$

7-3 CONDUCTING CYLINDER, DIELECTRIC CYLINDER, AND COATED CYLINDER

The main reason for us to have an eigenfunction expansion for $\overline{\overline{G}}_{eo}$ is for the construction of other kinds of electric dyadic Green functions. Thus, the function $\overline{\overline{G}}_{e1}$ for an infinite conducting cylinder of radius equal to "a" concentric with the z-axis can be found by the method of scattering superposition. We let

$$\overline{\overline{G}}_{e1}(\overline{R}, \overline{R}') = \overline{\overline{G}}_{e0}(\overline{R}, \overline{R}') + \overline{\overline{G}}_{1s}(\overline{R}, \overline{R}'), \tag{7.22}$$

where $\overline{\overline{G}}_{1s}(\overline{R}, \overline{R}')$ must have the form

$$\overline{\overline{G}}_{1s}(\overline{R}, \overline{R}') = \frac{i}{8\pi} \int_{-\infty}^{\infty} dh \sum_{n} \frac{2 - \delta_0}{\eta^2}$$
$$\cdot \left[a_\eta \overline{M}_\eta^{(1)}(h) \overline{M}'^{(1)}_\eta(-h) + b_\eta \overline{N}_\eta^{(1)} \overline{N}'^{(1)}_\eta(-h) \right]. \quad (7.23)$$

The choice of $\overline{M}^{(1)}$ and $\overline{N}^{(1)}$ as the field functions in $\overline{\overline{G}}_{1s}$ is dictated by the radiation condition that the scattered field must consist of outgoing waves, and the choice of $\overline{M}'^{(1)}$ and $\overline{N}'^{(1)}$ as the excitation functions is guided by the expression for $\overline{\overline{G}}_{eo}$ and the boundary condition that at $r = a$, $\overline{\overline{G}}_{e1}$ must satisfy the Dirichlet boundary condition which can be satisfied only if the excitation functions are the same as that of $\overline{\overline{G}}_{eo}$ for $r < r'$. To determine the unknown coefficients a_η and b_η we require, at $r = a$,

$$\hat{r} \times \left[\overline{M}_\eta(h) + a_\eta \overline{M}_\eta^{(1)}(h) \right] = 0 \quad (7.24)$$

and

$$\hat{r} \times \left[\overline{N}_\eta(h) + b_\eta \overline{N}_\eta^{(1)}(h) \right] = 0, \quad (7.25)$$

which yields

$$a_\eta = -\frac{[\partial J_n(x)]/\partial x}{[\partial H_n^{(1)}(x)]/\partial x}, \quad x = \eta a \quad (7.26)$$

$$b_\eta = -\frac{[J_n(x)]}{[H_n^{(1)}(x)]}, \qquad x = \eta a. \quad (7.27)$$

The expression for $\overline{\overline{G}}_{e2}$ can be obtained most expediently by taking advantage of the symmetrical property between $\overline{\overline{G}}_{e1}$ and $\overline{\overline{G}}_{e2}$. The net result is that if we interchange the role of a_η and b_η in $\overline{\overline{G}}_{e1}$, we have the expression for $\overline{\overline{G}}_{e2}$.

For a cylinder made of homogeneous isotropic material such as a dielectric cylinder, the relevant functions are $\overline{\overline{G}}_e^{(11)}$ and $\overline{\overline{G}}_e^{(21)}$ for a source placed in region 1 ($r > a$) and region 2 to be the interior region of the cylinder ($r \le a$). For clarity, we adopt the following notations for various parameters defined in the two regions; they are

$$k_1 = \omega \left(\mu_0 \epsilon_0\right)^{\frac{1}{2}}, \quad k_2 = \omega \left(\mu \epsilon\right)^{\frac{1}{2}}$$
$$\eta = \left(k_1^2 - h^2\right)^{\frac{1}{2}}, \quad \xi = \left(k_2^2 - h^2\right)^{\frac{1}{2}}, \quad (7.28)$$

where μ and ϵ denote, respectively, the permeability and the permittivity of the cylinder which may be complex. To construct the electric dyadic Green functions of the third kind, we let

$$\overline{\overline{G}}_e^{(11)} = \overline{\overline{G}}_{e0} + \overline{\overline{G}}_{es}^{(11)}, \quad r \ge a \quad (7.29)$$

$$\overline{\overline{G}}_e^{(21)} = \overline{\overline{G}}_{es}^{(21)}, \quad r \le a. \quad (7.30)$$

In order to satisfy the radiation condition at infinity and the boundary condition at the interface, the two scattering terms must have the following forms

$$\begin{aligned}\overline{\overline{G}}_{es}^{(11)}(\overline{R},\overline{R}') =& \frac{i}{8\pi}\int_{-\infty}^{\infty} dh \sum_n \frac{2-\delta_0}{\eta^2} \\ &\cdot\Big\{\Big[A_{\substack{e\\o}\eta}\overline{M}^{(1)}_{\substack{e\\o}\eta}(h) + B_{\substack{o\\e}\eta}\overline{N}^{(1)}_{\substack{o\\e}\eta}(h)\Big]\overline{M}'^{(1)}_{\substack{e\\o}\eta}(-h) \\ &+\Big[C_{\substack{e\\o}\eta}\overline{N}^{(1)}_{\substack{e\\o}\eta}(h) + D_{\substack{o\\e}\eta}\overline{M}^{(1)}_{\substack{o\\e}\eta}(h)\Big]\overline{N}'^{(1)}_{\substack{e\\o}\eta}(-h)\Big\}; \qquad (7.31)\\ \overline{\overline{G}}_{es}^{(21)}(\overline{R},\overline{R}') =& \frac{i}{8\pi}\int_{-\infty}^{\infty} dh \sum_n \frac{2-\delta_0}{\eta^2} \\ &\cdot\Big\{\Big[a_{\substack{e\\o}\xi}\overline{M}_{\substack{e\\o}\xi}(h) + b_{\substack{o\\e}\xi}\overline{N}_{\substack{o\\e}\xi}(h)\Big]\overline{M}'^{(1)}_{\substack{e\\o}\eta}(-h) \\ &+\Big[c_{\substack{e\\o}\xi}\overline{N}_{\substack{e\\o}\xi}(h) + d_{\substack{o\\e}\xi}\overline{M}_{\substack{o\\e}\xi}(h)\Big]\overline{N}'^{(1)}_{\substack{e\\o}\eta}(-h)\Big\}. \qquad (7.32)\end{aligned}$$

It is observed that we have restored the notations for the even and odd functions because the functions with coefficients $A_{\substack{e\\o}\eta}$ must combine with functions with coefficients $B_{\substack{o\\e}\eta}$ in that manner in order to satisfy the boundary condition at the interface; more specifically, the expanded version of a typical combination is

$$\begin{aligned}&\Big[A_{\substack{e\\o}\eta}\overline{M}^{(1)}_{\substack{e\\o}\eta}(h) + B_{\substack{o\\e}\eta}\overline{N}^{(1)}_{\substack{o\\e}\eta}\Big]\overline{M}'^{(1)}_{\substack{e\\o}\eta}(-h) \\ =& \Big[\overline{A}_{e\eta}\overline{M}^{(1)}_{e\eta}(h) + \overline{B}_{o\eta}\overline{N}^{(1)}_{o\eta}\Big]\overline{M}'^{(1)}_{e\eta}(-h) \\ &+ \Big[\overline{A}_{o\eta}\overline{M}^{(1)}_{o\eta}(h) + \overline{B}_{e\eta}\overline{N}^{(1)}_{e\eta}\Big]\overline{M}'^{(1)}_{o\eta}(-h)\end{aligned}$$

similarly for the other combinations. We choose the functions of the first kind as the field functions for $\overline{\overline{G}}_e^{(11)}$ in order to satisfy the radiation condition at infinity. The field functions for $\overline{\overline{G}}_e^{(21)}$ are so chosen because they are the solutions for the vector wave equation in region 2, and they must be finite in that region; that is,

$$\begin{aligned}\overline{M}_{\substack{e\\o}\xi}(h) &= \nabla\times\left[J_n(\xi r)\,{\cos\atop sin}\, n\phi e^{ihz}\hat{z}\right] \\ \overline{N}_{\substack{e\\o}\xi}(h) &= \frac{1}{k_2}\nabla\times\nabla\times\left[J_n(\xi r)\,{\cos\atop \sin}\, n\phi e^{ihz}\hat{z}\right],\end{aligned}$$

where $\xi = \left(k_2^2 - h^2\right)^{\frac{1}{2}}$. The inclusion of terms with coefficeints B_η, D_η, b_ξ, and d_ξ are necessary in order to satisfy the boundary condition at the interface, although these terms are absent in $\overline{\overline{G}}_{eo}$. In other words, for a dielectric cylinder, an incident TE mode will excite both a scattered TE mode and a scattered TM mode. Such a phenomenon does not occur for a conducting cylinder. The fact that the even $\overline{M}$ functions play similar roles as the odd $\overline{N}$ functions, or vice versa, is because they have the same angular functions of the ϕ-component. The boundary condition at the interface, $r = a$, requires

$$\hat{r} \times \left[\overline{\overline{G}}_e^{(11)}(\overline{R}, \overline{R}') - \overline{\overline{G}}_e^{(21)}(\overline{R}, \overline{R}')\right] = 0 \tag{7.33}$$

and

$$\hat{r} \times \left[\frac{\nabla \times \overline{\overline{G}}_e^{(11)}(\overline{R}, \overline{R}')}{\mu_0} - \frac{\nabla \times \overline{\overline{G}}_e^{(21)}(\overline{R}, \overline{R}')}{\mu}\right] = 0 \tag{7.34}$$

that enables us to determine the 16 scattering coefficients. The results are given in Table 7-1 in the form of 16 linear equations grouped in four sets. The square matrices $[D_1]$ and $[D_2]$ and the column matrices $[C_1], [C_2], [F_1]$, and $[F_2]$ are defined as follows:

$$[D_1] = \begin{bmatrix} 0 & \dfrac{-\eta^2 H_n(\eta a)}{k_1} & 0 & \dfrac{\xi^2 J_n(\xi a)}{k_2} \\ \dfrac{-\partial H_n(\eta a)}{\partial a} & \dfrac{\pm ihn H_n(\eta a)}{k_1 a} & \dfrac{\partial J_n(\xi a)}{\partial a} & \dfrac{\mp ihn J_n(\xi a)}{k_2 a} \\ \dfrac{-\eta^2 H_n(\eta a)}{\mu_0} & 0 & \dfrac{\xi^2 J_n(\xi a)}{\mu} & 0 \\ \dfrac{\mp ihn H_n(\eta a)}{\mu_0 a} & \dfrac{-k_1}{\mu_0}\dfrac{\partial H_n(\eta a)}{\partial a} & \dfrac{\pm ihn J_n(\xi a)}{\mu a} & \dfrac{k_2}{\mu}\dfrac{\partial J_n(\xi a)}{\partial a} \end{bmatrix}$$

$$[D_2] = \begin{bmatrix} \dfrac{-\eta^2 H_n(\eta a)}{k_1} & 0 & \dfrac{\xi^2 J_n(\xi a)}{k_2} & 0 \\ \dfrac{\pm ihn H(\eta a)}{k_1 a} & \dfrac{-\partial H_n(\eta a)}{\partial a} & \dfrac{\pm ihn J_n(\xi a)}{k_2 a} & \dfrac{\partial J_n(\xi a)}{\partial a} \\ 0 & \dfrac{-\eta^2 H_n(\eta a)}{\mu_0} & 0 & \dfrac{\xi^2 J_n(\xi a)}{\mu} \\ \dfrac{-k_1}{\mu_0}\dfrac{\partial H_n(\eta a)}{\partial a} & \dfrac{\pm ihn H_n(\eta a)}{\mu_0 a} & \dfrac{k_2}{\mu}\dfrac{\partial J_n(\xi a)}{\partial a} & \dfrac{\mp ihn J_n(\xi a)}{\mu a} \end{bmatrix}$$

$$[C_1] = \left[A_{\substack{e\\o}\eta}, B_{\substack{o\\e}\eta}, a_{\substack{e\\o}\xi}, b_{\substack{o\\e}\xi}\right]$$

$$[C_2] = \left[C_{\substack{e\\o}\eta}, D_{\substack{o\\e}\eta}, c_{\substack{e\\o}\xi}, d_{\substack{o\\e}\xi}\right]$$

$$[F_1] = \left[0, \frac{\partial J_n(\eta a)}{\partial a}, \frac{\eta^2 J_n(\eta a)}{\mu_0}, \frac{\pm ihn J_n(\eta a)}{\mu_0 a}\right]$$

$$[F_2] = \left[\frac{\eta^2 J_n(\eta a)}{k_1}, \frac{\pm ihn J_n(\eta a)}{k_1 a}, 0, \frac{k_1}{\mu_0}\frac{\partial J_n(\eta a)}{\partial a}\right].$$

TABLE 7-1 System of Equations for the Scattering Coefficients

$[D_1][C_1] = [F_1]$
$[D_2][C_2] = [F_2]$

In Table 7-1, $H_n^{(1)}$ is denoted simply by H_n.

For example, one typical equation of the above system of equations is

$$-\frac{\partial H_n(\eta a)}{\partial a}A_{e\eta}+\frac{ihnH_n(\eta a)}{k_1 a}B_{0\eta}+\frac{\partial J_n(\xi a)}{\partial a}a_{e\xi}$$
$$-\frac{ihnJ_n(\xi a)}{k_2 a}b_{0\xi}=\frac{\partial J_n(\eta a)}{\partial a}. \tag{7.35}$$

The same method can be applied to find the functions for a conducting cylinder coated by a layer of dielectric. These functions will be denoted by $\overline{\overline{G}}_e^{(11)}$ and $\overline{\overline{G}}_{e1}^{(21)}$ for a source in region 1, corresponding to the exterior region of the coated cylinder. Region 2 is within the layer ($b \geq r \geq a$) for a conducting cylinder of radius a with thickness of the layer equal to $b-a$. The subscript notation "e1" for the functions indicate that there is a Dirichlet boundary to be satisfied; hence they are electric dyadic Green functions of the first kind as well as of the third kind.

By the method of scattering superposition, we let

$$\overline{\overline{G}}_{e1}^{(11)}=\overline{\overline{G}}_{e0}+\overline{\overline{G}}_{es}^{(11)}$$
$$\overline{\overline{G}}_{e1}^{(21)}=\overline{\overline{G}}_{es}^{(21)};$$

then $\overline{\overline{G}}_{es}^{(11)}$ still has the form as (7.31), but $\overline{\overline{G}}_{es}^{(21)}$ or $\overline{\overline{G}}_{e1}^{(21)}$ must have the form

$$\begin{aligned}\overline{\overline{G}}_{es}^{(21)}(\overline{R},\overline{R}')=&\frac{i}{8\pi}\int_{-\infty}^{\infty}dh\sum_n\frac{2-\delta_0}{\eta^2}\\&\cdot\Big\{\Big[a_{\substack{e\\o}\xi}\overline{M}_{\substack{e\\o}\xi}(h)+a'_{\substack{e\\o}\xi}\overline{M}_{\substack{e\\o}\xi}^{(1)}(h)+b_{\substack{o\\e}\xi}\overline{N}_{\substack{o\\e}\xi}(h)\\&\qquad+b'_{\substack{o\\e}\xi}\overline{N}_{\substack{o\\e}\xi}^{(1)}(h)\Big]\overline{M}'^{(1)}_{\substack{e\\o}\eta}(-h)\\&+\Big[c_{\substack{e\\o}\xi}\overline{N}_{\substack{e\\o}\xi}(h)+c'_{\substack{e\\o}\xi}\overline{N}_{\substack{e\\o}\xi}^{(1)}(h)\\&\qquad+d_{\substack{o\\e}\xi}\overline{M}_{\substack{o\\e}\xi}(h)+d'_{\substack{o\\e}\xi}\overline{M}'_{\substack{o\\e}\xi}(h)\Big]\overline{N}'^{(1)}_{\substack{e\\o}\eta}(0h)\Big\}.\end{aligned} \tag{7.36}$$

Eight terms with coefficients a', b', c', and d', both even and odd, are involved because the functions of the first kind must also be included in the dielectric layer. In addition to the boundary conditions of the form stated by (7.33) and (7.34), now applying to $r=b$, the function $\overline{\overline{G}}_{e1}^{(21)}$ must satisfy the Dirichlet boundary condition at $r=a$, the site of the conducting cylinder. These boundary conditions enable us to determine the 24 scattering coefficients in this formulation, resulting in a quite complicated system of equations, the result of which will not be given here. In Chap. 11, a similar but simpler problem of plane stratified media resting on a conducting ground plane will be formulated and solved.

7-4 ASYMPTOTIC EXPRESSION

To find the far-zone field of a radiating source in the presence of a perfectly conducting cylinder or the far-zone field of an aperture antenna on the surface of the cylinder, we need only the asymptotic expression for $\overline{\overline{G}}_{e1}$ or $\overline{\overline{G}}_{e2}$. This expression can be found by the method of saddle-point integration. Let us consider the function of the first kind. Assuming the ηr is large compared to unity, the Hankel function in $\overline{M}^{(1)}$ and $\overline{N}^{(1)}$ can be approximated by its asymptotic expression; that is,

$$H_n^{(1)}(\eta r) \simeq \left(\frac{2}{\pi\eta r}\right)^{\frac{1}{2}} (-i)^{n+\frac{1}{2}}\, e^{i\eta r}.$$

The functions $\overline{M}^{(1)}$ and $\overline{N}^{(1)}$, therefore, become

$$\overline{M}^{(1)}_{\substack{e\\o}n\eta}(h) \simeq (-i)^{n+\frac{3}{2}}\eta \left(\frac{2}{\pi\eta r}\right) e^{i(\eta r+hz)} \begin{matrix}\cos\\ \sin\end{matrix}\, n\phi\hat{\phi} \tag{7.37}$$

$$\overline{N}^{(1)}_{\substack{e\\o}n\eta}(h) \simeq (-i)^{n+\frac{1}{2}}\frac{\eta}{k} \left(\frac{2}{\pi\eta r}\right)^{\frac{1}{2}} e^{i(\eta r+hz)} \begin{matrix}\cos\\ \sin\end{matrix}\, n\phi\,(-h\hat{r}+\eta\hat{z}). \tag{7.38}$$

The approximate expression for $\overline{\overline{G}}_{e1}$, using (7.21) and (7.23) with the functions $\overline{M}^{(1)}$ and $\overline{N}^{(1)}$ replaced by (7.37) and (7.38), can be written in the form

$$\begin{aligned}\overline{\overline{G}}_{e1}(\overline{R},\overline{R}') \simeq & \frac{i}{4\pi}\int_{-\infty}^{\infty} dh \sum_n \frac{2-\delta_0}{\eta}\frac{1}{(2\pi\eta r)^{\frac{1}{2}}}(-i)^{n+\frac{1}{2}} e^{i(\eta r+hz)} \\ & \cdot\left\{ -i \begin{matrix}\cos\\ \sin\end{matrix}\, n\phi\hat{\phi}\left[\overline{M}'_{\substack{e\\o}n\eta}(-h) + a_{\substack{e\\o}n}\overline{M}^{(1)}_{\substack{e\\o}n\eta}(-h)\right]\right. \\ & \left. + \frac{1}{k}\begin{matrix}\cos\\ \sin\end{matrix}\, n\phi\,(-h\hat{r}+\eta\hat{z})\left[\overline{N}'_{\substack{e\\o}n\eta}(-h) + b_{\substack{e\\o}n}\overline{N}^{(1)}_{\substack{e\\o}n\eta}(-h)\right]\right\}, \end{aligned} \tag{7.39}$$

where terms of the order equal to and higher than $(\eta r)^{-3/2}$ have been neglected.

We now change the cylindrical variables into spherical variables by letting

$$\begin{aligned} \eta &= k\sin\beta, & h &= k\cos\beta \\ R &= R\sin\theta, & z &= R\cos\theta; \end{aligned}$$

hence

$$\begin{aligned} \eta r + hz &= kR\cos(\theta-\beta) \\ -h\hat{r}+\eta\hat{z} &= k(-\hat{r}\cos\beta + \hat{z}\sin\beta) \\ dh &= -k\sin\beta d\beta. \end{aligned}$$

Applying the integral formula (1.99) to (7.39), we obtain

$$\begin{aligned}\overline{\overline{G}}_{e1}(\overline{R}, \overline{R}') \simeq & \frac{e^{ikR}}{4\pi kR\sin\theta} \sum_{n=0}^{\infty} (2-\delta_0)(-i)^{n+1} \begin{matrix}\cos\\ \sin\end{matrix} n\phi \\ & \cdot \left\{ \hat{\phi} \left[\overline{M}'_{\substack{e\\o}ns}(-k\cos\theta) + a_{\substack{e\\o}n} \overline{M}'^{(1)}_{\substack{e\\o}ns}(-k\cos\theta) \right] \right. \\ & \left. - i\hat{\theta} \left[\overline{N}'_{\substack{e\\o}ns}(-k\cos\theta) + a_{\substack{e\\o}n} \overline{N}'^{(1)}_{\substack{e\\o}ns}(-k\cos\theta) \right] \right\}, \end{aligned} \tag{7.40}$$

where $s = k\sin\theta$, corresponding to the values of η evaluated at $\beta = \theta$. Equation (7.40), of course, is valid only when $kR \gg 1$. The singularity at $\theta = 0$, in practice, does not exist because θ never goes to zero due to the finite radius of the cylinder. Conversely, when $kR' \gg 1$, corresponding to a remote source, we can find the asymptotic expression for $\overline{\overline{G}}_{e1}$. The result would provide for us the implicit solution of the field due to a plane wave incident on a conducting cylinder. We leave this problem as an exercise in Appendix C, where the readers are requested to recover some of the classical formulas based on the present technique. As far as the actual field is concerned, it is of course necessary to use the formula

$$\overline{E}(\overline{R}) = i\omega\mu_0 \iiint \overline{\overline{G}}_{e1}(\overline{R}, \overline{R}') \cdot \overline{J}(\overline{R}')\, dV' \tag{7.41}$$

to find the field for a current source, or

$$\overline{E}(\overline{R}) = \iint \nabla \times \overline{\overline{G}}_{e2}(\overline{R}, \overline{R}') \cdot [\hat{n}' \times \overline{E}(\overline{R}')]\, dS'$$

for an aperture field source, where $\hat{n}'$ denotes the outward normal to the cylindrical surface.

8

Perfectly Conducting Elliptical Cylinder

Vector wave functions in an elliptical cylinder coordinate system are generated when the elliptical cylinder scalar wave functions are used. Once the orthogonal properties of these functions are known, we can find the eigenfunction expansion of the free-space dyadic Green function without any difficulty. The function of the first and second kind can then be constructed by the method of scattering superposition. Because both the angular functions and the radial functions depend on the wave number, it is impossible to find an orthogonal expansion for the functions of the third kind. Problems involving a dielectric elliptical cylinder, therefore, cannot be formulated in the same manner as those involving a dielectric circular cylinder.

8-1 VECTOR WAVE FUNCTIONS IN AN ELLIPTICAL CYLINDER COORDINATE SYSTEM

The scalar wave equation in an elliptical cylinder coordinate system can be written in the form

$$\frac{1}{\beta^2}\left(\frac{\partial^2\psi}{\partial u^2}+\frac{\partial^2\psi}{\partial v^2}\right)+\frac{\partial^2\psi}{\partial z^2}+\kappa^2\psi=0, \tag{8.1}$$

where

$$\beta=c\left(\cos h^2 u-\cos^2 v\right)^{\frac{1}{2}}.$$

The variables u, v, and z and the parameter c are defined in Sec. 1-1 and are illustrated in Fig. 1-3. Equation (8.1) is derived by using (A.50), (A.51), and (A.55) of Appendix A. The eigenfunctions associated with (8.1), called the elliptical cylinder wave functions, have been discussed very thoroughly by Stratton [1941, pp. 375–387]. We shall follow very closely Stratton's presentation with, however, some minor changes of notation in order to conform with our previous notation. These changes will be obvious to the readers when the two texts are compared. In order to give a full treatment of the orthogonal property of the vector wave functions in elliptical cylinder coordinate systems, a brief review of the scalar wave functions is necessary. According to Stratton, the eigenfunctions pertaining to (8.1) can be written in the form

$$\psi_{\substack{e\\o}m\lambda}(h) = S_{\substack{e\\o}m\lambda}(v)R_{\substack{e\\o}m\lambda}(u)e^{ihz}, \tag{8.2}$$

where

$$h^2 + \lambda^2 = \kappa^2.$$

The angular functions $S_{\substack{e\\o}m\lambda}$ and the radial functions $R_{\substack{e\\o}m\lambda}$ satisfy, respectively, the Mathieu equations

$$\frac{d^2 S_{\substack{e\\o}m\lambda}(v)}{dv^2} + \left(b_{\substack{e\\o}m} - c^2\lambda^2\cos^2 v\right) S_{\substack{e\\o}m\lambda}(v) = 0 \tag{8.3}$$

$$\frac{d^2 R_{\substack{e\\o}m\lambda}(u)}{du^2} - \left(b_{\substack{e\\o}m} - c^2\lambda^2\cos h^2 u\right) R_{\substack{e\\o}m\lambda}(u) = 0. \tag{8.4}$$

The eigenvalues b_{em} or b_{om} form a denumerable set such that the corresponding angular functions will be periodic functions of v. We require the angular functions to be periodic with respect to v so that the field represented by these functions would be a single-valued function of position. These periodic angular functions can be represented by either the cosine series in the case of even functions or the sine series in the case of odd functions. They are

$$S_{em\lambda}(v) = \sum_n{}' D_n^m \cos nv, \quad m = 1, 2, 3, \ldots \tag{8.5}$$

and

$$S_{om\lambda}(v) = \sum_n{}' F_n^m \sin nv, \quad m = 1, 2, 3, \ldots, \tag{8.6}$$

where the primed summation is to be extended over even values of n if m is even and odd values of n if m is odd. The coefficients D_n^m and F_n^m are so normalized that

$$\sum_n{}' D_n^m = 1 \text{ or } S_{em\lambda}(0) = 1 \tag{8.7}$$

$$\sum_n{}' nF_n^m = 1 \text{ or } \frac{d}{dv} S_{om\lambda}(v)\big|_{v=0} = 1. \tag{8.8}$$

Using (8.3) and (8.4), we can show that S_{em} and S_{om} form a complete orthogonal set; that is,

$$\int_0^{2\pi} S_{em\lambda}(v) S_{om'\lambda}(v) dV = 0 \tag{8.9}$$

$$\int_0^{2\pi} S_{em\lambda}(v) S_{em'\lambda}(v) dV = \begin{cases} 0, & m \neq m' \\ I_{em\lambda}, & m = m', \end{cases} \tag{8.10}$$

with

$$I_{em\lambda} = \pi \sum_n{}' (1 + \delta_0) (D_n^m)^2 ,$$

$$\int_0^{2\pi} S_{om\lambda}(v) S_{om'\lambda}(v) dV = \begin{cases} 0, & m \neq m' \\ I_{om\lambda}, & m = m', \end{cases} \tag{8.11}$$

with

$$I_{om\lambda} = \pi \sum_n{}' (F_n^m)^2 .$$

The radial functions which are solutions of (8.4) and finite at the origin can be written in the form of the series of Bessel functions. They are

$$R_{em\lambda}(u) = \left(\frac{\pi}{2}\right)^{\frac{1}{2}} \sum_n{}' (i)^{m-n} D_n^m J_n (c\lambda \cosh u) \tag{8.12}$$

$$R_{om\lambda}(u) = \left(\frac{\pi}{2}\right)^{\frac{1}{2}} \tanh u \sum_n{}' (i)^{n-m} F_n^m J_n (c\lambda \cosh u) . \tag{8.13}$$

We have omitted the superscript 1 as used by Stratton in the designation of these functions. Two other radial functions which would represent outgoing waves in the $e^{-i\omega t}$ system involved the Hankel function of the first kind. They are given by the series

$$R_{em\lambda}^{(1)}(u) = \left(\frac{\pi}{2}\right)^{\frac{1}{2}} \sum_n{}' (i)^{m-n} D_n^m H_n^{(1)} (c\lambda \cosh u) \tag{8.14}$$

$$R_{om\lambda}^{(1)}(u) = \left(\frac{\pi}{2}\right)^{\frac{1}{2}} \tanh u \sum_n{}' (i)^{n-m} F_n^m H_n^{(1)} (c\lambda \cosh u) . \tag{8.15}$$

In Stratton's original work, these functions are designated by $R^{(3)}_{\substack{e\\0}m\lambda}$. It is hoped that these minor changes of notation will not cause any inconvenience to readers. For the determination of the normalization constant for the vector wave function, we need the expansion formulas of the elliptical wave functions in terms of the circular wave functions. They are given by Stratton [1941, p. 386].

$$R_{em\lambda}(u) S_{em\lambda}(v) = \left(\frac{\pi}{2}\right)^{\frac{1}{2}} \sum_n{}' (i)^{n-m} D_n^m \cos n\phi J_n (\lambda r) \tag{8.16}$$

and

$$R_{om\lambda}(u) S_{om\lambda}(v) = \left(\frac{\pi}{2}\right)^{\frac{1}{2}} \sum_n{}' (i)^{n-m} F_n^m \sin n\phi J_n (\lambda r) . \tag{8.17}$$

This brief review is essential for our discussion of the corresponding vector wave functions. These functions are defined by

$$\overline{M}_{\substack{e\\o}m\lambda}(h) = \nabla \times \left[\psi_{\substack{e\\o}m\lambda}(h)\hat{z}\right] \tag{8.18}$$

$$\overline{N}_{\substack{e\\o}m\lambda}(h) = \frac{1}{\kappa}\nabla \times \overline{M}_{\substack{e\\o}m\lambda}(h), \tag{8.19}$$

where $\psi_{\substack{e\\o}m\lambda}(h)$ is defined by (8.2). They are, of course, solutions of the vector wave equations in elliptical cylinder coordinate systems. The complete expressions of these functions are given by

$$\overline{M}_{\substack{e\\o}m\lambda}(h) = \frac{1}{\beta}\left(R_{\substack{e\\o}m\lambda}\frac{\partial S_{\substack{e\\o}m\lambda}}{\partial v}\hat{u} - S_{\substack{e\\o}m\lambda}\frac{\partial R_{\substack{e\\o}m\lambda}}{\partial u}\hat{v}\right)e^{ihz} \tag{8.20}$$

$$\begin{aligned}\overline{N}_{\substack{e\\o}m\lambda}(h) = &\frac{1}{\kappa\beta}\left(ihS_{\substack{e\\o}m\lambda}\frac{\partial R_{\substack{e\\o}m\lambda}}{\partial u}\hat{u}\right.\\ &\left.+ihR_{\substack{e\\o}m\lambda}\frac{\partial S_{\substack{e\\o}m\lambda}}{\partial v}\hat{v} + \beta\lambda^2 R_{\substack{e\\o}m\lambda}S_{\substack{e\\o}m\lambda}\hat{z}\right)e^{ihz},\end{aligned} \tag{8.21}$$

where $\beta = c\left(\cosh^2 u - \cos^2 v\right)^{\frac{1}{2}}$.

The orthogonal properties of these functions are stated by the following equations:

$$\iiint \overline{M}_{\substack{e\\o}m\lambda}(h) \cdot \overline{N}_{\substack{e\\o}m'\lambda'}(-h')\,dV = 0 \tag{8.22}$$

$$\begin{aligned}&\iiint \overline{M}_{\substack{e\\o}m\lambda}(h) \cdot \overline{M}_{\substack{e\\o}m'\lambda'}(-h')\,dV\\ &\quad = \begin{cases} 0, & m \neq m' \\ \pi^2\lambda I_{\substack{e\\o}m\lambda}\delta\left(h-h'\right)\delta\left(\lambda-\lambda'\right), & m = m' \end{cases}\end{aligned} \tag{8.23}$$

$$\begin{aligned}&\iiint \overline{N}_{\substack{e\\o}m\lambda}(h) \cdot \overline{N}_{\substack{e\\o}m'\lambda'}\left(-h'\right)dV\\ &\quad = \begin{cases} 0, & m \neq m' \\ \pi^2\lambda I_{\substack{e\\o}m\lambda}\delta\left(h-h'\right)\delta\left(\lambda-\lambda'\right), & m = m', \end{cases}\end{aligned} \tag{8.24}$$

where $I_{\substack{e\\o}m\lambda}$ is defined by (8.10) and (8.11).

These relations are analogous to (7.1) through (7.3) except that the normalization constants are different. The fine structure, as characterized by the two delta functions in the h- and λ-domain, is exactly the same. The orthogonal relations can easily be proved because of (8.9) through (8.11). It is sufficient for us to prove (8.23) for the case $m = m'$. The proof of (8.24) is given as an exercise. We consider, for example, the integral

$$I = \iiint \overline{M}_{em\lambda}(h) \cdot \overline{M}_{em\lambda'}(-h')\,dV,$$

where the volume of integration is extended through the entire space. Using (8.2) and the differential relation that

$$\begin{aligned} dV &= h_1 h_2 h_3 \, du \, dv \, dz \\ &= \beta^2 du \, dv \, dz, \end{aligned}$$

we have

$$\begin{aligned} I = \int_0^\infty \int_0^{2\pi} \int_{-\infty}^\infty \Bigg[& R_{em\lambda} R_{em\lambda'} \frac{\partial S_{em\lambda}}{\partial v} \frac{\partial S_{em\lambda'}}{\partial v} \\ & + S_{em\lambda} S_{em\lambda'} \frac{\partial R_{em\lambda}}{\partial u} \frac{\partial R_{em\lambda'}}{\partial u} \Bigg] e^{i(h-h')z} du \, dv \, dz. \end{aligned} \tag{8.25}$$

To simplify (8.25) we first make use of the relations that

$$\begin{aligned} \int_0^{2\pi} \frac{\partial S_{em\lambda}}{\partial v} \frac{\partial S_{em\lambda'}}{\partial v} dv &= \frac{\partial S_{em\lambda}}{\partial v} S_{em\lambda'} \Bigg|_0^{2\pi} - \int_0^{2\pi} S_{em\lambda'} \frac{\partial^2 S_{em\lambda}}{\partial v^2} dv \\ &= \int_0^{2\pi} \left(b_{em} - c^2 \lambda^2 \cos^2 v \right) S_{em\lambda} S_{em\lambda'} \, dv \end{aligned}$$

and

$$\begin{aligned} \int_0^\infty \frac{\partial R_{em\lambda}}{\partial u} \frac{\partial R_{em\lambda'}}{\partial u} du &= R_{em\lambda'} \frac{\partial R_{em\lambda}}{\partial u} \Bigg|_0^\infty - \int_0^\infty R_{em\lambda'} \frac{\partial^2 R_{em\lambda}}{\partial u^2} du \\ &= - \int_0^\infty \left(b_{em} - c^2 \lambda^2 \cosh^2 u \right) R_{em\lambda} R_{em\lambda'} \, du. \end{aligned}$$

Thus (8.25) can be written in the form

$$I = \int_0^\infty \int_0^{2\pi} \int_{-\infty}^\infty \lambda^2 \beta^2 R_{em\lambda} R_{em\lambda'} S_{em\lambda} S_{em\lambda'} e^{i(h-h')z} du \, dv \, dz.$$

Using (8.16), and changing the domain of integration for the elliptical cylinder system to the circular cylinder system with the relation that

$$\beta^2 du \, dv = r \, dr \, d\phi,$$

we obtain

$$\begin{aligned} I =& \pi^2 \delta(h - h') \int_0^\infty \int_0^{2\pi} \lambda^2 \left[\sum_n {}' (i)^{m-n} D_n^m(\lambda) J_n(\lambda r) \cos n\phi \right] \\ & \cdot \left[\sum_n {}' (i)^{m-n} D_n^m(\lambda') J_n(\lambda' r) \cos n\phi \right] r \, dr \, d\phi. \end{aligned} \tag{8.26}$$

We have put into evidence that the coefficients D_n^m are functions of λ. As a result of the trigonometric functions and the integral representation of the weighted two-dimensional delta function, we obtain

$$I = (1+\delta_0)\,\pi^3\lambda\delta\,(h-h')\,\delta\,(\lambda-\lambda')\sum_n{}'\,[D_n^m(\lambda)]^2$$
$$= \pi^2\lambda I_{en\lambda}\delta\,(h-h')\,\delta\,(\lambda-\lambda'), \tag{8.27}$$

where $I_{en\lambda}$ is defined by (8.10). The proofs of the remaining relations in (8.23) and (8.24) are similar. Once the orthogonal properties of the vector wave functions are known, the way to the eigenfunction expansion of the free-space dyadic Green function is straightforward.

8-2 THE ELECTRIC DYADIC GREEN FUNCTION OF THE FIRST KIND

Following the procedure described in Sec. 7-2 for the perfectly conducting cylinder, we shall first derive the eigenfunction expansion of the free-space magnetic dyadic Green function. We let

$$\nabla\times\left[\bar{\bar{I}}\delta(\bar{R}-\bar{R}')\right] = \int_0^\infty d\lambda\int_{-\infty}^{\infty} dh$$
$$\sum_m\left[\bar{N}_{\substack{e\\o}m\lambda}(h)\bar{A}_{\substack{e\\o}m\lambda}(h)\bar{M}_{\substack{e\\o}m\lambda}(h)\bar{B}_{\substack{e\\o}m\lambda}(h)\right]. \tag{8.28}$$

The unknown vector coefficients $\bar{A}$ and $\bar{B}$ are determined by taking the anterior scalar product of (8.28) with $\bar{M}_{\substack{e\\o}m'\lambda'}(-h')$ and $\bar{N}_{\substack{e\\o}m'\lambda'}(-h')$ in turn and integrating the resulting equation through the entire space. As a result of (8.22) through (8.24) we obtain

$$\bar{A}_{\substack{e\\o}m\lambda}(h) = \frac{1}{\pi^2\lambda I_{\substack{e\\o}m\lambda}}\nabla'\times\bar{N}'_{\substack{e\\o}m\lambda}(-h)$$
$$= \frac{\kappa}{\pi^2\lambda I_{\substack{e\\o}m\lambda}}\bar{M}'_{\substack{e\\o}m\lambda}(-h) \tag{8.29}$$

and

$$\bar{B}_{\substack{e\\o}m\lambda} = \frac{1}{\pi^2\lambda I_{\substack{e\\o}m\lambda}}\nabla'\times\bar{M}'_{\substack{e\\o}m\lambda}(-h)$$
$$= \frac{\kappa}{\pi^2\lambda I_{\substack{e\\o}m\lambda}}\bar{N}'_{\substack{e\\o}m\lambda}(-h). \tag{8.30}$$

The expression for $\bar{\bar{G}}_{mo}(\bar{R},\bar{R}')$ can then be written in the form

$$\bar{\bar{G}}_{mo}(\bar{R},\bar{R}') = \int_0^\infty d\lambda\int_{-\infty}^{\infty} dh\sum_m\frac{\kappa}{\pi^2\lambda\,(\kappa^2-k^2)\,I_{\substack{e\\o}m\lambda}}$$
$$\cdot\left[\bar{N}_{\substack{e\\o}m\lambda}(h)\bar{M}'_{\substack{e\\o}m\lambda}(-h)+\bar{M}_{\substack{e\\o}m\lambda}(h)\bar{N}'_{\substack{e\\o}m\lambda}(-h)\right]. \tag{8.31}$$

To construct the electric dyadic Green function of the first kind the λ integration can be evaluated in a closed form by using the same technique as described in

Sec. 7-2 that yields

$$\overline{\overline{G}}{}^{\pm}_{mo}(\overline{R}, \overline{R}') = \frac{ik}{2\pi} \int_{-\infty}^{\infty} dh \sum_{m} \frac{1}{\eta^2 I_{\substack{e\\o}mn}}$$
$$\cdot \begin{cases} \overline{N}^{(1)}_{\eta}(h)\overline{M}'_{\eta}(-h) + \overline{M}^{(1)}_{\eta}(h)\overline{N}'_{\eta}(-h), & u > u' \\ \overline{N}_{\eta}(h)\overline{M}'^{(1)}_{\eta}(-h) + \overline{M}_{\eta}(h)\overline{N}'^{(1)}_{\eta}(-h), & u < u', \end{cases} \tag{8.32}$$

where a condensed notation has been used; for example,

$$\overline{N}^{(1)}_{\eta}(h) = \overline{N}^{(1)}_{\substack{e\\o}mn}(h) = \frac{1}{k}\nabla \times \nabla \times \left[R^{(1)}_{\substack{e\\o}mn}(u)S_{\substack{e\\o}mn}(v)e^{ihz}\hat{z}\right],$$

with $\eta = \left(k^2 - h^2\right)^{\frac{1}{2}}$. The function of the first kind with superscript (1) is defined with respect to the radial function of the first kind as defined by (8.14) with λ replaced by η. Knowing $\overline{\overline{G}}_{m0}$, we can obtain $\overline{\overline{G}}_{eo}$ by the formula

$$\overline{\overline{G}}_{eo}(\overline{R}, \overline{R}') = \frac{1}{k^2}\Big[- \hat{u}\hat{u}\delta(\overline{R} - \overline{R}')$$
$$+ \left(\nabla \times \overline{\overline{G}}{}^{+}_{mo}\right) U\left(u - u'\right) + \left(\nabla \overline{\overline{G}}{}^{-}_{mo}\right) U\left(u' - u\right)\Big],$$

which gives

$$\overline{\overline{G}}_{eo}(\overline{R}, \overline{R}') = -\frac{1}{k^2}\hat{u}\hat{u}\delta(\overline{R} - \overline{R}')$$
$$+ \frac{i}{2\pi} \int_{-\infty}^{\infty} dh \sum_{m} \frac{1}{\eta^2 I_{\substack{e\\o}mn}}$$
$$\begin{cases} \overline{M}^{(1)}_{\eta}(h)\overline{M}'_{\eta}(-h) + \overline{N}^{(1)}_{\eta}(h)\overline{N}'_{\eta}(-h), & u > u' \\ \overline{M}_{\eta}(h)\overline{M}'^{(1)}_{\eta}(-h) + \overline{N}_{\eta}(h)\overline{N}'^{(1)}_{\eta}(-h), & u < u'. \end{cases} \tag{8.33}$$

By applying the method of scattering superposition, we can find $\overline{\overline{G}}_{e1}$ for a perfectly conducting elliptical cylinder placed in space as a scattering body. The major axis and the minor axis of the cylinder are defined by

$$a = c \cosh u_0$$
$$b = c \sinh u_0.$$

The equation of the elliptical surface is, therefore, subscribed by

$$\frac{x^2}{a^2} + \frac{y^2}{b^2} = 1$$

or

$$\frac{x^2}{c^2 \cosh^2 u_0} + \frac{y^2}{c^2 \sinh^2 u_0} = 1. \tag{8.34}$$

Omitting the details, we find

$$\begin{aligned}\overline{\overline{G}}_{e1}(\overline{R},\overline{R}') = &-\frac{1}{k^2}\hat{u}\hat{u}\delta(\overline{R}-\overline{R}')\\ &+\frac{i}{2\pi}\int_{-\infty}^{\infty} dh \sum_m \frac{1}{\eta^2 I_{\substack{e\\o}mn}}\\ &\cdot\begin{cases}\overline{M}_\eta^{(1)}(h)\left[\overline{M}_\eta'(-h)+\alpha_\eta \overline{M}_\eta'^{(1)}(-h)\right]\\ \quad +\overline{N}_\eta^{(1)}(h)\left[\overline{N}_\eta'(-h)+\beta_\eta \overline{N}_\eta'^{(1)}(-h)\right], & u>u'\\ \left[\overline{M}_\eta(h)+\alpha_\eta \overline{M}_\eta^{(1)}(h)\right]\overline{M}_\eta'^{(1)}(-h)\\ \quad +\left[\overline{N}_\eta(h)+\beta_\eta \overline{N}_\eta^{(1)}(h)\right]\overline{N}_\eta'^{(1)}(-h), & u<u',\end{cases}\end{aligned} \tag{8.35}$$

where

$$\alpha_\eta = \alpha_{\substack{e\\o}m\eta} = -\frac{[\partial R_\eta(u_0)]/\partial u_0}{[\partial R_\eta^{(1)}(u_0)]/\partial u_0} \tag{8.36}$$

$$\beta_\eta = \beta_{\substack{e\\o}m\eta} = \frac{-R_\eta(u_0)}{R_\eta^{(1)}(u_0)} \tag{8.37}$$

with $\eta = (k^2-h^2)^{\frac{1}{2}}$ and

$$\begin{aligned} I_{\substack{e\\o}m\eta} &= \int_0^{2\pi} S^2_{\substack{e\\o}m\eta}(v)dv\\ &= \begin{cases}\pi \sum_n'(1+\delta_0)(D_n^m)^2\\ \pi\sum_n'(F_n^m)^2.\end{cases}\end{aligned} \tag{8.38}$$

To find the asymptotic expression for (8.35), we substitute in (8.35) the asymptotic values of the radial functions $R^{(1)}_{\substack{e\\o}m\eta}$ assuming $\eta \cosh u$ to be large. The problem is straightforward and is assigned as an exercise.

Once the function of the first kind is known, the function of the second kind can be found by making use of the symmetry relation between these two functions. As was mentioned at the beginning of the chapter, it is impossible to find an orthogonal expansion for the function of the third kind associated with a dielectric elliptical cylinder because both the angular functions and the radial functions depend on the wave number, which has different values for the two regions interior and exterior to the cylinder. The mixed boundary condition, therefore, cannot be satisfied if we use expressions analogous to (7.31) and (7.32) for the dielectric cylinder. As far as we know, the dyadic Green function of the third kind has been found so far only for the layered media, the circular cylinder, and the sphere. We shall treat the layered media and the dielectric sphere in the subsequent chapters.

9

Perfectly Conducting Wedge and the Half Sheet

In his book on optics [1954, p. 266], Sommerfeld remarked that the generalization of the edge diffraction problem to three dimensions is directly possible only for scalar or acoustic problems. The master did not reveal the reason why this cannot be done for vector or electromagnetic problems. It was this remark that inspired the present author in 1954 to resolve the problem using the then relatively new technique of the dyadic Green functions. This material now forms the main body of this chapter. This chapter starts with a derivation of the electric dyadic Green function of the first kind for a perfectly conducting wedge of arbitrary wedge angle. The function for a half sheet is just a limiting case when the wedge angle approaches zero. Radiation from dipoles, both electric and magnetic, and from apertures in the presence of a half sheet, are treated in detail. The examples illustrate quite adequately the versatility of the dyadic Green function technique. The numerical results obtained for the half sheet may also be of interest to engineers designing antennas that are similar to the ones considered here.

9-1 DYADIC GREEN FUNCTIONS FOR A PERFECTLY CONDUCTING WEDGE

For reasons that will become clear later, we judiciously choose the boundary of a perfectly conducting wedge corresponding to $\phi = 0$ and $\phi = 2\pi - \phi_0$ as shown in Fig. 9-1, where ϕ_0 denotes the angle of the wedge.

The vector wave function which will be used in the eigenfunction expansion of the dyadic Green functions of the first and second kind pertaining to the wedge are defined by

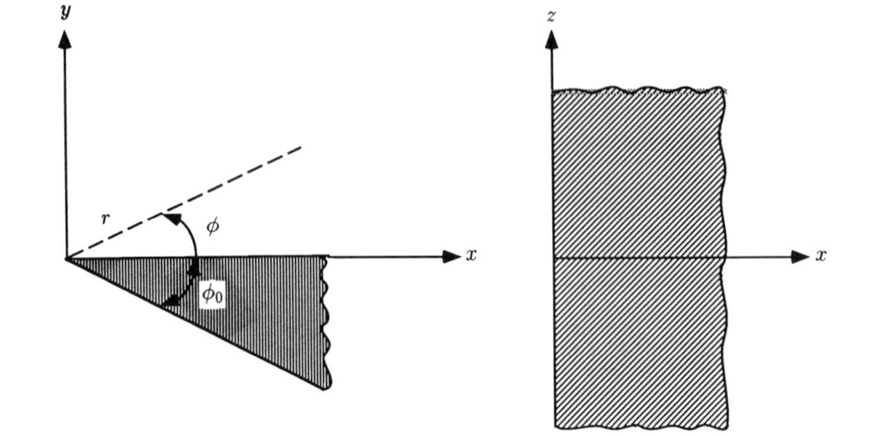

Fig. 9-1 Cross-sectional views of a perfectly conducting wedge

$$\overline{M}_{\substack{e\\o}\nu\lambda}(h) = \nabla\left[J_\nu(\lambda r)\genfrac{}{}{0pt}{}{\cos}{\sin}\nu\phi e^{ihz}\hat{z}\right]$$

$$\overline{N}_{\substack{e\\o}\nu\lambda}(h) = \frac{1}{\kappa}\nabla\times\nabla\times\left[J_\nu(\lambda r)\genfrac{}{}{0pt}{}{\cos}{\sin}\nu\phi e^{ihz}\hat{z}\right],$$

where

$$\kappa^2 = \lambda^2 + h^2$$

$$\nu = \frac{n}{(2-\phi_0/\pi)},\ n = 0, 1, 2, 3, \ldots.$$

The complete expressions of these functions are given by

$$\overline{M}_{\substack{e\\o}\nu\lambda}(h) = \left[\mp\frac{\nu}{r}J_\nu(\lambda r)\genfrac{}{}{0pt}{}{\sin}{\cos}\nu\phi\hat{r} - \frac{\partial J_\nu(\lambda r)}{\partial r}\genfrac{}{}{0pt}{}{\cos}{\sin}\nu\phi\hat{\phi}\right]e^{ihz} \tag{9.1}$$

$$\begin{aligned}\overline{N}_{\substack{e\\o}\nu\lambda}(h) = \frac{1}{\kappa}\Bigg[&ih\frac{\partial J_\nu(\lambda r)}{\partial r}\genfrac{}{}{0pt}{}{\cos}{\sin}\nu\phi\hat{r} \mp \frac{ih\nu}{r}J_\nu(\lambda r)\genfrac{}{}{0pt}{}{\sin}{\cos}\nu\phi\hat{\phi}\\ &+\lambda^2 J_\nu(\lambda r)\genfrac{}{}{0pt}{}{\cos}{\sin}\nu\phi\hat{z}\Bigg]e^{ihz}.\end{aligned} \tag{9.2}$$

These functions satisfy the boundary bonditions that on the surface of the wedge, corresponding to $\phi = 0$ and $\phi = 2\pi - \phi_0$,

$$\begin{aligned}\hat{n}\times\overline{M}_e &= 0\\ \hat{n}\times\overline{N}_o &= 0\\ \hat{n}\times\nabla\times\overline{M}_o &= 0\\ \hat{n}\times\nabla\times\overline{N}_e &= 0.\end{aligned}$$

This shows why we have chosen the wedge geometry with respect to the coordinate system as shown in Fig. 9-1 so that the required vector wave functions would have a relatively simple form. Unlike the case for a circular cylinder, the construction of the functions of the first and the second kind for the wedge will be pursued directly using these functions. It would be much more complicated in this case to find the free-space dyadic Green function first and then to apply the method of scattering superposition to determine the functions of the first and second kind.

The vector functions defined by (9.1) and (9.2) have the following orthogonal relations:

$$\iiint \overline{M}_{e\nu\lambda}(h) \cdot \overline{N}_{o\nu'\lambda'}(-h')\, dV = 0 \tag{9.3}$$

$$\iiint \overline{M}_{e\nu\lambda}(h) \cdot \overline{M}_{e\nu'\lambda'}(-h')\, dV = \begin{cases} 0, \nu \neq \nu' \\ (1+\delta_0)\,\pi\,(2\pi - \phi_0)\,\lambda\delta\,(\lambda - \lambda')\,\delta\,(h - h'),\ \nu = \nu' \end{cases} \tag{9.4}$$

$$\iiint \overline{N}_{o\nu\lambda}(h) \cdot \overline{N}_{o\nu'\lambda'}(-h')\, dV = \begin{cases} 0, & \nu \neq \nu' \\ (1+\delta_0)\,\pi\,(2\pi - \phi_0)\,\lambda\delta\,(\lambda - \lambda')\,\delta\,(h - h'), & \nu = \nu', \end{cases} \tag{9.5}$$

where the domain of integration is extended through the space exterior to the wedge. The derivation of (9.3) to (9.5) follows the same procedure as described in Sec. 7-1. It will not be repeated here. According to the method of $\overline{\overline{G}}_m$, we let

$$\nabla \times \left[\overline{\overline{I}}\delta(\overline{R} - \overline{R}')\right] = \int_0^\infty d\lambda \int_0^\infty dh \sum_\nu \left[\overline{N}_{e\nu\lambda}(h)\overline{A}_{e\nu\lambda}(h) + \overline{M}_{o\nu\lambda}(h)\overline{B}_{o\nu\lambda}(h)\right]. \tag{9.6}$$

As a result of the orthogonal relationships between $\overline{N}_e$ and $\overline{M}_0$, we find

$$\overline{A}_{e\nu\lambda}(h) = \frac{2-\delta_0}{2\pi\,(\pi - \phi_0)\,\lambda}\nabla' \times \overline{N}'_{e\nu\lambda}(-h) = \frac{(2-\delta_0)\kappa}{2\pi(\pi - \phi_0)\lambda}\overline{M}'_{e\nu\lambda}(-h) \tag{9.7}$$

$$\overline{B}_{0\nu\lambda}(h) = \frac{2-\delta_0}{2\pi(\pi - \phi_0)\lambda}\nabla' \times \overline{M}_{0\nu\lambda}(-h) = \frac{(2-\delta_0)\kappa}{2\pi(\pi - \phi_0)\lambda}\overline{N}'_{0\nu\lambda}(-h). \tag{9.8}$$

The expression for $\overline{\overline{G}}_{m2}$ can then be written in the form

$$\overline{\overline{G}}_{m2}(\overline{R}, \overline{R}') = \int_0^\infty d\lambda \int_{-\infty}^\infty dh \sum_\nu \frac{(2-\delta_0)\kappa}{2\pi(\pi - \phi_0)\lambda(\kappa^2 - k^2)} \cdot \left[\overline{N}_{e\nu\lambda}(h)\overline{M}'_{e\nu\lambda}(-h) + \overline{M}_{0\nu\lambda}(h)\overline{N}'_{0\nu\lambda}(-h)\right]. \tag{9.9}$$

The λ-integration can be evaluated with the aid of the residue theorem in the λ-plane. That yields

$$\begin{aligned}\overline{\overline{G}}_{m2}^{\pm}(\overline{R},\overline{R}') = & \frac{ik}{4(2\pi-\phi_0)}\int_{-\infty}^{\infty}dh\sum_{\nu}\frac{2-\delta_0}{\eta^2} \\ & \cdot\begin{cases}\overline{N}_{e\nu\eta}^{(1)}(h)\overline{M}'_{e\nu\eta}(-h)+\overline{M}_{0\nu\eta}^{(1)}(h)\overline{N}'_{0\nu\eta}(-h), & r>r' \\ \overline{N}_{e\nu\eta}(h)\overline{M}'^{(1)}_{e\nu\eta}(-h)+\overline{M}_{0\nu\eta}(h)\overline{N}'^{(1)}_{0\nu\eta}(-h), & r<r',\end{cases}\end{aligned} \tag{9.10}$$

where

$$\eta = (k^2-h^2)^{\frac{1}{2}}.$$

The vector wave functions of the first kind are defined with respect to the Hankel function of the first kind with order ν. They are

$$\overline{M}_{0\nu\eta}^{(1)}(h) = \nabla\times\left[H_\nu^{(1)}(\eta r)\sin\nu\phi e^{ihz}\hat{z}\right]$$

$$\overline{N}_{e\nu\eta}^{(1)}(h) = \frac{1}{k}\nabla\times\nabla\times\left[H_\nu^{(1)}(\eta r)\cos\nu\phi e^{ihz}\hat{z}\right].$$

The primed functions in (9.10) are defined with respect to (r', ϕ', z'). The function $\overline{\overline{G}}_{e1}$ can now be obtained in the usual manner that is given by

$$\begin{aligned}\overline{\overline{G}}_{e1}(\overline{R},\overline{R}) = & -\frac{1}{k^2}\hat{r}\hat{r}\delta(\overline{R}-\overline{R}') \\ & +\frac{i}{4(2\pi-\phi_0)}\int_{-\infty}^{\infty}dh\sum_{\nu}\frac{2-\delta_0}{\eta^2} \\ & \cdot\begin{cases}\overline{M}_{e\nu\eta}^{(1)}(h)\overline{M}'_{e\nu\eta}(-h)+\overline{N}_{0\nu\eta}^{(1)}(h)\overline{N}'_{0\nu\eta}(-h), & r>r' \\ \overline{M}_{e\nu\eta}(h)\overline{M}'^{(1)}_{e\nu\eta}(-h)+\overline{N}_{0\nu\eta}(h)\overline{N}'^{(1)}_{0\nu\eta}(-h), & r>r',\end{cases}\end{aligned} \tag{9.11}$$

where

$$\overline{M}_{e\nu\eta}^{(1)}(h) = \nabla\times\left[H_\nu^{(1)}(\eta r)\cos\nu\phi e^{ihz}\hat{z}\right]$$

$$\overline{N}_{0\nu\eta}^{(1)}(h) = \frac{1}{k}\nabla\times\nabla\times\left[H_\nu^{(1)}(\eta r)\sin\nu\phi e^{ihz}\hat{z}\right],$$

with

$$\nu = \frac{n}{(2-\phi_0/\pi)}, n = 0, 1, 2, \ldots$$

$$\eta = (k^2-h^2)^{\frac{1}{2}}.$$

By making use of the symmetric relations between the functions of the first and the second kind, we can find the expression for $\overline{\overline{G}}_{e2}(\overline{R},\overline{R}')$, which is given by replacing the even $\overline{M}$ functions and the odd $\overline{N}$ functions in (9.11), respectively, by the odd $\overline{M}$ functions and the even $\overline{N}$ functions. Like the case of a dielectric elliptical cylinder, there is no way of constructing the dyadic Green function of the third kind for a dielectric wedge because of the lack of proper orthogonal sets of vector wave functions.

9-2 THE HALF SHEET

When the angle of a wedge approaches zero, the wedge degenerates into a half sheet. The function of the first kind is thus obtained by putting $\phi_0 = 0$ in (9.11), which yields

$$
\begin{aligned}
\overline{\overline{G}}_{e1}(\overline{R}, \overline{R}') = &- \frac{1}{k^2}\hat{r}\hat{r}\delta(\overline{R} - \overline{R}') \\
&+ \frac{i}{8\pi}\int_{-\infty}^{\infty} dh \sum \frac{2-\delta_0}{\eta^2} \\
&\cdot \begin{cases} \overline{M}^{(1)}_{e\nu\eta}(h)\overline{M}'_{e\nu\eta}(-h) + \overline{N}^{(1)}_{0\nu\eta}(h)\overline{N}'_{0\nu\eta}(-h), & r > r' \\ \overline{M}_{e\nu\eta}(h)\overline{M}'^{(1)}_{e\nu\eta}(-h) + \overline{N}_{0\nu\eta}(h)\overline{N}'^{(1)}_{0\nu\eta}(-h), & r < r', \end{cases}
\end{aligned} \tag{9.12}
$$

and from the symmetrical relationship between $\overline{\overline{G}}_{e1}$ and $\overline{\overline{G}}_{e2}$, we have

$$
\begin{aligned}
\overline{\overline{G}}_{e2}(\overline{R}, \overline{R}') = &- \frac{1}{k^2}\hat{r}\hat{r}\delta(\overline{R} - \overline{R}') \\
&+ \frac{i}{8\pi}\int_{-\infty}^{\infty} dh \sum_{\nu} \frac{2-\delta_0}{\eta^2} \\
&\cdot \begin{cases} \overline{M}^{(1)}_{0\nu\eta}(h)\overline{M}'_{0\nu\eta}(-h) + \overline{N}^{(1)}_{e\nu\eta}(h)\overline{N}'_{e\nu\eta}(-h), & r > r' \\ \overline{M}_{0\nu\eta}(h)\overline{M}'^{(1)}_{0\nu\eta}(-h) + \overline{N}_{e\nu\eta}(h)\overline{N}'^{(1)}_{e\nu\eta}(-h), & r < r', \end{cases}
\end{aligned} \tag{9.13}
$$

where

$$
\begin{aligned}
\nu &= \frac{n}{2}, n = 0, 1, 2, \ldots \\
\eta &= (k^2 - h^2)^{\frac{1}{2}}.
\end{aligned}
$$

It is seen that only Bessel functions and Hankel functions of half order and integer order are involved in these expressions. Once the expressions for $\overline{\overline{G}}_{e1}$ and $\overline{\overline{G}}_{e2}$ are known, the field from current elements with known distributions or apertures with known field distributions can be calculated by evaluating some integrals. In general, because of the Fourier integrals involved in the expressions for the dyadic Green functions, we are unable to evaluate these integrals in closed forms. However, for certain restricted ranges of the parameters, corresponding to the far-zone range, most of the integrals can be evaluated in a closed form in terms of tabulated functions. For convenience, the problems to be considered will be divided into four categories:

1. Radiation from electric dipoles in the presence of a half sheet
2. Radiation from magnetic dipoles in the presence of a half sheet

3. Radiation from slots cut in a half sheet
4. Diffraction of a plane wave by a half sheet

9-3 RADIATION FROM ELECTRICAL DIPOLES IN THE PRESENCE OF A HALF SHEET

9-3.1 Longitudinal Electric Dipole

Let us consider an infinitesimal electric dipole with a current moment c oriented along the direction of the longitudinal axis of the half sheet, and located at $(a, \phi_0, 0)$ as shown in Fig. 9-2. The electric current density can be written as

$$\bar{J}(\bar{R}') = c\frac{\delta(r'-a)\delta(\phi'-\phi_0)\delta(z'-0)}{r'}\hat{z}. \tag{9.14}$$

The delta functions have a weighting factor $1/r'$ so that

$$\iiint \bar{J}(\bar{R}')\,dV' = \iiint \bar{J}(\bar{R}')r'\,dr'\,d\phi'\,dz' = c\hat{z}.$$

Substituting (9.12) and (9.14) into

$$\bar{E}(\bar{R}) = i\omega\mu_0 \iiint \bar{\bar{G}}_{e1}(\bar{R},\bar{R}')\cdot\bar{J}(\bar{R}')\,dV',$$

we obtain, for $r > a$,

$$\bar{E}(\bar{R}) = \frac{-\omega\mu_0 c}{4\pi k}\int_{-\infty}^{\infty} dh \sum_{\nu=0}^{\infty} \bar{N}^{(1)}_{0\nu\eta}(h)\bar{J}_\nu(\eta a)\sin\nu\phi_0.$$

The TM modes are not excited in this case because $\bar{M}'_{e\nu\eta}$ is transverse to the longitudinal axis. For numerical calculation, we shall investigate only the z-component of the electric field which is given by

$$E_z(\bar{R}) = \frac{-\omega\mu_0 c}{4\pi k^2}\int_{-\infty}^{\infty} \eta^2 e^{ihz} \sum_{n=1,2,\ldots} \sin\left(\frac{n\phi_0}{2}\right) \sin\left(\frac{n\phi}{2}\right)\cdot J_{\frac{n}{2}}(\eta a)H^{(1)}_{\frac{n}{2}}(\eta r)\,dh, \tag{9.15}$$

where we have already converted ν into $n/2$. The Fourier integral contained in (9.15) cannot be evaluated in a closed form for arbitrary values of r and z. However, when the point of observation is far away from the origin, the integration can be done by means of the saddle-point method of integration. The procedure is the same as the one described in Sec. 7-4. Without repeating the details,

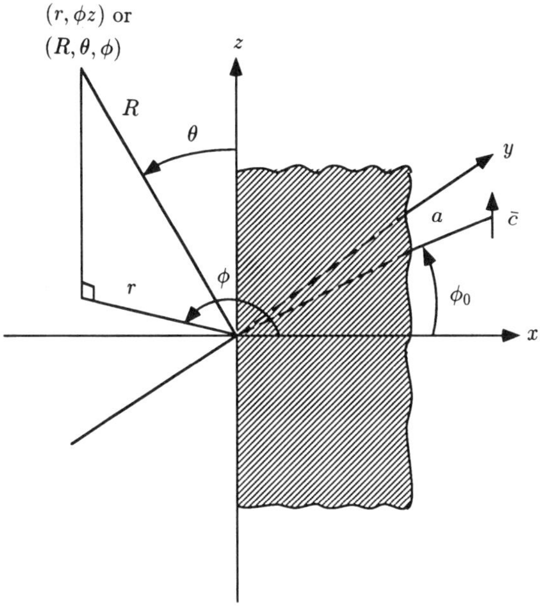

Fig. 9-2 A longitudinal electric dipole located at $(a, \phi_0, 0)$ in the presence of a half sheet

we obtain

$$E_z(\overline{R}) = \frac{i\omega\mu c e^{ikR}}{2\pi R} \sin^2\theta \sum_{n=1}^{\infty} (-i)^{\frac{n}{2}} \sin\left(\frac{n\phi_0}{2}\right) \sin\left(\frac{n\phi}{2}\right) J_{\frac{n}{2}}\ (ka\sin\theta), \tag{9.16}$$

where (R, θ, ϕ) denotes the point of observation in the spherical coordinate system as shown in Fig. 9-2. The series contained in (9.16) can be transformed into a function involving some Fresnel integrals by applying an expansion theorem due to Hargreaves [1918]. The analysis is found in Morse and Feshbach [1953, p. 1386]. To transform (9.16) into the desirable form, we let

$$S(\rho, \phi) = \frac{1}{2} \sum_{n=0}^{\infty} (2 - \delta_0)(-i)^{\frac{n}{2}} \cos\frac{n\phi}{2} J_{\frac{n}{2}}\ (\rho). \tag{9.17}$$

The integral representation of (9.17) according to Hargreaves's theorem, is given by

$$S(\rho, \phi) = (\pi)^{-\frac{1}{2}} e^{-i(\rho\cos\phi + \pi/4)} \int_{-\infty}^{(2\rho)^{\frac{1}{2}} \cos\frac{\phi}{2}} e^{iS^2} dS. \tag{9.18}$$

The integral contained in (9.18) can be expressed in terms of the Fresnel integral functions defined by

$$C(x) = \int_0^x \frac{\cos t}{(2\pi t)^{\frac{1}{2}}}\, dt, \quad S(x) = \int_0^x \frac{\sin t}{(2\pi t)^{\frac{1}{2}}}\, dt. \tag{9.19}$$

A short table of these functions is found in Watson [1922, p. 744]. By a change of the variable of integration, (9.18) can be written in the form

$$S(\rho,\phi) = 2^{-\frac{1}{2}} e^{-i(\rho\cos\phi + \pi/4)} \left\{ \frac{1}{2}(1+i) \pm [C(\rho+\rho\cos\phi) + iS(\rho+\rho\cos\phi)] \right\}, \tag{9.20}$$

where the $\pm$ sign in (9.20) is chosen according to the appropriate range of ϕ:

$$+ \text{ sign for } \pi \geq \phi \geq 0$$
$$- \text{ sign for } 2\pi \geq \phi \geq \pi.$$

Regarding $S(\rho,\phi)$ as a basic function in this work, we can transform (9.16) into an expression containing two of these functions with different arguments owing to the simple trigonometric identity that

$$2\sin\left(\frac{n\phi_0}{2}\right)\sin\left(\frac{n\phi}{2}\right) = \cos\frac{n(\phi-\phi_0)}{2} - \cos\frac{n(\phi+\phi_0)}{2}.$$

The final expression for $E_z(\overline{R})$ is given by

$$E_z(\overline{R}) = \frac{i\omega\mu_0 c e^{ikR}}{4\pi R}\sin^2\theta \cdot [S(ka\sin\theta, \phi-\phi_0) - S(ka\sin\theta, \phi+\phi_0)]. \tag{9.21}$$

In the principal plane, corresponding to $\theta = \pi/2$, the pattern of the electric field is described by

$$F_1(\phi) = S(ka, \phi-\phi_0) - S(ka, \phi+\phi_0). \tag{9.22}$$

By using (9.20) instead of (9.17) we can determine very accurately the numerical values of the pattern function. It should be mentioned that the pattern function (9.22) which we obtained here is equivalent to the one obtained by Harrington [1953] for a line source in the presence of a half sheet. In fact, the complex integral involved in Harrington's work based on the method of the Weiner-Hopf integral equation can readily be reduced to the Fresnel integral.

9-3.2 Horizontal Electrical Dipole

The orientation and the location of the dipole for this case is shown in Fig. 9-3, which displays only a cross-sectional view in the plane $z = 0$.

In the principal plane, corresponding to $\theta = \pi/2$ or $z = 0$, the far-zone electric field has only a ϕ-component. Its expression is given by

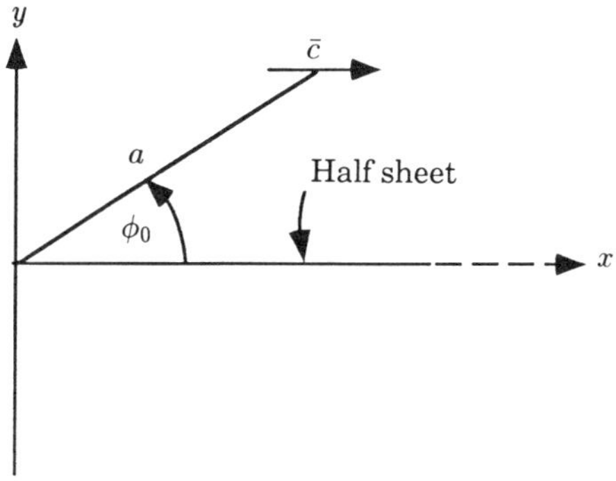

Fig. 9-3 A horizontal electric dipole located at $(a, \phi_0, 0)$

$$E_\phi\left(R, \frac{\pi}{2}, \phi\right) = \frac{\omega\mu_0 c e^{ikR}}{4\pi k a R} \cdot \left[-\cos\phi_0 \frac{\partial}{\partial\phi} \sum_{n=0}^{\infty} (2-\delta_0)(-i)^{\frac{n}{2}} J_{\frac{n}{2}}(\rho) \sin\frac{n\phi_0}{2} \sin\frac{n\phi}{2} + \rho\sin\phi_0 \frac{\partial}{\partial\rho} \sum_{n=0}^{\infty} (2-\delta_0)(-i)^{\frac{n}{2}} J_{\frac{n}{2}}(\rho) \cos\frac{n\phi_0}{2} \cos\frac{n\phi}{2} \right]_{\rho=ka}.$$

The two series contained in the brackets can be expressed in terms of the S-function defined by (9.17). Thus we have

$$\frac{\partial}{\partial\phi} \sum_{n=0}^{\infty} (2-\delta_0)(-i)^{\frac{n}{2}} J_{\frac{n}{2}}(\rho) \sin\frac{n\phi_0}{2} \sin\frac{n\phi}{2} = \frac{\partial}{\partial\phi} \left[S(\rho, \phi-\phi_0) - S(\phi, \phi+\phi_0) \right]$$

and

$$\frac{\partial}{\partial\rho} \sum_{n=0}^{\infty} (2-\delta_0)(-i)^{\frac{n}{2}} J_{\frac{n}{2}}(\rho) \cos\frac{n\phi_0}{2} \cos\frac{n\phi}{2} = \frac{\partial}{\partial\rho} \left[S(\rho, \phi-\phi_0) - S(\phi, \phi+\phi_0) \right].$$

By taking the partial derivatives of the S-functions as represented by (9.18) or (9.20) and simplifying the result, we obtain

$$E_\phi\left(R, \frac{\pi}{2}, \phi\right) = \frac{-i\omega\mu_0 c e^{ikR}}{4\pi R} \cdot \left\{ \sin\phi \left[S(\rho, \phi-\phi_0) - S(\rho, \phi+\phi_0) \right] + \left(\frac{2}{\pi\rho}\right)^{\frac{1}{2}} e^{i(\rho+\pi/4)} \sin\frac{\phi_0}{2} \sin\frac{\phi}{2} \right\}_{\rho=ka}. \tag{9.23}$$

The pattern function in this case is therefore described by

$$F_2(\phi) = \sin\phi\,[S(ka, \phi-\phi_0) - S(ka, \phi+\phi_0)] + \left(\frac{2}{\pi ka}\right)^{\frac{1}{2}} e^{i(ka+\pi 4)} \sin\frac{\phi_0}{2}\sin\frac{\phi}{2}. \tag{9.24}$$

9-3.3 Vertical Electric Dipole

The orientation and the location of the dipole for this case are shown in Fig. 9-4. The far-zone electric field in the principal plane again has only a ϕ-component. Its expression is given by

$$E_\phi\left(R, \frac{\pi}{2}, \phi\right) = \frac{i\omega\mu_0 c e^{ikR}}{4\pi R}\left\{\cos\phi\,[S(ka, \phi-\phi_0) + S(ka, \phi+\phi_0)] + \left(\frac{2}{\pi ka}\right)^{\frac{1}{2}} e^{i(ka+\pi/4)} \cos\frac{\phi_0}{2}\cos\frac{\phi}{2}\right\}. \tag{9.25}$$

The pattern function in the principal plane is therefore represented by

$$F_3(\phi) = \cos\phi\,[S(ka, \phi-\phi_0) + S(ka, \phi+\phi_0)] + \left(\frac{2}{\pi ka}\right)^{\frac{1}{2}} e^{i(ka+\pi/4)} \cos\frac{\phi_0}{2}\cos\frac{\phi}{2}. \tag{9.26}$$

Some sample calculations based on (9.22), (9.24), and (9.26), after being normalized with respect to the maximum value of each individual pattern, are plotted in Figs. 9-5, 9-6, and 9-7. More patterns are found in the original edition of this book [Tai, 1971].

It should be mentioned that the problem of the diffraction of a dipole field by a perfectly conducting half sheet has also been investigated by Senior [1953]. The relationship between the present formulation and the one based on the method of potentials is discussed by Bowman, Senior, and Uslenghi [1969].

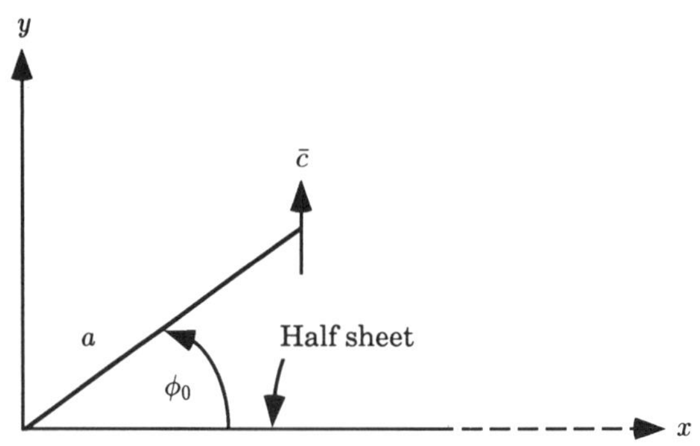

Fig. 9-4 A vertical electric dipole located at $(a, \phi_0, 0)$

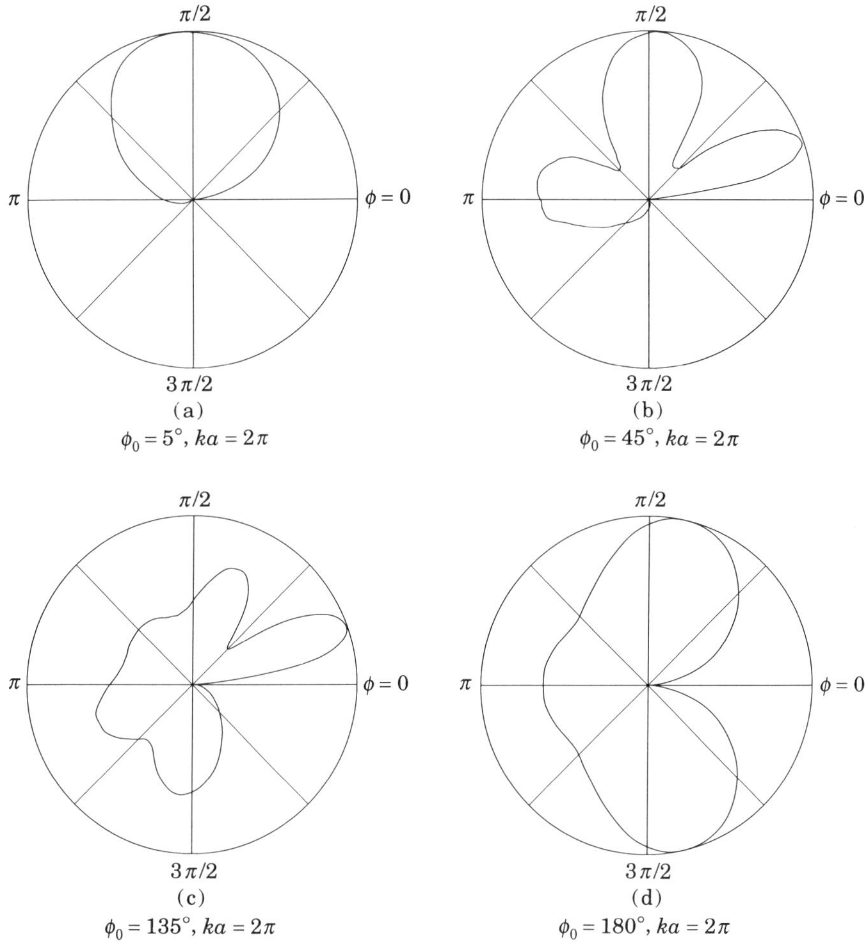

Fig. 9-5 Radiation pattern of a longitudinal dipole placed in the front of a half sheet

9-4 RADIATION FROM MAGNETIC DIPOLES IN THE PRESENCE OF A HALF SHEET

In principle, the electric field for any current distribution in the presence of a conducting body can be obtained by using the formula

$$\overline{E}(\overline{R}) = i\omega\mu_0 \iiint \overline{\overline{G}}_{e1}(\overline{R}, \overline{R}') \cdot \overline{J}(\overline{R}')dV'.$$

In practice, for currents in the form of small current loops or magnetic dipoles it is more convenient to use an alternative formula. If we introduce an equivalent magnetization vector $\overline{M}$ such that $\nabla \times \overline{J} = k^2\overline{M}$, then the equation for $\overline{H}$ becomes

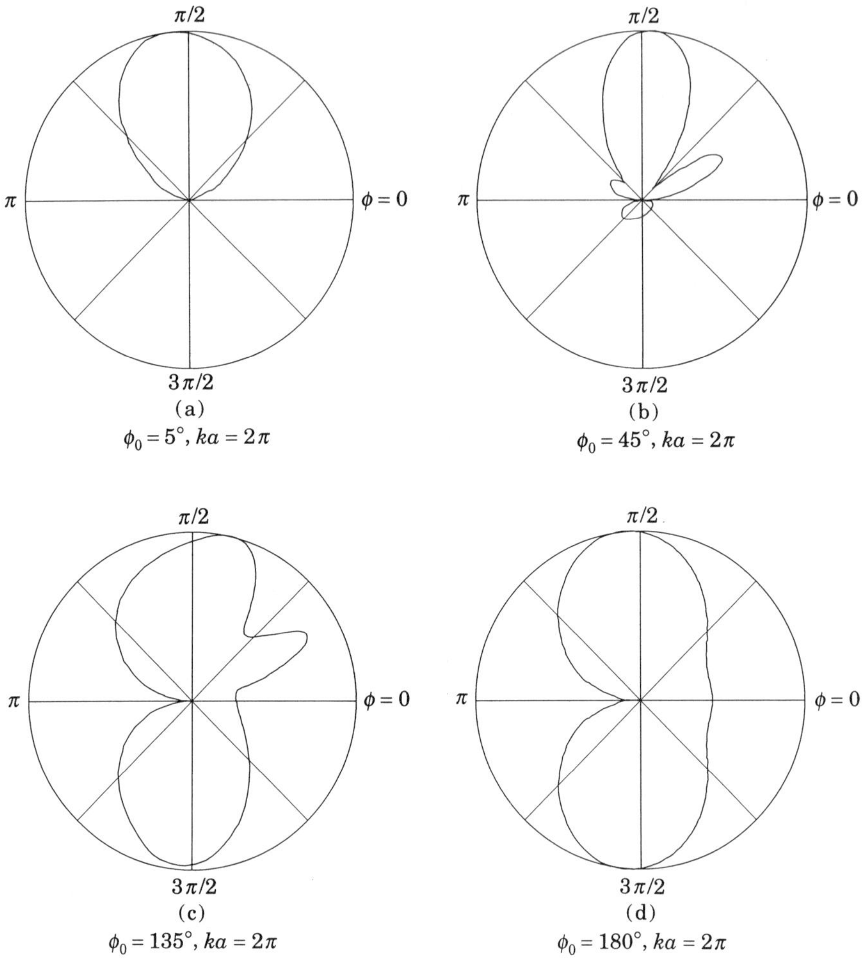

Fig. 9-6 Radiation pattern of a horizontal dipole placed in the front of a half sheet

$$\nabla \times \nabla \times \overline{H} - k^2 \overline{H} = k^2 \overline{M}. \tag{9.27}$$

By integrating this equation with the aid of $\overline{\overline{G}}_{e2}(\overline{R}, \overline{R}')$, we obtain

$$\overline{H}(\overline{R}) = k^2 \iiint \overline{\overline{G}}_{e2}(\overline{R}, \overline{R}') \cdot \overline{M}(\overline{R}')\, dV'. \tag{9.28}$$

Hence, in the region outside of a current source,

$$\begin{aligned} \overline{E}(\overline{R}) &= \frac{i}{\omega\epsilon_0} \nabla \times \overline{H}(\overline{R}) \\ &= i\omega\mu_0 \iiint \nabla \times \overline{\overline{G}}_{e2}(\overline{R}, \overline{R}') \cdot \overline{M}(\overline{R}')\, dV'. \end{aligned} \tag{9.29}$$

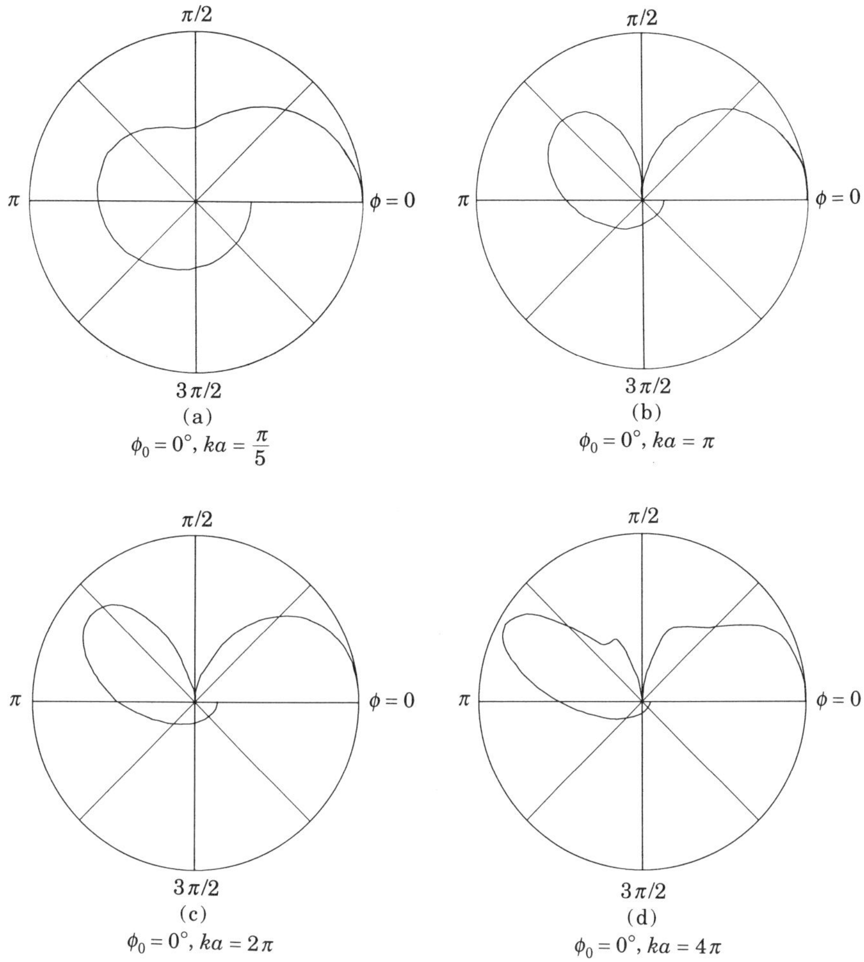

Fig. 9-7 Radiation pattern of a vertical dipole placed in the front of a half sheet

It can be shown that this equation can also be obtained from (4.176) with some transformations and by letting $\nabla \times \overline{J} = k^2\overline{M}$. In the case of a half sheet, for $\overline{R} \neq \overline{R}'$,

$$\nabla \times \overline{\overline{G}}_{e2}(\overline{R}, \overline{R}') = \frac{ik}{8\pi} \int_{-\infty}^{\infty} dh \sum_{\nu} \frac{2 - \delta_0}{\eta^2} \cdot \begin{cases} \overline{N}^{(1)}_{0\nu\eta}(h)\overline{M}'_{0\nu\eta}(-h) + \overline{M}^{(1)}_{e\nu\eta}(h)\overline{N}'_{e\nu\eta}(-h), & r > r' \\ \overline{N}_{0\nu\eta}(h)\overline{M}'^{(1)}_{0\nu\eta}(-h) + \overline{M}_{e\nu\eta}(h)\overline{N}'^{(1)}_{e\nu\eta}(-h), & r > r'. \end{cases} \tag{9.30}$$

For an infinitesimal magnetic dipole located at $r = a, \phi = \phi_0, z = 0$ with a dipole moment $\bar{m}$ pointed in the $\hat{x}_i$-direction we have

$$\overline{M} = \bar{m}\delta(\overline{R} - \overline{R}') = m\hat{x}_i\delta(\overline{R} - \overline{R}'),$$

where $m = IA$ is the magnetic dipole moment of a small current loop with area A and current I; then

$$\overline{E}(\overline{R}) = i\omega\mu_0 m\nabla \times \overline{\overline{G}}_{e2}(\overline{R}, \overline{R}') \cdot \hat{x}_i. \tag{9.31}$$

Following the same analysis as the case of an electric dipole, we find that the far-zone electric field in the horizontal plane ($z = 0$) is given by the following expressions for three orientations of the magnetic dipole.

1. Longitudinal magnetic dipole, $\hat{x}_i = \hat{z}$

$$E_\phi\left(R, \frac{\pi}{2}, \phi\right) = \frac{\omega\mu_0 m e^{ikR}}{4\pi R}\left[S(ka, \phi, \phi_0) - S(ka, \phi + \phi_0)\right] \tag{9.32}$$

2. Horizontal magnetic dipole, $\bar{x}_i = \hat{x}$

$$E_z\left(R, \frac{\pi}{2}, \phi\right) = \frac{\omega\mu_0 k m e^{ikR}}{4\pi R}\left\{\sin\phi\left[S(ka, \phi, \phi_0) + S(ka, \phi + \phi_0)\right]\right.$$
$$\left. + \left(\frac{2}{\pi ka}\right)^{\frac{1}{2}} e^{i(ka+\pi/4)} \cos\frac{\phi_0}{2}\cos\frac{\phi}{2}\right\}. \tag{9.33}$$

3. Vertical magnetic dipole, $\hat{x}_i = \hat{y}$

$$E_z\left(R, \frac{\pi}{2}, \phi\right) = \frac{-\omega\mu_0 k m e^{ikR}}{4\pi R}\left\{\cos\phi\left[S(ka, \phi - \phi_0) - S(ka, \phi + \phi_0)\right]\right.$$
$$\left. - \left(\frac{2}{\pi ka}\right)^{\frac{1}{2}} e^{i(ka+\pi/4)} \sin\frac{\phi_0}{2}\sin\frac{\phi}{2}\right\}. \tag{9.34}$$

9-5 SLOTS CUT IN A HALF SHEET

To find the electric field due to radiating slots cut in a half sheet, we can use the integral expression (4.174) which is given by

$$\overline{E}(\overline{R}) = -\int_{S_d} \nabla \times \overline{\overline{G}}_{e2}(\overline{R}, \overline{R}') \cdot \left[\hat{n}' \times \overline{E}(\overline{R}')\right] dS', \tag{9.35}$$

where $\hat{n}' = -\hat{y}$ for the problem in consideration. This expression can also be obtained from (9.29) if we assume $\overline{M}$ to be distributed on a surface with a surface magnetization vector $\overline{M}_s$; then (9.29) becomes

$$\overline{E}(\overline{R}) = i\omega\mu_0 \iint_{S_d} \nabla \times \overline{\overline{G}}_{e2}(\overline{R}, \overline{R}') \cdot \overline{M}_s(\overline{R}')dS'. \tag{9.36}$$

By letting $i\omega\mu_0\overline{M}_s(\overline{R}') = -\hat{n}' \times \overline{E}(\overline{R}')$, (9.36) becomes the same as (9.35).

In applying (9.35) to radiating apertures on a half sheet, we shall distinguish two types of excitations, designated as "one sided" and "two sided." In practice, the one-sided excitation corresponds to an aperture radiation from a waveguide terminated on a half sheet. If the opening on the sheet is excited by a two-wire transmission line, we have a two-sided excitation. The two cases are illustrated in Fig. 9-8.

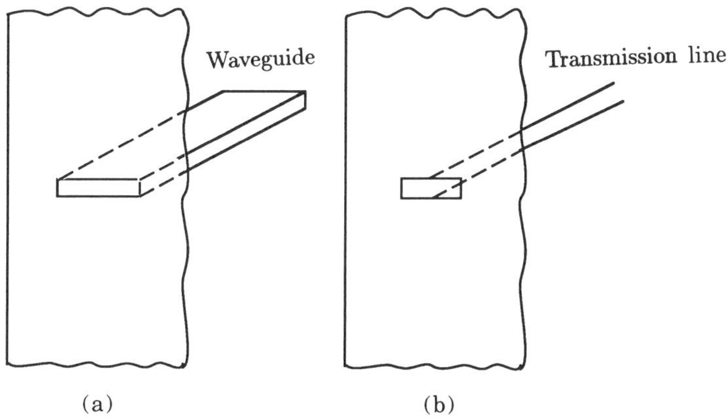

Fig. 9-8 (a) One-sided excitation (b) Two-sided excitation

For numerical calculation, we shall consider only apertures in the form of narrow slots. This simplifies our analysis considerably. The types of slots to be studied are shown in Fig. 9-9, where the polarization of the aperture electric field is indicated by the arrows.

Fig. 9-9 (a) Longitudinal slot (b) Horizontal slot

9-5.1 Longitudinal Slot

The field distribution for an infinitesimally narrow longitudinal slot can be written in the form

$$\overline{E}(\overline{R}') = f(z')\delta(r-a)\hat{r},$$

where $f(z')$ is a given function of z', depending on the actual field distribution along the slot. In the case of a so-called half-wave resonant slot, we let

$$f(z') = V_0 \cos kz', \quad \frac{\lambda}{4} \geq z' \geq -\frac{\lambda}{4}.$$

Since we are going to discuss the far-zone field in the principal plane, the exact knowledge of $f(z')$ is not needed. We merely demand that it is an even function of z'. For a one-sided slot, we assume that the opening is facing the $\hat{y}$ direction. In this case, the radiation pattern is the same as that of a longitudinal magnetic dipole placed at $r = a$ and $\phi_0 = 0$. According to (9.32) with $\phi_0 = 0$ the field pattern is given by

$$F_1(\phi) = S(ka, \phi). \tag{9.37}$$

For a two-sided excitation, the pattern function would be

$$F_2(\phi) = S(ka, \phi) - S(ka, 2\pi - \phi). \tag{9.38}$$

9-5.2 Horizontal Slot

In general it is much more difficult to analyze the field pattern due to a horizontal slot of finite length. For an infinitesimally narrow slot, we let

$$\overline{E}(\overline{R}') = f(r')\delta(z' - 0)\hat{z}; \tag{9.39}$$

then the far-zone electric field in the principal plane for a one-sided slot is represented by the expression

$$E_z\left(R, \frac{\pi}{2}, \phi\right) = \frac{e^{ikR}}{4\pi R} \sum_{n=1}^{\infty} n(-i)^{\frac{n}{2}} \sin\frac{n\phi}{2} \int f(r') J_{\frac{n}{2}}(kr') \frac{dr'}{r'}. \tag{9.40}$$

This expression is obtained by substituting (9.39) into (9.35) with the Fourier integral simplified by the method of saddle-point integration. The radial integral contained in (9.40) depends strongly on the specific form of the function $f(r')$. In general, it cannot be evaluated in a closed form. If we assume that the slot is infinitesimally short, then $f(r')$ can be replaced by $\delta(r' - a)$ where a denotes the location of the slot from the edge of the half sheet. In that case we have

$$\begin{aligned} E_z\left(R, \frac{\pi}{2}, \phi\right) &= \frac{e^{ikR}}{4\pi aR} \sum_{n=1}^{\infty} n(-i)^{\frac{n}{2}} \sin\frac{n\phi}{2} J_{\frac{n}{2}}(ka) \\ &= -\frac{e^{ikR}}{4\pi aR} \frac{\partial}{\partial\phi} S(ka, \phi). \end{aligned} \tag{9.41}$$

The pattern function in this case is the same as that of a horizontal magnetic dipole, pointed in the x-direction, placed at the surface of the half sheet, that is, by putting $\phi_0 = 0$ in (9.33). A convenient form of the pattern function to be used for numerical calculation is

$$F_1(\phi) = \frac{1}{ika}\frac{\partial}{\partial\phi}S(ka,\phi) = \sin\phi S(ka,\phi) + \frac{e^{i(ka+\pi/4)}}{(2\pi ka)^{\frac{1}{2}}}\sin\frac{\phi}{2}. \tag{9.42}$$

For a two-sided infinitesimal horizontal slot, the pattern function can most conveniently be written in the form

$$F_2(\phi) = |\sin\phi|\, e^{9ka\cos\phi}\,[C(x) + iS(x)]_{x=ka(1+\cos\phi)} + \frac{ie^{ika}}{(\pi ka)^{\frac{1}{2}}}\sin\frac{\phi}{2}. \tag{9.43}$$

Numerical calculations based on (9.37), (9.38), (9.42), and (9.43) for several different values of ka are plotted in Figs. 9-10 to 9-13.

It is of some practical interest to examine in detail the radiation pattern of a slot when it is located far away from the edge of a half sheet. A typical pattern for a longitudinal slot placed at a distance corresponding to $ka = 30$ is shown in Fig. 9-14. The locations of the maxima and minima of such a pattern can be ascertained in a relatively simple manner. In the case of a one-sided longitudinal slot, the extreme values are determined by the equation

$$\frac{\partial\,|S(ka,\phi)|}{\partial\phi} = 0. \tag{9.44}$$

Using the explicit expressions for $S(ka,\phi)$ represented by (9.20), we find that (9.44) yields

$$\frac{0.5 + C(x)}{0.5 + S(x)} = -\tan x, \quad x = ka(1+\cos\phi). \tag{9.45}$$

The graphical solution for the roots of (9.45) is shown in Fig. 9-15. Denoting these roots by x_m, it is seen from the display that the approximate solutions for these roots are given by

$$x_m = (4m-1)\frac{\pi}{4}, \; m = 1, 2, \ldots$$
$$\cos\phi_m = (4m-1)\frac{\pi}{4ka} - 1, \; m = 1, 2, \ldots. \tag{9.46}$$

When m is odd, it gives the angular position of a minimum. Even values of m provide the positions of the maxima. Since the magnitude of $|S(ka,\phi_1)|$ and that of $|S(ka,\pi)|$ are independent of ka, their ratio is also independent of ka and it is given by

$$\frac{|S(ka,\pi)|}{|S(ka,\phi_1)|} = 0.428. \tag{9.47}$$

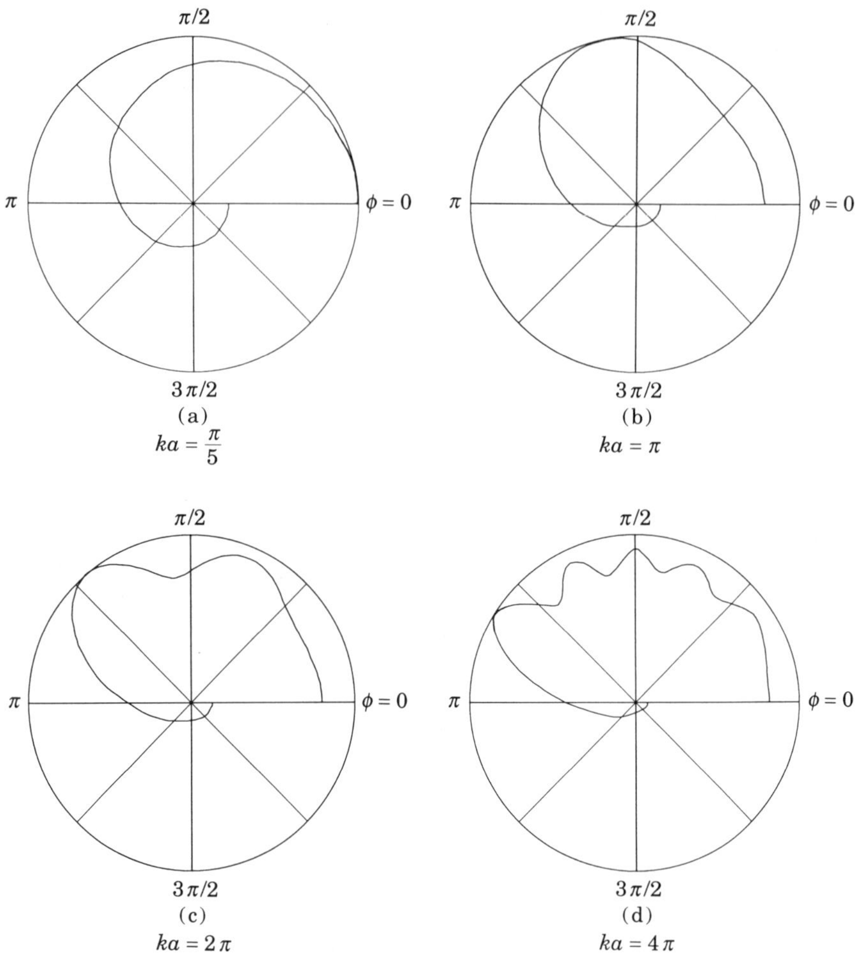

Fig. 9-10 Radiation pattern of a one-sided longitudinal slot

This number can be used as a measure of the rate of decay of the field intensity from its peak value to the value observed at the grazing direction. Equation (9.46) shows that when ka is large, ϕ_1 approaches π, but there is always a sharp decay of the field as the point of observation changes from the lit region to the shadow region. Such a spill-over effect must be taken into consideration in the design of a flush-mounted antenna over a flat surface of finite dimension.

Other problems of the type considered here can be formulated, such as a quarter-wave electric monopole attached to the edge of a half sheet, and a quarter-wave notch antenna cut in a half sheet. These problems were discussed in the first edition of this book, and they will be omitted in this edition to restrain the size of this new edition.

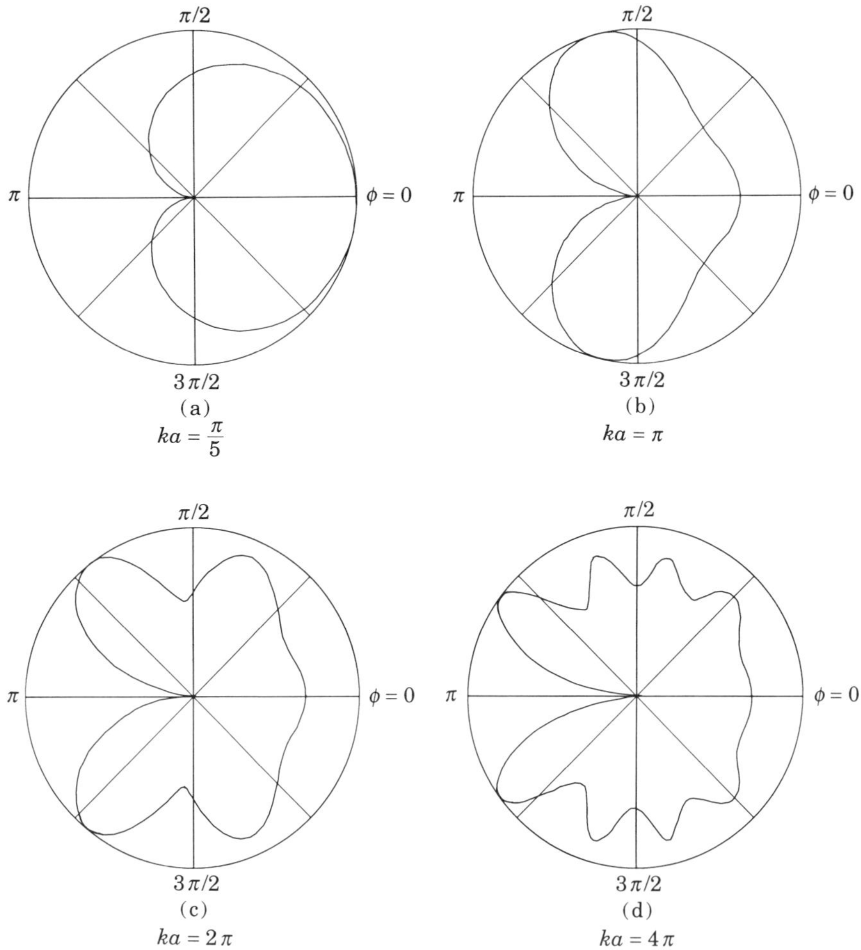

Fig. 9-11 Radiation pattern of a two-sided longitudinal slot

9-6 DIFFRACTION OF A PLANE WAVE BY A HALF SHEET

The conventional method of treating the diffraction of a plane electromagnetic wave by a simple geometrical body is to expand the plane wave in terms of the proper vector wave functions pertaining to the body and then find the appropriate vector wave functions for the diffracted field such that both the radiation condition and the boundary conditions at the surface of the body are satisfied. An alternative method, which is usually more clumsy, is to determine the far-zone field of an electric dipole placed in the neighborhood of the diffracting body; then by means of the reciprocity theorem the solution for the diffraction

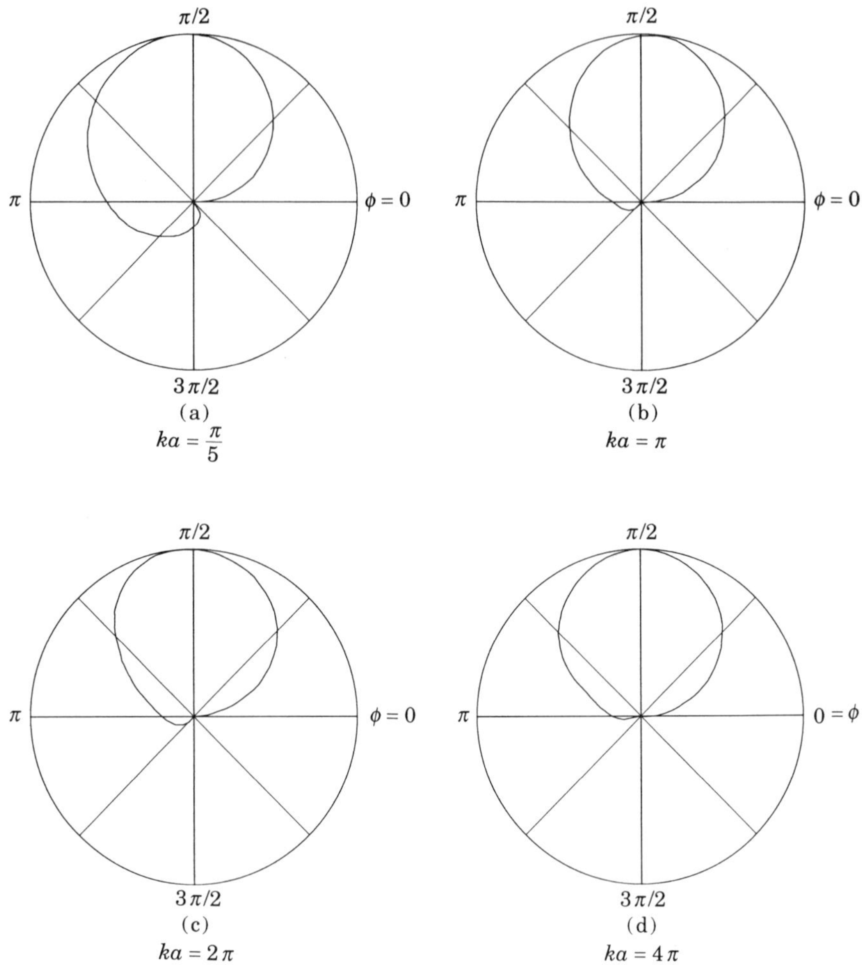

Fig. 9-12 Radiation pattern of a one-sided horizontal slot

of a plane electromagnetic wave by the same body can be found. In the second method we must have at our disposal the far-zone field of a dipole with an arbitrary orientation with respect to the diffracting body. The dyadic Green function technique actually comprises both these formulations. Without applying the reciprocity theorem, let us examine the diffraction problem for a half sheet as an independent problem within the framework of our general formulation.

We consider an electric dipole which is perpendicular to the radial vector $\bar{R}_0$ as shown in Fig. 9-16. As $\bar{R}_0$ recedes to infinity from the origin, the primary field of the dipole would degenerate into a plane wave.

The problem shown in Fig. 9-16 is similar to those treated in Sec. 9-3 except that we demand a certain orientation of the dipole with respect to its position

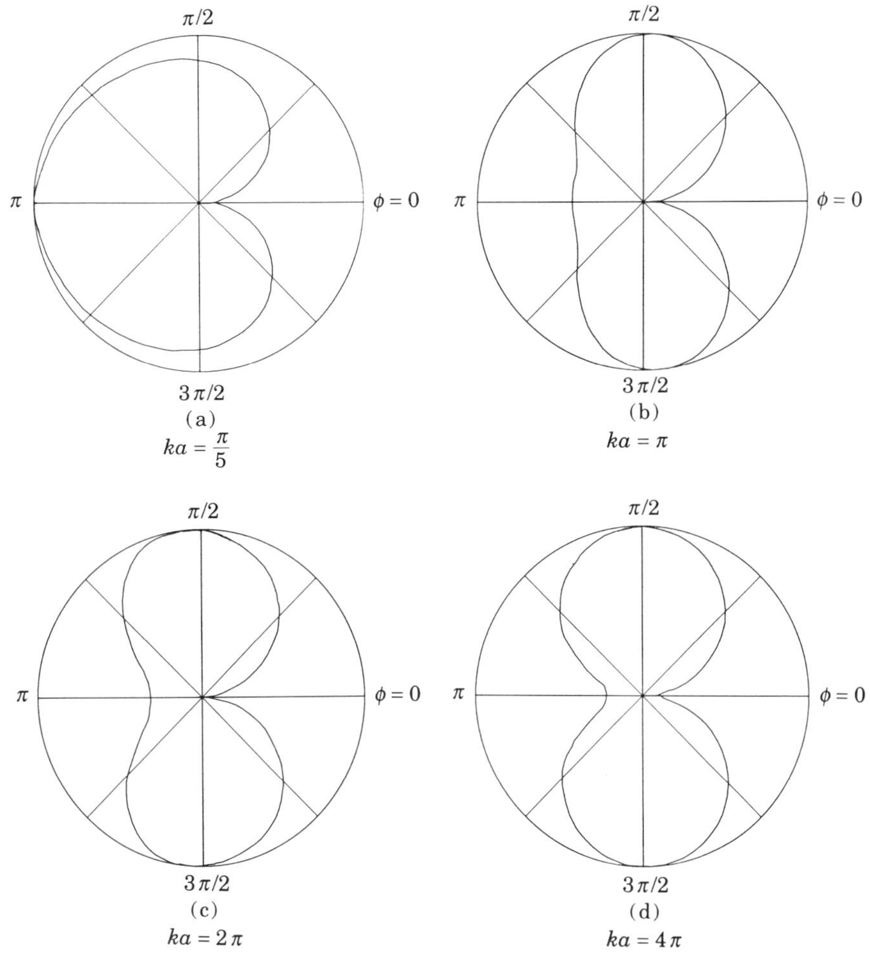

Fig. 9-13 Radiation pattern of a two-sided horizontal slot

vector. To identify the plane wave field with the field of a dipole measured at a large distance, we go back to (3.87). According to that formula the far-zone of a field of a dipole with current moment $\bar{c}$ as observed in a broadside direction can be written as

$$\overline{E} = \frac{i\omega\mu_0\bar{c}e^{ikR_d}}{4\pi R_d}, \tag{9.48}$$

where R_d is the perpendicular distance measured from the dipole. If we let

$$R_d = R_0 + d$$

and assume $d \ll R_0$, then

$$\overline{E} = \left(\frac{i\omega\mu_0 ce^{ikR_0}}{4\pi R_0}\right)\hat{c}e^{ikd} = \left(\frac{i\omega\mu_0 ce^{ikR_0}}{4\pi R_0}\right)\hat{c}e^{i\bar{k}\cdot\overline{R}}, \tag{9.49}$$

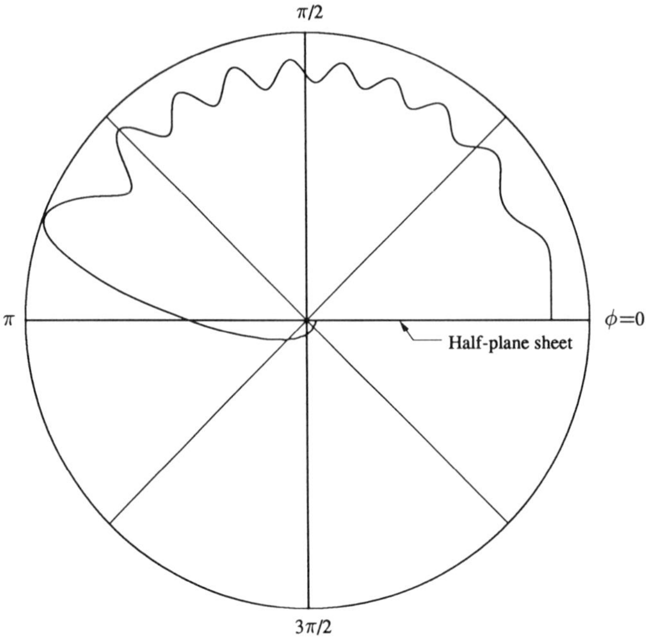

Fig. 9-14 Radiation pattern of a longitudinal slot placed at $ka = 30$

where the relations between R_d, R_0, d, and $\overline{R}$ are shown graphically in Fig. 9-17. When R_0 is very large, we designate the quantity inside the brackets of (9.49) as the amplitude of an incident plane wave propagating in the direction $\hat{k}$; that is,

$$\overline{E} = E_0 \hat{c} e^{i\overline{k}\cdot\overline{R}}, \tag{9.50}$$

where

$$E_0 = \frac{i\omega\mu_0 c e^{ikR_0}}{4\pi R_0}.$$

We now consider the asymptotic solution of the electric field resulting from

$$\overline{E}(\overline{R}) = i\omega\mu_0 \iiint \overline{\overline{G}}_{e1}(\overline{R}, \overline{R}') \cdot \overline{J}(\overline{R}')\, dV', \tag{9.51}$$

with $\overline{\overline{G}}_{e1}(\overline{R}, \overline{R}')$ given by (9.12) under the condition that $r < r'$. The analysis is very similar to the one covered in Sec. 9-3 except that the roles of $\overline{R}$ and $\overline{R}'$ are now interchanged. For illustration, let us treat the case that

$$\overline{J}(\overline{R}') = c\frac{\delta\,(r' - R_0)\,\delta\,(\phi - \phi_0)\,\delta\,(z' - 0)}{r'}\hat{z};$$

then

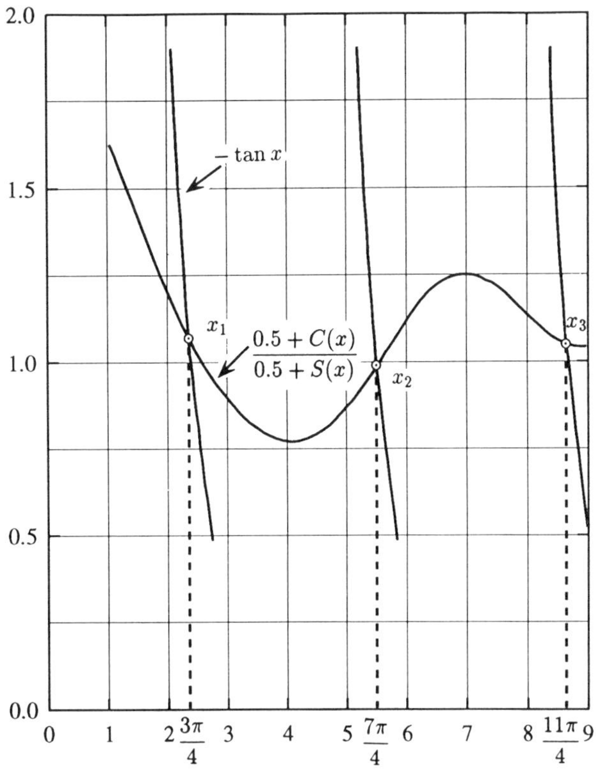

Fig. 9-15 Roots of the equation $\frac{0.5+C(x)}{0.5+S(x)} = -\tan x$

$$\overline{E}(\overline{R}) = -\frac{\omega\mu_0 c}{8\pi k}\int_{-\infty}^{\infty}\sum_{\nu=0}^{\infty}\lambda^2\overline{N}_{0\nu\eta}(h)H_\nu^{(1)}(\lambda R_0)\sin\nu\phi_0\,dh, \tag{9.52}$$

where

$$\eta^2 = k^2 - h^2$$
$$\nu = \frac{n}{2}, n = 1, 2, \ldots.$$

For large values of kR_0, the asymptotic expression for (9.52) is given by

$$\overline{E}(\overline{R}) = \frac{i\omega\mu_0 c e^{ikR_0}}{2\pi R_0}\sum_{n=1}^{\infty}(-i)^{\frac{n}{2}}\sin\frac{n\phi_0}{2}\sin\frac{n\phi}{2}J_{\frac{n}{2}}(kr)\hat{z}. \tag{9.53}$$

Of course, if we invoke the reciprocity theorem, (9.53) can be obtained from (9.21) by putting $\theta = \pi/2$ in that equation, and substituting for R and a, respectively, by R_0 and r. In view of (9.51), (9.53) can be converted to

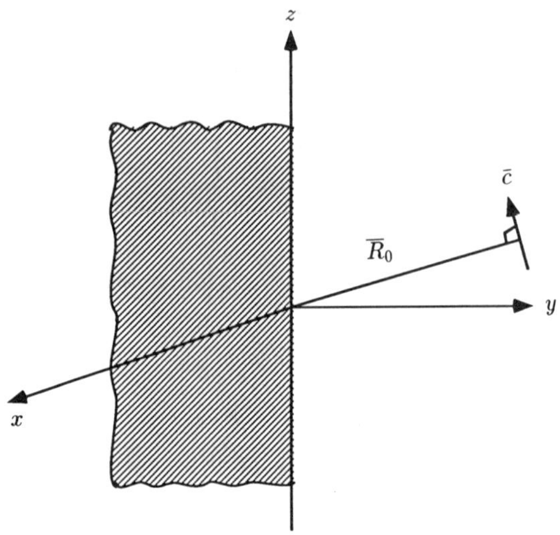

Fig. 9-16 An electric dipole oriented in a direction perpendicular to a radial vector $\overline{R}_0$

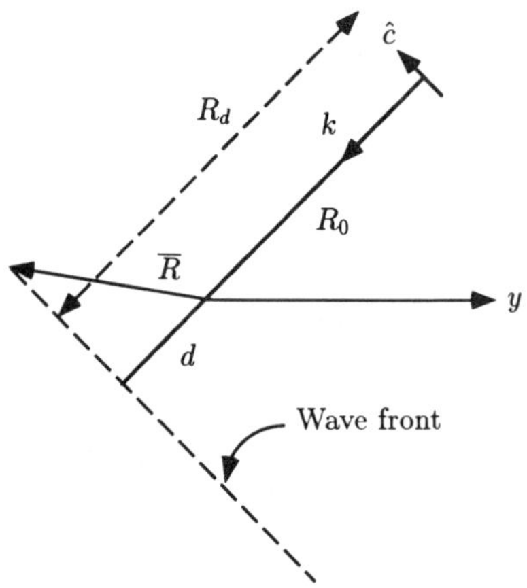

Fig. 9-17 Relations between various position vectors

$$
\begin{aligned}
E_z(\overline{R}) &= \frac{i\omega\mu_0 c e^{ikR_0}}{2\pi R_0} \sum_{n=1}^{\infty} (-i)^{\frac{n}{2}} \sin\frac{n\phi_0}{2} \sin\frac{n\phi}{2} J_{\frac{n}{2}}(kr) \\
&= E_0 \left[S(kr, \phi - \phi_0) - S(kr, \phi + \phi_0) \right].
\end{aligned}
\tag{9.54}
$$

This represents the total electric field resulting from the incidence of a plane wave

$$\overline{E}_i = E_0 e^{-ikr\cos(\phi-\phi_0)}\hat{z}$$

on the half sheet. Because our method of approach is different from Sommerfeld's original formulation [Sommerfeld, 1954, p. 249], our final expression also appears different in form although they are equivalent. In addition to Sommerfeld's original and masterful treatment of this subject, the problem under consideration was also covered very thoroughly by Baker and Copson [1950], particularly from the point of view of geometrical theory of diffraction. For completeness, we shall repeat some of these discussions but based on our formulation. For large values of x, the Fresnel integrals defined by (9.19) have the asymptotic form

$$C(x) + iS(x) = \frac{1}{2}(1+i) - \frac{ie^{ix}}{\sqrt{2\pi x}}, \quad x \gg 1.$$

The asymptotic formula for the function $S(\rho, \phi)$ defined by (9.20) is therefore given by

$$S(\rho, \pi) = \begin{cases} e^{-i\rho\cos\phi} - \frac{e^{i(\rho+\frac{\pi}{4})}}{2\sqrt{\pi\rho(1+\cos\phi)}}, & \pi > \phi \geq 0 \\ \frac{e^{i(\rho+\pi/4)}}{2\sqrt{\pi\rho(1+\cos\phi)}}, & 2\pi \geq \phi > \pi. \end{cases} \tag{9.55}$$

It is understood that (9.55) is valid only if $\rho(1 + \cos\phi) \gg 1$. Thus it cannot be applied to the region where ϕ is near π. To discuss the asymptotic expression for (9.54), we shall divide the region of observation into three distinct zones as shown in Fig. 9-18.

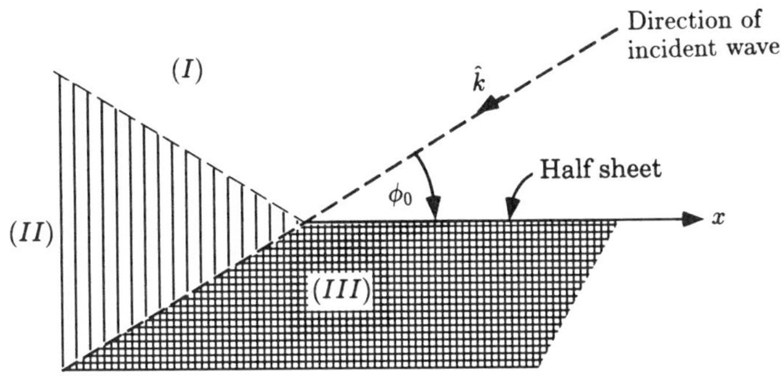

Fig. 9-18 Three distinct zones

(I) Reflection Zone; $\phi < \pi - \phi_0$

In this zone

$$S(kr, \phi - \phi_0) = e^{-ikr\cos(\phi-\phi_0)} - \frac{e^{i(kr+\pi/4)}}{2\sqrt{2\pi kr}\cos\left(\frac{\phi-\phi_0}{2}\right)}$$

$$S(kr, \phi + \phi_0) = e^{-i(kr\cos(\phi+\phi_0)} - \frac{e^{i(kr+\pi 4)}}{2\sqrt{2\pi kr}\cos\left(\frac{\phi+\phi_0}{2}\right)},$$

provided that $kr\,[1+\cos(\phi+\phi_0)] \gg 1$. The inequality implies that ϕ cannot be too close to $\pi-\phi_0$. Substituting these two expressions into (9.54) we obtain

$$E_z(\overline{R}) = E_0\left[e^{-ikr\cos(\phi-\phi_0)} - e^{-ikr\cos(\phi+\phi_0)} + \frac{e^{i(kr+\frac{\pi}{4})}}{2\sqrt{2\pi kr}} \cdot\left(\frac{1}{\cos\frac{\phi+\phi_0}{2}} - \frac{1}{\cos\frac{\phi-\phi_0}{2}}\right)\right]. \tag{9.56}$$

The first term in the above expression is simply the incident wave. The second is the same as the reflected wave from a perfectly conducting full-sheet. The remaining term is the diffracted field attributed to the edge of the half sheet. It has the form of a cylindrical wave emerging from the edge.

(II) Interference Zone: $\pi-\phi_0 < \phi < \pi+\phi_0$

In this zone

$$S(kr,\phi-\phi_0) = e^{-ikr\cos(\cos\,-\,\cos_0)} - \frac{e^{i(kr+\pi 4)}}{2\sqrt{2\pi kr}\cos\frac{\phi-\phi_0}{2}}$$

$$S(kr,\phi+\phi_0) = \frac{e^{i(kr+\frac{\pi}{4})}}{2\sqrt{2\pi kr}\left|\cos\frac{\phi+\phi_0}{2}\right|} = -\frac{e^{i(kr+\frac{\pi}{4})}}{2\sqrt{s\pi kr}\cos\frac{\phi+\phi_0}{2}},$$

provided that $kr\,[1+\cos(\phi+\pm\phi_0)] \gg 1$. Thus ϕ cannot be too close to $\pi-\phi_0$ or $\pi+\phi_0$. Substituting these expressions into (9.54), we obtain

$$E_z(\overline{R}) = E_0\left[e^{-ikr\cos(\phi-\phi_0)} + \frac{e^{i(kr+\frac{\pi}{4})}}{2\sqrt{2\pi kr}}\left(\frac{1}{\cos\frac{\phi+\phi_0}{2}} - \frac{1}{\cos\frac{\phi-\phi_0}{2}}\right)\right]. \tag{9.57}$$

We have only the direct wave and the diffracted cylindrical wave in this zone.

(III) Shadow Zone; $\pi+\phi_0 < \phi$

In this zone

$$S(kr,\phi-\phi_0) = \frac{e^{i(kr+\frac{\pi}{4})}}{2\sqrt{2\pi kr}\left|\cos\frac{\phi-\phi_0}{2}\right|} = -\frac{e^{i(kr+\frac{\pi}{4})}}{2\sqrt{2\pi kr}\cos\frac{\phi-\phi_0}{2}}$$

$$S(kr,\phi+\phi_0) = \frac{e^{i(kr+\frac{\pi}{4})}}{2\sqrt{2\pi kr}\left|\cos\frac{\phi+\phi_0}{2}\right|} = -\frac{e^{i(kr+\frac{\pi}{4})}}{2\sqrt{2\pi kr}\cos\frac{\phi+\phi_0}{2}},$$

provided that ϕ is not near $\pi+\phi_0$. In this case, we have

$$E_z(\overline{R}) = \frac{e^{i(kr+\frac{\pi}{4})}}{2\sqrt{2\pi kr}}\left(\frac{1}{\cos\frac{\phi+\phi_0}{2}} - \frac{1}{\cos\frac{\phi-\phi_0}{2}}\right). \tag{9.58}$$

Equations (9.56) to (9.58) are identical to the ones previously given by Sommerfeld and Baker and Copson.

As we have emphasized, these asymptotic expressions are valid approximations only if ϕ is not near $\pi - \phi_0$ or $\pi + \phi_0$. To be more specific, we can define two parabolic contours such that

$$kr\,[1 + \cos(\phi + \phi_0)] = K$$
$$kr\,[1 + \cos(\phi - \phi_0)] = K,$$

where K is a positive constant equal to or greater than 10. Then (9.56)–(9.58) are good approximations when the point of observation (r, ϕ) lies outside of these contours. Figure 9-19 shows two typical contours corresponding to $K = 4\pi$ so that

$$\frac{r}{\lambda}\,[1 + \cos(\phi \pm \phi_0)] = 2.$$

The three distinct zones are now more clearly defined in this figure. When the point of observation lies inside the parabolic regions, we must use the exact expressions as described by (9.54) to evaluate the field. Near the edge of the sheet, it is sufficient to keep the leading terms of the series expansion of the Fresnel integral given by

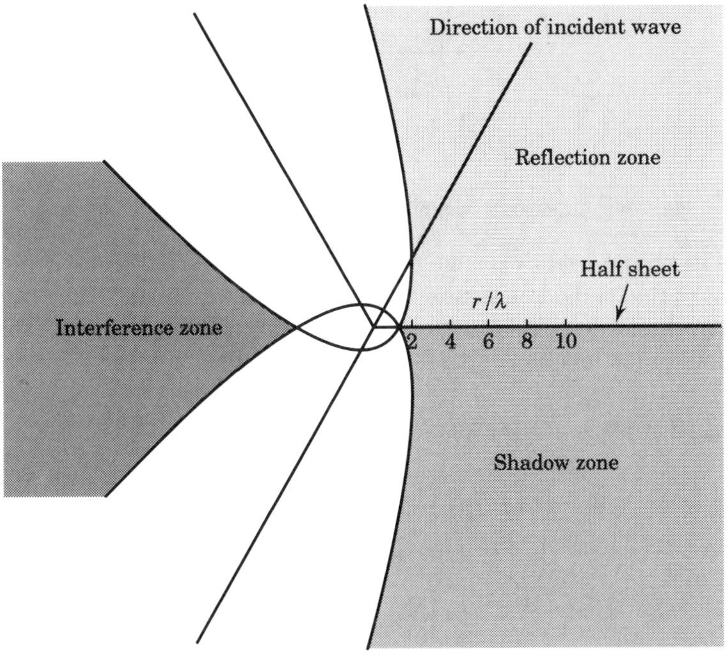

Fig. 9-19 Contours defining the various regions of the asymptotic solutions

$$C(x) + iS(x) = \frac{2x}{\pi}^{\frac{1}{2}} \sum_{n=0}^{\infty} \frac{(ix)^n}{n!(2n+1)} \tag{9.59}$$

for the numerical calculation of the field intensity. The so-called "edge condition" that characterizes the behavior of an electromagnetic field in the neighborhood of the sharp edge of a conductor can be investigated by this approach.

9-7 CIRCULAR CYLINDER AND HALF SHEET

The dyadic Green functions which we have derived for the half sheet can be generalized to include the effect of a cylinder, conducting or dielectric, mounted at the edge of the sheet as shown in Fig. 9-20.

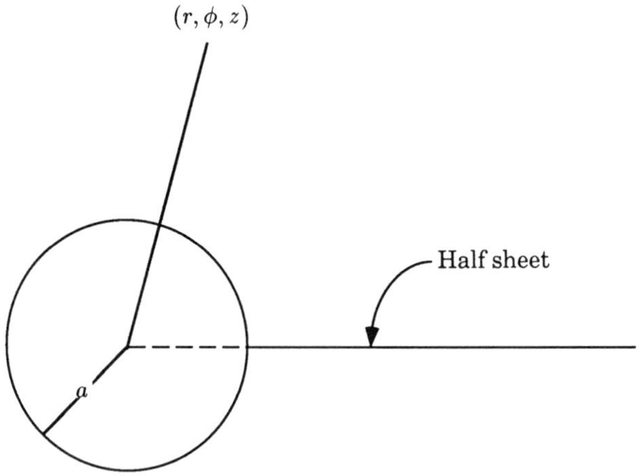

Fig. 9-20 A composite body made of a circular cylinder and a half sheet

For illustration, let us assume the cylinder to be perfectly conducting, then by means of the method of scattering superposition we can find the function of the first or the second kind for the composite body without difficulty. The final expression for the function of the first kind is given by

$$\begin{aligned}\overline{\overline{G}}_{e1}(\overline{R}, \overline{R}') = &-\frac{1}{k^2}\hat{r}\hat{r}\delta(\overline{R} - \overline{R}') + \frac{i}{8\pi}\int_{-\infty}^{\infty} dh \sum_{\nu=0}^{\infty} \frac{2-\delta_0}{\eta^2} \\ &\cdot \begin{cases} \overline{M}^{(1)}_{e\nu\eta}(h)\left[\overline{M}'_{e\nu\eta}(-h) + \alpha\overline{M}'^{(1)}_{e\nu\eta}(-h)\right] \\ \left[\overline{M}_{e\nu\eta}(h) + \alpha\overline{M}^{(1)}_{e\nu\eta}(h)\right]\overline{M}'^{(1)}_{e\nu\eta}(-h) \end{cases} \\ &\begin{rcases} +\overline{N}^{(1)}_{o\nu\eta}(h)\left[\overline{N}'_{o\nu\eta}(-h) + \beta\overline{N}'^{(1)}_{o\nu\eta}(-h)\right] \\ +\left[\overline{N}_{o\nu\eta}(h) + \beta\overline{N}^{(1)}_{o\nu\eta}(h)\right]\overline{N}'^{(1)}_{o\nu\eta}(-h) \end{rcases} \begin{matrix} r > r' \\ r < r', \end{matrix}\end{aligned} \tag{9.60}$$

where

$$\alpha = -\frac{[\partial J_\nu(x)]/\partial x}{[\partial H_\nu^{(1)}(x)]/\partial x}, \quad x = \eta a$$
$$\beta = -J_\nu(\eta a)/H_\nu^{(1)}(\eta a)$$
$$\nu = \frac{n}{2}, \quad n = 0, 1, 2, \ldots$$
$$\eta^2 = k^2 - h^2.$$

It is obvious that the same technique can be applied to composite bodies made of a conducting wedge and a cylinder and a cylindrical waveguide partitioned by a conducting wedge. In the next chapter we shall discuss the same for a body made of a sphere and a conducting cone.

10

Spheres and Perfectly Conducting Cones

The construction of the dyadic Green functions pertaining to spheres, perfectly conducting or dielectric, and perfectly conducting cones follows the same procedure used for the cylinder, except that the spherical and the conical vector wave functions are used. While the rectangular and the cylindrical vector wave functions are closely related to the Hertzian potentials, the spherical and the conical ones are akin to the Debye potentials. Once the orthogonal relations of these functions are known, the remaining task is straightforward.

In order to appreciate the elegance of the Ohm-Rayleigh method which we have used so far in finding various dyadic Green functions, we shall also present in this chapter an alternative, but rather tedious, algebraic method of deriving the eigenfunction expansion of the free-space dyadic Green function outside the source region. This method avoids the necessity of discussing the orthogonal properties of the vector wave functions, but its execution depends heavily upon our recognition of the intricate recurrence relations between various types of vector wave functions which are all generated from the same scalar wave functions but with different piloting vectors. The singular term, however, cannot be extracted from this method.

10-1 EIGENFUNCTION EXPANSION OF THE FREE-SPACE DYADIC GREEN FUNCTIONS

The eigenfunctions which are solutions of the scalar wave equation $\nabla^2\psi+\kappa^2\psi = 0$ in spherical coordinate systems and finite at origin can be written in the form

$$\psi_{\substack{e\\o}mn}(\kappa) = j_n(\kappa R) P_n^m(\cos\theta) \frac{\cos}{\sin} m\phi, \tag{10.1}$$

where $j_n(x)$ denotes the spherical Bessel function of order n which satisfies the differential equation

$$\frac{d^2}{dx^2}[x j_n(x)] + \left[1 - \frac{n(n+1)}{x^2}\right] x j_n(x) = 0.$$

The spherical Bessel function is related to the half-order cylindrical Bessel function by

$$j_n(x) = \left(\frac{\pi}{2x}\right)^{\frac{1}{2}} J_{n+1/2}(x). \tag{10.2}$$

Later on, we need the spherical Hankel function of the first kind, denoted by $h_n^{(1)}(x)$, which is related to the cylindrical Hankel function of the first kind in the same way. That is,

$$h_n^{(1)}(x) = \left(\frac{\pi}{2x}\right)^{\frac{1}{2}} H_{n+1/2}^{(1)}(x). \tag{10.3}$$

The function $P_n^m(\cos\theta)$ in (10.1) denotes the associated Legendre functions of order (n, m). It satisfies the differential equation

$$\frac{1}{\sin\theta}\frac{d}{d\theta}\left[\sin\theta \frac{dP_n^m(\cos\theta)}{d\theta}\right] + \left[n(n+1) - \frac{m^2}{\sin^2\theta}\right] P_n^m(\cos\theta) = 0. \tag{10.4}$$

A very thorough description of the general properties of the spherical Bessel functions and the associated Legendre functions is found in Stratton [1941, pp. 399–411] and will not be repeated here. For problems involving a sphere, n and m are integers. Later on when we apply these functions to conical structures, n, in general, will be fractional. In that case we shall use a different notation. Until then, it is understood that n as well as m represent integers. As was first shown by Stratton [p. 415], two sets of spherical vector wave functions can be defined which are solutions of the vector equation $\nabla \times \nabla \times \overline{F} - \kappa^2 \overline{F} = 0$. They are

$$\overline{M}_{\substack{e\\o}mn}(\kappa) = \nabla \times \left(\psi_{\substack{e\\o}mn} \overline{R}\right) \tag{10.5}$$

$$\overline{N}_{\substack{e\\o}mn}(\kappa) = \frac{1}{\kappa} \nabla \times \nabla \times \left(\psi_{\substack{e\\o}mn} \overline{R}\right). \tag{10.6}$$

These functions, like the other vector wave functions introduced before, satisfy the symmetrical relations

$$\overline{N}_{\substack{e\\o}mn}(\kappa) = \frac{1}{\kappa} \nabla \times \overline{M}_{\substack{e\\o}mn}(\kappa) \tag{10.7}$$

$$\overline{M}_{\substack{e\\o}mn}(\kappa) = \frac{1}{\kappa} \nabla \times \overline{N}_{\substack{e\\o}mn}(\kappa) \tag{10.8}$$

The complete expressions of these functions are given by

$$\overline{M}_{\substack{e\\o}mn}(\kappa) = \mp \frac{m}{\sin\theta} j_n(\kappa R) P_n^m(\cos\theta) \begin{matrix}\sin\\ \cos\end{matrix} m\phi\hat{\theta} - j_n(\kappa R)\frac{\partial P_n^m(\cos\theta)}{\partial\theta} \begin{matrix}\cos\\ \sin\end{matrix} m\phi\hat{\phi} \tag{10.9}$$

$$\overline{N}_{\substack{e\\o}mn}(\kappa) = \frac{n(n+1)}{\kappa R} j_n(\kappa R) P_n^m(\cos\theta) \begin{matrix}\cos\\ \sin\end{matrix} m\phi\hat{R} + \frac{1}{\kappa R}\frac{\partial}{\partial R}\left[R j_n(\kappa R)\right]\left[\frac{\partial P_n^m(\cos\theta)}{\partial\theta} \begin{matrix}\cos\\ \sin\end{matrix} m\phi\hat{\theta} \mp \frac{m}{\sin\theta} P_n^m(\cos\theta) \begin{matrix}\sin\\ \cos\end{matrix} m\phi\hat{\phi}\right]. \tag{10.10}$$

It should be pointed out that the θ- and ϕ-components of these functions have the radial function, either

$$j_n(\kappa R) \text{ or } \frac{1}{\kappa R}\frac{\partial}{\partial R}\left[R j_n(\kappa R)\right],$$

as their constituent. This characteristic has an important bearing on the boundary conditions which must be satisfied by an electromagnetic field at the surface of a sphere. The two sets of spherical vector wave functions thus defined have the following orthogonal relations:

$$\iiint \overline{M}_{\substack{e\\o}mn}(\kappa) \cdot \overline{N}_{\substack{e\\o}m'n'}(\kappa')\, dV = 0 \tag{10.11}$$

$$\iiint \overline{M}_{\substack{e\\o}mn}(\kappa) \cdot \overline{M}_{\substack{e\\o}m'n'}(\kappa')\, dV = \begin{cases} 0, & m \neq m', n \neq n' \\ \frac{(1+\delta_0)\pi^2 n(n+1)(n+m)!}{\kappa^2(2n+1)(n-m)!}\delta(\kappa-\kappa'), & m = m', n = n' \end{cases} \tag{10.12}$$

$$\delta_0 = \begin{cases} 1, m = 0 \\ 0, m \neq 0 \end{cases}$$

$$\iiint \overline{N}_{\substack{e\\o}mn}(\kappa) \cdot \overline{N}_{\substack{e\\o}m'n'}(\kappa') dV = \begin{cases} 0, & m \neq m', n \neq n' \\ \frac{(1+\delta_0)\pi^2 n(n+1)(n+m)!}{\kappa^2(2n+1)(n-m)!}\delta(\kappa-\kappa'), & m = m', n = n'. \end{cases} \tag{10.13}$$

The domain of integration is understood to be through the entire space. The proof of these relations for $m \neq m'$ and $n \neq n'$ is relatively simple because of the orthogonal properties of the trigonometric functions and those of the associated Legendre functions. It is sufficient for us to demonstrate the cases where $m = m'$ and $n = n'$. We consider, for example, the integral

$$I = \iiint \overline{N}_{emn}(\kappa) \cdot \overline{N}_{emn}(\kappa')dV$$
$$= \int_{-}^{\infty} \int_{0}^{\pi} \int_{0}^{2\pi} \overline{N}_{emn}(\kappa) \cdot \overline{N}_{emn}(\kappa')R^2 \sin\theta \, dR \, d\theta \, d\phi. \qquad (10.14)$$

Substituting the explicit expression for $\overline{N}_{emn}$ as given by (10.10) into (10.14) and performing the ϕ-integration we obtain

$$1 = \frac{(1+\delta_0)\pi}{\kappa\kappa'} \int_0^{\infty} \int_0^{\infty} \left\{ n^2(n+1)^2 j_n(\kappa R) j_n(\kappa' R) \left[P_n^m(\cos\theta)\right]^2 \right.$$
$$\left. + \frac{\partial}{\partial R}\left[R j_n(\kappa R)\right] \frac{\partial}{\partial R}\left[R j_n(\kappa' R)\right] \left[\left(\frac{\partial P_n^m}{\partial \theta}\right)^2 + \left(\frac{m P_n^m}{\sin\theta}\right)^2 \right] \right\} \sin\theta \, dR \, d\theta.$$

It is known [Stratton, 1941, pp. 403, 417] that

$$\int_0^{\pi} \left[P_n^m(\cos\theta)\right]^2 \sin\theta \, d\theta = \frac{2}{2n+1} \frac{(n+m)!}{(n-m)!}$$
$$\int_0^{\pi} \left[\left(\frac{\partial P_n^m}{\partial \theta}\right)^2 + \left(\frac{n P_n^m}{\sin\theta}\right)^2 \right] \sin\theta \, d\theta$$
$$= \frac{2n(n+1)}{2n+1} \frac{(n+m)!}{(n-m)!}.$$

Thus, after the integration with respect to θ, we have

$$I = \frac{2(1+\delta_0)\pi n(n+1)}{\kappa\kappa'(2n+1)} \frac{(n+m)!}{(n-m)!} \int_0^{\infty} \left\{ n(n+1) j_n(\kappa R) j_n(k' R) \right.$$
$$\left. + \frac{\partial}{\partial R}\left[R j_n(\kappa R)\right] \frac{\partial}{\partial R}\left[R j_n(\kappa' R)\right] \right\} dR. \qquad (10.15)$$

From the recurrence for the Bessel functions, we can deduce the following formulas for the spherical Bessel functions:

$$\frac{d}{dx}\left[x j_n(x)\right] = \frac{x}{2n+1}\left[(n+1) j_{n-1}(x) - n j_{n+1}(x)\right] \qquad (10.16)$$
$$j_n(x) = \frac{x}{2n+1}\left[j_{n-1}(x) + j_{n+1}(x)\right]. \qquad (10.17)$$

Using these relations, the integrand in (10.15) can be changed to

$$\frac{R^2}{2n+1}\left[(n+1) j_{n-1}(\kappa R) j_{n-1}(\kappa' R) + n j_{n+1}(\kappa R) j_{n+1}(\kappa' R)\right].$$

In view of the integral representation of the weighted delta function in the three-dimensional case as stated by (1.84), we have

$$\int_0^\infty R^2 j_{n-1}(\kappa R) j_{n-1}(\kappa' R) dR = \frac{\pi\delta(\kappa-\kappa')}{2\kappa^2}$$

and

$$\int_0^\infty R^2 j_{n+1}(\kappa R) j_{n+1}(\kappa' R) dR = \frac{\pi\delta(\kappa-\kappa')}{2\kappa^2}.$$

Hence (10.15) reduces to

$$I = (1+\delta_0)\pi^2 \frac{n(n+1)}{(2n+1)} \frac{(n+m)!}{(n-m)!} \frac{\delta(\kappa-\kappa')}{\kappa^2}, \tag{10.18}$$

which is the normalization factor found in (10.13). The proof for (10.12) follows the same procedure with some minor variations in detail. Knowing the proper vector wave functions to be used and their orthogonal properties, we can find the eigenfunction expansion of the free-space dyadic Green function by means of the Ohm-Rayleigh method in the same manner as described in Chapter 7. The only difference is that we now have two discrete sets of eigenvalues, m and n, and one set of continuous eigenvalues, κ. Thus we let

$$\begin{aligned} &\nabla \times \left[\bar{\bar{I}}\delta(\bar{R}-\bar{R}')\right] \\ &= \int_0^\infty d\kappa \sum_{m,n} \left[\bar{N}_{\substack{e\\o}mn}(\kappa)\bar{A}_{\substack{e\\o}mn}(\kappa) + \bar{M}_{\substack{e\\o}mn}(\kappa)\bar{B}_{\substack{e\\o}mn}(\kappa)\right] \end{aligned} \tag{10.19}$$

where n starts with unity and m starts with zero because for $n = 0$ and $m = 0$ the functions $\bar{M}$ and $\bar{N}$ are null. By taking the anterior scalar product of (10.19) with $\bar{M}_{\substack{e\\o}m'n'}(\kappa')$ and $\bar{N}_{\substack{e\\o}m'n'}(\kappa')$ in turn and integrating the resultant equations in the entire space, we find as a result of the orthogonal properties of the spherical vector wave functions that

$$\begin{aligned} \bar{A}_{\substack{e\\o}mn}(\kappa) &= \frac{C_{mn}}{2\pi^2}\kappa^2\nabla' \times \bar{N}'_{\substack{e\\o}mn}(\kappa) \\ &= \frac{C_{mn}}{2\pi^2}\kappa^3 \bar{M}'_{\substack{e\\o}mn}(\kappa) \\ \bar{B}_{\substack{e\\o}mn}(\kappa) &= \frac{C_{mn}}{2\pi^2}\kappa^2\nabla' \times \bar{M}'_{\substack{e\\o}mn}(\kappa) \\ &= \frac{C_{mn}}{2\pi^2}\kappa^3 \bar{N}'_{\substack{e\\o}mn}(\kappa), \end{aligned}$$

where

$$C_{mn} = (2-\delta_0)\frac{2n+1}{n(n+1)}\frac{(n-m)!}{(n+m)!} \tag{10.20}$$

and the primed functions are defined with respect to (R', θ', ϕ'), the coordinates of the position vector $\bar{R}'$. Equation (10.19) then becomes

$$\begin{aligned} &\nabla \times \left[\bar{\bar{I}}\delta(\bar{R}-\bar{R}')\right] \\ &= \frac{1}{2\pi^2}\int_0^\infty d\kappa \sum_{m,n} C_{mn}\kappa^3 \left[\bar{N}(\kappa)\bar{M}'(\kappa) + \bar{M}(\kappa)\bar{N}'(\kappa)\right]. \end{aligned} \tag{10.21}$$

A condensed notation for the vector wave functions has been used now to simplify the writing by dropping the subscript $_{\substack{e\\o}mn}$. According to the method of $\overline{\overline{G}}_m$, we have

$$\overline{\overline{G}}_{m0}(\overline{R}, \overline{R}') = \frac{1}{2\pi^2}\int_0^\infty d\kappa \sum_{m,n} \frac{C_{mn}\kappa^3}{\kappa^2 - k^2}\left[\overline{N}(\kappa)\overline{M}'(\kappa) + \overline{M}(\kappa)\overline{N}'(\kappa)\right]. \tag{10.22}$$

By writing the dyadics such as $\overline{N}(\kappa)\overline{M}'(\kappa)$ in an operational form

$$\overline{N}(\kappa)\overline{M}'(\kappa) = \overline{\overline{T}}_\kappa\left[j_n(\kappa R) j_n(\kappa R')\right],$$

we have

$$\int_0^\infty \frac{\kappa^3}{\kappa^2 - k^2}\overline{\overline{T}}_\kappa\left[j_n(\kappa R) j_n(\kappa R')\right] d\kappa = \frac{i\pi k^2}{2}\begin{cases} \overline{N}^{(1)}(k)\overline{M}'(k), & R > R' \\ \overline{N}(k)\overline{M}'^{(1)}(k), & R < R', \end{cases}$$

where $\overline{N}^{(1)}(k)$ means that the function is defined with respect to the spherical Hankel function of the first kind; that is,

$$\overline{N}^{(1)}(k) = \overline{N}^{(1)}_{\substack{e\\o}mn}(k) = \frac{1}{k}\nabla \times \nabla \times \left[h_n^{(1)}(kR) P_n^m(\cos\theta) \begin{matrix}\cos\\ \sin\end{matrix} m\phi \overline{R}\right]$$

and similarly for $\overline{M}^{(1)}$. By doing the same reduction for $\overline{M}(\kappa)\overline{N}'(\kappa)$, we obtain finally the eigenfunction expansion for $\overline{\overline{G}}_{m0}$, namely,

$$\overline{\overline{G}}{}^{\pm}_{m0}(\overline{R}, \overline{R}') = \frac{ik^2}{4\pi}\sum_{m,n} C_{mn}\begin{cases} \overline{N}^{(1)}(k)\overline{M}'(k) + \overline{M}^{(1)}(k)\overline{N}'(k), & R > R' \\ \overline{N}(k)\overline{M}'^{(1)}(k) + \overline{M}(k)\overline{N}'^{(1)}(k), & R < R'. \end{cases} \tag{10.23}$$

The function is discontinuous at $R = R'$. The corresponding expression for $\overline{\overline{G}}_{eo}$ is then given by

$$\overline{\overline{G}}_{eo}(\overline{R}, \overline{R}') = -\frac{1}{k^2}\hat{R}\hat{R}\delta(\overline{R} - \overline{R}') + \frac{ik}{4\pi}\sum_{m,n} C_{mn}\begin{cases} \overline{M}^{(1)}(k)\overline{M}'(k) + \overline{N}^{(1)}(k)\overline{N}'(k), & R > R' \\ \overline{M}(k)\overline{M}'^{(1)}(k) + \overline{N}(k)\overline{N}'^{(1)}(k), & R < R'. \end{cases} \tag{10.24}$$

This expression for $\overline{\overline{G}}_{eo}$ will later be used to construct other kinds of electric dyadic Green functions for a spherical body. Before doing so, we would like to describe an alternative approach that can be used to find $\overline{\overline{G}}_{eo}$ outside the source region, that is without the singular term $-\hat{R}\hat{R}\delta(\overline{R} - \overline{R}')k^2$. The analysis is rather tedious, but it does not need the Hankel transform, and it also shows the intimate relations between various types of spherical vector wave functions and many recurrence formulas of the associated Legendre functions not found in existing books.

10-2 AN ALGEBRAIC METHOD OF FINDING $\overline{\overline{G}}_{eo}$ WITHOUT THE SINGULAR TERM

To prepare for the conversion of various vector functions that will be involved in this method, we shall first introduce other types of spherical vector wave functions in addition to the two standard ones defined by (10.5) and (10.6). According to the general discussion of the vector wave functions in the beginning of Sec. 5-1, functions of the type

$$\overline{M}^{(c)} = \nabla \times (\psi \hat{c})$$
$$\overline{N}^{(c)} = \frac{1}{k} \nabla \times \nabla \times (\psi \hat{c})$$

are solutions of the vector wave equation as long as ψ is a solution of the scalar wave equation $\nabla^2 \psi + k^2 \psi = 0$ and the piloting vector $\hat{c}$ is a constant vector. If we identify $\hat{c}$ to be $\hat{x}$ or $\hat{y}$ or $\hat{z}$ and ψ to be the eigenfunction of the scalar wave equation in spherical coordinate systems, that is,

$$\psi = \psi_{\substack{e \\ o}mn}(k) = j_n(kR) P_n^m(\cos\theta) \frac{\cos}{\sin} m\phi,$$

then we can define six more spherical vector wave functions as follows:

$$\overline{M}^{(c)}_{\substack{e \\ o}mn}(k) = \frac{1}{k} \nabla \times \left(\psi_{\substack{e \\ o}mn} \hat{c} \right) \tag{10.25}$$

$$\overline{N}^{(c)}_{\substack{e \\ o}mn}(k) = \frac{1}{k^2} \nabla \times \nabla \times \left(\psi_{\substack{e \\ o}mn} \hat{c} \right), \tag{10.26}$$

with $\hat{c} = \hat{x}$, $\hat{y}$, and $\hat{z}$, respectively. For clarity, we shall call these functions the spherical wave functions of the c-type. The additional constant $1/k$ which is included in these functions makes them of the same dimension as that of $\overline{M}$ or $\overline{N}$ defined by (10.5) and (10.6). By converting the unit vectors $\hat{x}$, $\hat{y}$, and $\hat{z}$ into the unit vectors in the spherical coordinate system with the aid of Table 1-2, we can find the explicit expressions for these functions. We shall now demonstrate some interesting relations between these types of functions and the standard or the radial spherical vector wave functions. These identities are the key relations to be used later in our derivation of the free-space dyadic Green's function based on the algebraic method. We will not prove all these relations in the text, but will give one example in detail. It will show quite clearly our approach in the analysis. We consider, for example, the function

$$\overline{M}^{(x)}_{emn}(k) = \frac{1}{k} \nabla \times (\psi_{emn'} \hat{x}), \tag{10.27}$$

with

$$\psi_{emn} = j_n(kR) P_n^m(\cos\theta) \cos m\phi.$$

Substituting into (10.27)

$$\hat{x} = \sin\theta\cos\phi\hat{R} + \cos\theta\cos\phi\hat{\theta} - \sin\phi\hat{\phi},$$

we obtain

$$\begin{aligned}\overline{M}^{(x)}_{emn}(k) =& \frac{1}{kR}\left\{\frac{-1}{\sin\theta}\left[\frac{\partial}{\partial\theta}(\sin\theta\sin\phi\psi_{emn}) + \frac{\partial}{\partial\phi}(\cos\theta\cos\phi\psi_{emn})\right]\hat{R}\right.\\ &+\left[\frac{1}{\sin\theta}\frac{\partial}{\partial\phi}(\sin\theta\cos\phi\psi_{emn}) + \frac{\partial}{\partial R}(R\sin\phi\psi_{emn})\right]\hat{\theta}\\ &\left.+\left[\frac{\partial}{\partial R}(R\cos\theta\cos\phi\psi_{emn}) - \frac{\partial}{\partial\theta}(\sin\theta\cos\phi\psi_{emn})\right]\hat{\phi}\right\}. \qquad (10.28)\end{aligned}$$

Let us now study the composition of the radial component of this function first, which is given by

$$\begin{aligned}\hat{R}\cdot\overline{M}^{(x)}_{emn} =& \frac{-j_n(\rho)}{\rho\sin\theta}\left[\sin\phi\cos m\phi\frac{d}{d\theta}(\sin\theta P_n^m)\right.\\ &\left.+\cos\theta P_n^m\frac{d}{d\phi}(\cos\phi\cos m\phi)\right], \qquad (10.29)\end{aligned}$$

where ρ denotes kR. Using the sum and difference formulas for the trigonometric functions, we can write (10.29) in the form

$$\begin{aligned}\hat{R}\cdot\overline{M}^{(x)}_{emn} =& \frac{-j_n(\rho)}{2\rho\sin\theta}\left\{\left[\frac{d}{d\theta}(\sin\theta P_n^m) - (m+1)\cos\theta P_n^m\right]\sin(m+1)\phi\right.\\ &\left.-\left[\frac{d}{d\theta}(\sin\theta P_n^m) + (m-1)\cos\theta P_n^m\right]\sin(m-1)\phi\right\}\\ =& \frac{-j_n(\rho)}{2\rho}\left[\left(\frac{dP_n^m}{d\theta} - m\frac{\cos\theta}{\sin\theta}P_n^m\right)\sin(m+1)\phi\right.\\ &\left.-\left(\frac{dP_n^m}{d\theta} + m\frac{\cos\theta}{\sin\theta}P_n^m\right)\sin(m-1)\phi\right]. \qquad (10.30)\end{aligned}$$

For further reduction we use the recurrence relations that

$$\frac{dP_n^m}{d\theta} \pm m\frac{\cos\theta}{\sin\theta}P_n^m = \begin{cases}(n-m+1)(n+m)P_n^{m-1}\\ -P_n^{m+1}\end{cases}$$

which can be obtained by combining the following two formulas found in Stratton [p. 401]:

$$\frac{1}{2}\left[(n-m+1)(n+m)P_n^{m-1} \pm P_n^{m+1}\right] = \begin{cases}\frac{m\cos\theta}{\sin\theta}P_n^m\\ \frac{dP_n^m}{d\theta}.\end{cases}$$

Thus (10.30) can be written in the form

$$\begin{aligned}\hat{R}\cdot\overline{M}^{(x)}_{emn} =& \frac{j_n(\rho)}{2\rho}\left[P_n^{m+1}\sin(m+1)\phi\right.\\ &\left.+(n-m+1)(n+m)P_n^{m-1}\sin(m-1)\phi\right]. \qquad (10.31)\end{aligned}$$

In view of (10.10), this is recognized as the radial component of two odd $\overline{N}$ functions with orders $(n, m+1)$ and $(n, m-1)$, respectively, or

$$\begin{aligned}\hat{R}\cdot\overline{M}^{(x)}_{emn}(k) =&\hat{R}\cdot\frac{1}{2n(n+1)}\left[\overline{N}_{o(m+1)n}(k)\right.\\&\left.+(n+m)(n-m+1)\overline{N}_{0(m-1)n}(k)\right].\end{aligned} \tag{10.32}$$

The identifications of the composition of the θ-component or the ϕ-component of (10.28) is much more intricate. We consider the θ-component which is given by

$$\begin{aligned}\hat{\theta}\cdot\overline{M}^{(x)}_{emn}(k) =&\frac{1}{\rho}\left[j_n P_n^m\frac{d}{d\phi}(\cos\phi\cos m\phi)\right.\\&\left.+\frac{d}{d\rho}(\rho j_n)P_n^m\sin\phi\cos m\phi\right].\end{aligned} \tag{10.33}$$

A reduction of the trigonometric functions yields

$$\begin{aligned}\hat{\theta}\cdot\overline{M}^{(x)}_{emn}(k) =&\frac{P_n^m}{2\rho}\left\{\left[\frac{d(\rho j_n)}{d\rho}-(m+1)j_n\right]\sin(m+1)\phi\right.\\&\left.-\left[\frac{d(\rho j_n)}{d\rho}+(m-1)j_n\right]\sin(m-1)\phi\right\}.\end{aligned} \tag{10.34}$$

This is obviously not equal to the θ-component of the two odd functions contained in (10.32). To reduce it further we need the following recurrence relations for the associated Legendre functions and the spherical Bessel functions:

$$P_n^m = \frac{1}{(2n+1)\sin\theta}\left[P_{n+1}^{m+1}-P_{n-1}^{m+1}\right] \tag{10.35}$$

$$\begin{aligned}P_n^m =&\frac{1}{(2n+1)\sin\theta}\left[-(n-m+2)(n-m+1)P_{n+1}^{m-1}\right.\\&\left.+(n+m)(n+m-1)P_{n-1}^{m-1}\right]\end{aligned} \tag{10.36}$$

$$\frac{dP_n^{m+1}}{d\theta} =\frac{n(n+1)}{(2n+1)\sin\theta}\left[\frac{n-m}{n+1}P_{n+1}^{m+1}-\frac{n+m+1}{n}P_{n-1}^{m+1}\right] \tag{10.37}$$

$$\frac{dP_n^{m-1}}{d\theta} =\frac{n(n+1)}{(2n+1)\sin\theta}\left[\frac{n-m+2}{n+1}P_{n+1}^{m-1}-\frac{n+m-1}{n}P_{n-1}^{m-1}\right] \tag{10.38}$$

$$\frac{j_n}{\rho} =\frac{(\rho j_n)'}{(n+1)\rho}+\frac{j_{n+1}}{n+1} \tag{10.39}$$

$$\frac{j_n}{\rho} =\frac{-(\rho j_n)'}{n\rho}+\frac{j_{n-1}}{n}, \tag{10.40}$$

where $(\rho j_n)'$ denotes the derivatives of ρj_n. Equation (10.35) is one of the relations listed by Stratton [p. 401]. Equations (10.36)–(10.40) can be derived by a proper combination of the pertinent ones found in his list. Equations (10.39) and (10.40) can be derived from (10.16) and (10.17). To reduce (10.34) to the desired recognizable form, the algebra is long and tedious and is outlined below. Using (10.35), (10.36), (10.39), and (10.40), we first change (10.34) to

$$
\begin{aligned}
\hat{\theta}\cdot\overline{M}^{(x)}_{emn} =& \frac{1}{2(2n+1)}\left\{\frac{P^{m+1}_{n+1}}{\sin\theta}\left[\frac{n-m}{n+1}\frac{(\rho j_n)'}{\rho}-\frac{m+1}{n+1}j_{n+1}\right]\right.\\
&\left.-\frac{P^{m+1}_{n-1}}{\sin\theta}\left[\frac{n+m+1}{n}\frac{(\rho j_n)'}{\rho}-\frac{m+1}{n}j_{n-1}\right]\right\}\sin(m+1)\phi\\
&+\frac{1}{2(2n+1)}\left\{(n-m+2)(n-m+1)\frac{P^{m-1}_{n+1}}{\sin\theta}\left[\frac{n+m}{n+1}\frac{(\rho j_n)'}{\rho}\right.\right.\\
&\left.+\frac{m-1}{n+1}j_{n+1}\right]-(n+m)(n+m-1)\frac{P^{m-1}_{n-1}}{\sin\theta}\left[\frac{n-m+1}{n}\frac{(\rho j_n)'}{\rho}\right.\\
&\left.\left.+\frac{m-1}{n}j_{n-1}\right]\right\}\sin(m-1)\phi.
\end{aligned}
\tag{10.41}
$$

We rearrange it in the form

$$
\begin{aligned}
\hat{\theta}\cdot\overline{M}^{(x)}_{emn} =& \frac{1}{2(2n+1)}\left\{-\frac{m+1}{n+1}\frac{P^{m+1}_{n+1}}{\sin\theta}j_{n+1}+\frac{m+1}{n}\frac{P^{m+1}_{n-1}}{\sin\theta}j_{n-1}\right.\\
&\left.+\frac{(\rho j_n)'}{\rho\sin\theta}\left[\frac{n-m}{n+1}P^{m+1}_{n+1}-\frac{n+m+1}{n}P^{m+1}_{n-1}\right]\right\}\sin(m+1)\phi\\
&+\frac{1}{2(2n+1)}\left\{\frac{(m-1)(n-m+2)(n-m+1)}{(n+1)}\frac{P^{m-1}_{n+1}}{\sin\theta}j_{n+1}\right.\\
&-\frac{(m-1)(n+m)(n+m-1)}{n}\frac{P^{m-1}_{n-1}}{\sin\theta}j_{n-1}\\
&+(n+m)(n-m+1)\frac{(\rho j_n)'}{\rho\sin\theta}\left[\frac{(n-m+2)}{n+1}\frac{P^{m-1}_{n+1}}{\sin\theta}\right.\\
&\left.\left.-\frac{(n+m-1)}{n}\frac{P^{m-1}_{n-1}}{\sin\theta}\right]\right\}\sin(m-1)\phi.
\end{aligned}
\tag{10.42}
$$

In view of (10.37) and (10.38), it can be written in the form

$$
\begin{aligned}
\hat{\theta}\cdot\overline{M}^{(x)}_{emn} =& \frac{1}{2}\left\{\frac{m+1}{2n+1}\left[\frac{-1}{n+1}\frac{P_{n+1}^{m+1}}{\sin\theta}j_{n+1}+\frac{1}{n}\frac{P_{n-1}^{m+1}}{\sin\theta}j_{n-1}\right]\right.\\
&\left.+\frac{1}{n(n+1)}\frac{\partial P_n^{m+1}}{\partial\theta}\frac{(\rho j_n)'}{\rho}\right\}\sin(m+1)\phi\\
&+\frac{1}{2}\left\{\frac{m-1}{2n+1}\left[\frac{(n-m+2)(n-m+1)}{n+1}\frac{P_{n+1}^{m-1}}{\sin\theta}j_{n+1}\right.\right.\\
&\left.-\frac{(n+m)(n+m-1)}{n}\cdot\frac{P_{n-1}^{m-1}}{\sin\theta}j_{n-1}\right]\\
&\left.-\frac{(n+m)(n+m-1)}{n(n+1)}\frac{\partial P_n^{m-1}}{\partial\theta}\frac{(\rho j_n)'}{\rho}\right\}\cdot\sin(m-1)\phi. \qquad (10.43)
\end{aligned}
$$

Each term in (10.43) can now be identified with the θ-component of a standard spherical vector wave function. There are six of them. In fact, (10.43) is described more characteristically by

$$
\begin{aligned}
\hat{\theta}\cdot\overline{M}^{(x)}_{emn} =& \hat{\theta}\cdot\left\{\frac{1}{2n(n+1)}\left[\overline{N}_{0(m+1)n}\right.\right.\\
&\left.+(n+m)(n-m+1)\overline{N}_{0(m-1)n}\right]\\
&+\frac{1}{2(n+1)(2n+1)}\left[\overline{M}_{e(m+1)(n+1)}\right.\\
&\left.-(n-m+1)(n-m+2)\overline{M}_{e(m-1)(n+1)}\right]\\
&-\frac{1}{2n(2n+1)}\left[\overline{M}_{e(m+1)(n-1)}\right.\\
&\left.\left.-(n+m-1)(n+m)\overline{M}_{e(m-1)(n-1)}\right]\right\}. \qquad (10.44)
\end{aligned}
$$

It is noted that the two $\overline{N}$-functions in (10.44), including the coefficients, are the same as the ones appearing in (10.32). Since the standard $\overline{M}$-functions do not have a radial component, they are, therefore, not involved in (10.32). If we do the same study for the ϕ-component the result shows that it is of the same form as (10.44). Thus we have established a complete relationship between $\overline{M}^{(x)}_{emn}$ and the six radial vector wave functions that are given by (10.44) with $\left(\hat{\theta}\cdot\right)$ deleted. The relations between all the vector wave functions of the c-type and the standard ones have been found and they are tabulated in Sec. B-3 of Appendix B. Knowing these relations, we can proceed to rederive (10.24) based on the algebraic method. As was shown in Sec. 4-2, the free-space electric dyadic Green function can be written in the form

$$
\overline{\overline{G}}_{eo}(\overline{R},\overline{R}') = \overline{G}^{(x)}_{eo}(\overline{R},\overline{R}')\hat{x}+\overline{G}^{(y)}_{eo}(\overline{R},\overline{R}')\hat{y}+\overline{G}^{(z)}_{eo}(\overline{R},\overline{R}')\hat{z}, \qquad (10.45)
$$

where

$$\overline{\overline{G}}_{eo}^{(c)}(\overline{R}, \overline{R}') = \left(1 + \frac{1}{k^2}\nabla\nabla\cdot\right) G_0(\overline{R}, \overline{R}')\hat{c}, \quad \hat{c} = \hat{x}, \hat{y}, \hat{z} \tag{10.46}$$

$$G_0(\overline{R}, \overline{R}') = \frac{e^{ik|\overline{R}-\overline{R}'|}}{4\pi\left|\overline{R} - \overline{R}'\right|}\cdot \tag{10.47}$$

For $\overline{R} \neq \overline{R}'$, (10.46) is equivalent to

$$\overline{\overline{G}}_{eo}^{(c)}(\overline{R}, \overline{R}') = \frac{1}{k^2}\nabla \times \nabla \times \left[G_0(\overline{R}, \overline{R}')\hat{c}\right]. \tag{10.48}$$

The three-dimensional free-space scalar Green function defined by (10.47) has a series expansion given by

$$G_0(\overline{R}, \overline{R}') = \frac{ik}{4\pi}\sum_{m,n} D_{mn}P_n^m(\cos\theta)P_n^m(\cos\theta')\cos m\,(\phi - \phi') \cdot \begin{cases} h_n^{(1)}(kR)j_n(kR'), & R > R' \\ j_n(kR)h_n^{(1)}(kR'), & R < R', \end{cases} \tag{10.49}$$

where

$$D_{mn} = (2 - \delta_0)(2n + 1)\frac{(n - m)!}{(n + m)!}.$$

This expression is obtained by combining two expressions given by Stratton (Eq. 46, p. 408, and Eq. 87, p. 414). Substituting (10.49) into (10.46) we obtain for $R < R'$

$$G_0^{(c)}(\overline{R}, \overline{R}') = \frac{ik}{4\pi}\sum_{m,n} \overline{A}_{{}^e_o mn}\overline{N}^{(c)}_{{}^e_o mn}, \tag{10.50}$$

where

$$\overline{A}_{{}^e_o mn} = D_{mn}h_n^{(1)}(kR')P_n^m(\cos\theta')\,{}^{\cos}_{\sin}\, m\phi' \tag{10.51}$$

and $\overline{N}^{(c)}_{{}^e_o mn}$ is defined by (10.26). We can now express the $\overline{N}^{(c)}_{{}^e_o mn}$ functions in (10.50) in terms of the radial spherical vector wave functions according to the formulas given in Sec. B-3 of Appendix B. The result can be written in the form

$$\overline{\overline{G}}_{eo}^{(c)}(\overline{R}, \overline{R}') = \frac{ik}{4\pi}\sum_{n=1}^{\infty}\sum_{m=0}^{n}\left(\alpha^{(c)}_{{}^e_o mn}\overline{M}_{{}^e_o mn} + \beta^{(c)}_{{}^e_o mn}\overline{N}_{{}^e_o mn}\right). \tag{10.52}$$

The coefficients $\alpha^{(c)}_{{}^e_o mn}$ and $\beta^{(c)}_{{}^e_o mn}$ are all expressible in terms of $\overline{A}_{{}^e_o mn}$ of different orders. For example, we have

$$\alpha^{(x)}_{\substack{e\\o}mn} = \frac{-1}{2n(n+1)}\left[(1+\delta_1)A_{0(m-1)n}\right.$$
$$\left. + (n+m+1)(n-m)A_{0(m+1)n}\right]$$
$$\beta^{(x)}_{\substack{e\\o}mn} = \frac{1}{2n(2n-1)}\left[(1+\delta_1)A_{e(m-1)(n-1)}\right.$$
$$\left. - (n-m-1)(n-m)A_{e(m+1)(n-1)}\right],$$

where

$$\delta_1 = \begin{cases} 1, & m = 1 \\ 0, & m \neq 1. \end{cases}$$

Substituting (10.52) into (10.45), we find, after another long exercise, that the following identities are true:

$$\alpha^{(x)}_{\substack{e\\o}mn}\hat{x} + \alpha^{(y)}_{\substack{e\\o}mn}\hat{y} + \alpha^{(z)}_{\substack{e\\o}mn}\hat{z} = C_{mn}\overline{M}'^{(1)}_{\substack{e\\o}mn}$$
$$\beta^{(x)}_{\substack{e\\o}mn}\hat{x} + \beta^{(y)}_{\substack{e\\o}mn}\hat{y} + \beta^{(z)}_{\substack{e\\o}mn}\hat{z} = C_{mn}\overline{N}'^{(1)}_{\substack{e\\o}mn}.$$

The coefficient C_{mn} is defined by (10.20); hence for $R < R'$,

$$\overline{\overline{G}}_{eo}(\overline{R}, \overline{R}') = \frac{ik}{4\pi}\sum_{m,n} C_{mn}\left[\overline{M}(k)\overline{M}'^{(1)}(k) + \overline{N}(k)\overline{N}'^{(1)}(k)\right], \qquad (10.53)$$

which is the same as that part of (10.24) without the singular term as derived by the Ohm-Rayleigh method. In retrospect, the preceding exercises demonstrate very convincingly the elegance of the Ohm-Rayleigh method which bypasses all the complicated manipulations involved in the algebraic method. The singular term, of course, cannot be derived from the algebraic method based on $\overline{\overline{G}}_{eo}$ but can be found starting with $\overline{\overline{G}}_{mo}$.

10-3 PERFECTLY CONDUCTING AND DIELECTRIC SPHERES

Based on (10.24), we can construct the various kinds of dyadic Green functions associated with a sphere by means of the method of scattering superposition. Thus, for a perfectly conducting sphere of radius equal to a with its center located at the origin of the coordinate system, we would deal with either the function of the first kind or the function of the second kind. To find the function of the first kind we let

$$\overline{\overline{G}}_{e1}(\overline{R}, \overline{R}') = \overline{\overline{G}}_{eo}(\overline{R}, \overline{R}') + \overline{\overline{G}}_{1s}(\overline{R}, \overline{R}'), \qquad (10.54)$$

where $\overline{\overline{G}}_{eo}(\overline{R}, \overline{R}')$ is given by (10.24). In view of the composition of $\overline{\overline{G}}_{eo}$ the term representing the scattered part must have the form

$$\bar{\bar{G}}_{1s}(\bar{R},\bar{R}') = \frac{ik}{4\pi}\sum_{m,n} C_{mn}\left[a_n\overline{M}^{(1)}(k)\overline{M}'^{(1)}(k) + b_n\overline{N}^{(1)}(k)\overline{N}'^{(1)}(k)\right]. \tag{10.55}$$

The coefficients A and B are determined by applying the Dirichlet boundary condition to $\bar{\bar{G}}_{e1}$ at the surface of the conducting sphere that yields

$$a_n = \frac{-j_n(ka)}{h_n^{(1)}(ka)} \tag{10.56}$$

$$b_n = \frac{(d/d\rho)\left[\rho j_n(\rho)\right]}{(d/d\rho)\left[\rho h_n^{(1)}(\rho)\right]}, \quad \rho = ka. \tag{10.57}$$

Similarly, for the function of the second kind we have

$$\bar{\bar{G}}_{2s}(\bar{R},\bar{R}') = \bar{\bar{G}}_{eo}(\bar{R},\bar{R}') + \bar{\bar{G}}_{2s}(\bar{R},\bar{R}'). \tag{10.58}$$

By a consideration of the symmetrical relationship between $\bar{\bar{G}}_{e1}$ and $\bar{\bar{G}}_{e2}$ we obtain

$$\begin{aligned}\bar{\bar{G}}_{2s}(\bar{R},\bar{R}') =& \frac{ik}{4\pi}\sum_{m,n} C_{mn}\left[b_n\overline{M}^{(1)}(k)\overline{M}'^{(1)}(k)\right.\\ &\left.+ a_n\overline{N}^{(1)}(k)\overline{N}'^{(1)}(k)\right].\end{aligned} \tag{10.59}$$

The same two sets of coefficients, a_n and b_n, are involved, but their roles have been interchanged.

For the functions of the third kind associated with imperfectly conducting or dielectric spheres, it is better to adopt the following notation for the constitutive constant defined in the two regions, exterior and interior to the sphere. We let

$$\begin{aligned} k_1 &= \omega\sqrt{\mu_1\epsilon_1} \quad \text{(region 1, } r \geq a)\\ k_2 &= \omega\sqrt{\mu_2\epsilon_2} \quad \text{(region 2, } r \leq a).\end{aligned}$$

For a dielectric sphere placed in air $\epsilon_1 = \epsilon_0, \epsilon_2 = \epsilon, \mu_1 = \mu_2 = \mu_0$, where ϵ denotes the permittivity of the sphere. For generality, no restriction will be placed on these constants. The constant k which appeared in $\bar{\bar{G}}_{eo}$ as expressed by (10.24) must now be replaced by k_1. With this change of notation, we let

$$\begin{aligned}\bar{\bar{G}}_e^{(11)}(\bar{R},\bar{R}') &= \bar{\bar{G}}_{eo}(\bar{R},\bar{R}') + \bar{\bar{G}}_{es}^{(11)}(\bar{R},\bar{R}'), \quad R \geq a\\ \bar{\bar{G}}_e^{(21)}(\bar{R},\bar{R}') &= \bar{\bar{G}}_{es}^{(21)}(\bar{R},\bar{R}'), \qquad\qquad\qquad R \leq a.\end{aligned}$$

In conformity to the superscript notations introduced in Sec. 4-4, the preceding two equations imply that we are dealing with a source located in region 1. The scattered terms now can be represented by

$$\overline{\overline{G}}_{es}^{(11)}(\overline{R},\overline{R}') = \frac{ik_1}{4\pi}\sum_{m,n} C_{mn}\left[A_n\overline{M}^{(1)}(k_1)\overline{M}'^{(1)}(k_1) + B_n\overline{N}^{(1)}(k_1)\overline{N}'^{(1)}(k_1)\right] \tag{10.60}$$

$$\overline{\overline{G}}_{es}^{(21)}(\overline{R},\overline{R}') = \frac{ik_1}{4\pi}\sum_{m,n} C_{mn}\left[C_n\overline{M}^{(1)}(k_2)\overline{M}'^{(1)}(k_1) + D_n\overline{N}^{(1)}(k_2)\overline{N}'^{(1)}(k_1)\right]. \tag{10.61}$$

Applying the boundary condition at the surface of the sphere as required by the function of the third kind, namely,

$$\left.\begin{aligned} \hat{R}\times\overline{\overline{G}}_e^{(11)} &= \hat{R}\times\overline{\overline{G}}_e^{(21)} \\ \frac{1}{\mu_1}\hat{R}\times\nabla\times\overline{\overline{G}}_e^{(11)} &= \frac{1}{\mu_2}\hat{R}\times\nabla\times\overline{\overline{G}}_e^{(21)} \end{aligned}\right\}, \quad R = a,$$

we find that the coefficients $A, B, C,$ and D must satisfy the following system of equations.

$$j_n(\rho_1) + A_n h_n^{(1)}(\rho_1) = C_n j_n(\rho_2) \tag{10.62}$$

$$\frac{k_1}{\mu_1}\left\{\frac{[\rho_1 j_n(\rho_1)]'}{\rho_1} + A_n\frac{[\rho_1 h_n^{(1)}(\rho_1)]'}{\rho_1}\right\} = \frac{k_2}{\mu_2}\left\{C_n\frac{[\rho_2 j_n(\rho_2)]'}{\rho_2}\right\} \tag{10.63}$$

$$\frac{[\rho_1 j_n(\rho_1)]'}{\rho_1} + B_n\frac{[\rho_1 h_n^{(1)}(\rho_1)]'}{\rho_1} = D_n\frac{[\rho_2 j_n(\rho_2)]'}{\rho_2} \tag{10.64}$$

$$\frac{k_1}{\mu_1}[j_n(\rho_1) + B_n h_n^{(1)}(\rho_1)] = D_n\left(\frac{k_2}{\mu_2}\right)j_n(\rho_2), \tag{10.65}$$

where

$$\rho_1 = k_1 a, \rho_2 = k_2 a$$

and the prime denotes the derivative of the function. It is a relatively simple matter to solve these equations. By comparing the functions of the third kind for the dielectric cylinder and for the dielectric sphere, we see that the spherical case is considerably simpler. The presence of the sphere does not introduce the coupling between the *TE* modes ($\overline{M}$-functions) and the *TM* modes ($\overline{N}$-functions) as manifested by the dielectric cylinder. We must, however, remember that the cylindrical vector wave functions are generated with the piloting vector pointed in the z-direction while the piloting vector of the spherical vector wave functions is $\overline{R}$. Their characteristics are entirely different.

In regard to the applications, the examples which we used for the half-sheet are equally valid for the sphere, perhaps, with more varieties because of the availability of the functions of the third kind. We shall, however, treat in some detail only the problem of radiation from a horizontal dipole in the presence of a sphere. Historically, the problem of a vertical dipole, electric or magnetic, and a sphere was a subject studied by many authors after the turn of the century in connection with radiowave propagation over a spherical earth. The corresponding problem involving a horizontal dipole, however, was not resolved until many decades later by Fock [1965] and Nomura [1951]. We shall now reformulate this problem using the dyadic Green function technique and show the identity between our result and the original one obtained by Nomura. We shall also recover Mie's solution for the diffraction of a plane wave by a sphere [Mie, 1908; Stratton, 1941, p. 563] from the asymptotic solution of the horizontal dipole problem. For simplicity, we assume the sphere to be perfectly conducting. Because of the available formula for the function of the third kind, the treatment of an imperfectly conducting or dielectric sphere is no more difficult.

Figure 10-1 shows the geometry of the problem under consideration. For an infinitesimal horizontal electric dipole with current moment c pointed in the x-direction and located at $R' = b, \theta' = 0, \phi' = 0$, we let

$$\bar{J}(\bar{R}') = c\frac{\delta(R'-b)\delta(\theta'-0)\delta(\phi'-0)}{b^2\sin\theta'}\hat{x}.$$

The electric field produced by this dipole in the presence of a perfectly conducting sphere with radius equal to a is then given by

$$\begin{aligned}\bar{E}(\bar{R}) &= i\omega\mu_0\iiint \bar{\bar{G}}_{e1}(\bar{R},\bar{R}')\cdot\bar{J}(\bar{R}')dV' \\ &= i\omega\mu_0\bar{\bar{G}}_{e1}(\bar{R},\bar{R}')\cdot\hat{x},\quad \bar{R}' = (b,0,0). \qquad (10.66)\end{aligned}$$

Using the expression for $\bar{\bar{G}}_{e1}$ given by (10.24), (10.55), and (10.56), we obtain

$$\begin{aligned}\bar{E}(\bar{R}) = &\frac{-k\omega\mu_0 c}{4\pi}\sum_{n=1}^{\infty}\frac{2n+1}{n(n+1)} \\ &\cdot\left\{\begin{matrix}\left[j_n(\rho_b)+a_nh_n^{(1)}(\rho_b)\right]\bar{M}^{(1)}(k) \\ h_n^{(1)}(\rho_b)\left[\bar{M}(k)+a_n\bar{M}^{(1)}(k)\right] \\ +\frac{1}{\rho_b}\left([\rho_bj_n(\rho_b)]'+b_n\left[\rho_bh_n^{(1)}(\rho_b)\right]'\right)\bar{N}^{(1)}(k) \\ +\frac{[\rho_bj_n(\rho_b)]'}{\rho_b}\left[\bar{N}(k)+b_n\bar{N}^{(1)}(k)\right]'\end{matrix}\right\}_{R<b}^{R>b}, \qquad (10.67)\end{aligned}$$

where

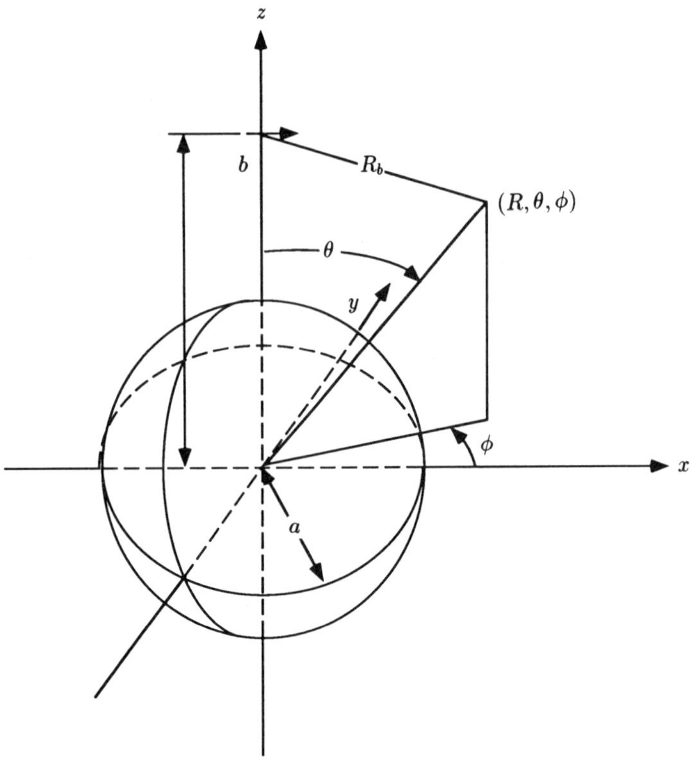

Fig. 10-1 A horizontal dipole in the presence of a perfectly conducting sphere

$$a_n = \frac{-j_n(\rho_a)}{h_n^{(1)}(\rho_a)}$$

$$b_n = \frac{-\left[\rho_a j_n(\rho_a)\right]'}{\left[\rho_a h_n^{(1)}(\rho_a)\right]'}$$

$$\rho_a = ka, \quad \rho_b = kb.$$

This expression is different in form from the one obtained by Nomura. He formulated the problem by the method of potential functions. As reviewed in Sec. 3-5 the primary or the incident field of the dipole can be expressed in the form

$$\overline{E}_i(\overline{R}) = \nabla \times \nabla \times (\Pi_i \hat{x})$$

$$\Pi_i = \frac{i\omega\mu_0 c e^{ikR_b}}{4\pi k^2 R_b}.$$

The free-space scalar Green function contained in Π_i, according to (10.49) for $\theta' = 0$ and $\phi' = 0$, has the series expansion

$$\frac{e^{ikR_b}}{4\pi R_b} = \frac{ik}{4\pi}\sum_{n=0}^{\infty}(2n+1)P_n(\cos\theta)\begin{cases} h_n^{(1)}(kR)j_n(kb), & R > b \\ j_n(kR)h_n^{(1)}(kb), & R < b. \end{cases} \tag{10.68}$$

Thus the incident field can be expressed in terms of a series of vector wave functions of the x-type defined by (10.26) with $m = 0$,

$$E_i(\overline{R}) = \frac{-k\omega\mu_0 c}{4\pi}\sum_{n=0}^{\infty}(2n+1)\begin{cases} j_n(kb)\overline{N}_{e0n}^{(x,1)}, & R > b \\ h_n^{(1)}(kb)\overline{N}_{e0n}^{(x)}, & R < b, \end{cases} \tag{10.69}$$

where $\overline{N}^{(x,1)}$ denotes a spherical vector wave function of the x-type using $h_n^{(1)}(kR)$ in the generating function. Nomura found the solution for the secondary or the scattered field by letting

$$\overline{E}_s(\overline{R}) = \frac{-k\omega\mu_0 c}{4\pi}\left[\sum_{n=0}^{\infty} c_n \overline{N}_{e0n}^{(x,1)}(k) + \sum_{n=1}^{\infty} d_n \overline{N}_{e1n}^{(1)}(k)\right]. \tag{10.70}$$

The coefficients c_n and d_n are determined by applying the boundary condition that $\hat{R}\times\left(\overline{E}_i + \overline{E}_s\right) = 0$ at the surface of the sphere which yields

$$c_n = -\frac{(2n+1)h_n^{(1)}(\rho_b)}{h_n^{(1)}(\rho_a)}$$

$$d_n = \frac{i}{\left[\rho_a h_n^{(1)}(\rho_a)\right]'}\left[\frac{h_{n-1}^{(1)}(\rho_b)}{h_{n-1}^{(1)}(\rho_a)} - \frac{h_{n+1}^{(1)}(\rho_b)}{h_{n+1}^{(1)}(\rho_a)}\right].$$

From the composition of (10.70), it is seen that the scattered field is derived partly from a Hertzian potential and partly from a Debye potential while the incident field is entirely derivable from a Hertzian potential. In view of our discussion of the relations between the functions of the x-type and the standard type, we can transform Nomura's formula into ours in the following manner. By taking the curl of Eq. $B.21$ of Appendix B and letting $m = 0$, we obtain

$$\begin{aligned} \overline{N}_{e0n}^{(x)} = & \frac{1}{2n(n+1)}\left[\overline{M}_{o1n} + n(n+1)\overline{M}_{o(-1)n}\right] \\ & + \frac{1}{2(n+1)(2n+1)}\left[\overline{N}_{e1(n+1)} - (n+1)(n+2)\overline{N}_{e(-1)(n+1)}\right] \\ & - \frac{1}{2n(2n+1)}\left[\overline{N}_{e1(n-1)} - (n-1)n\overline{N}_{e(-1)(n-1)}\right]. \end{aligned} \tag{10.71}$$

When m is negative, we use the formula [Sommerfeld, 1949, p. 129]

$$P_n^{-m} = (-1)^m\frac{(n-m)!}{(n+m)!}P_n^m$$

to obtain

$$\overline{M}_{\substack{e\\o}(-m)n} = \pm(-1)^m \frac{(n-m)!}{(n+m)!}\overline{M}_{\substack{e\\o}mn} \tag{10.72}$$

$$\overline{N}_{\substack{e\\o}(-m)n} = \pm(-1)^m \frac{(n-m)!}{(n+m)!}\overline{N}_{\substack{e\\o}mn}; \tag{10.73}$$

hence (10.71) can be simplified to read

$$\overline{N}^{(x)}_{e0n} = \frac{1}{n(n+1)}\overline{M}_{01n} + \frac{1}{(n+1)(2n+1)}\overline{N}_{e1(n+1)} - \frac{1}{n(2n+1)}\overline{N}_{e1(n-1)}.$$

By substituting this into (10.70) and changing the summation indices so that they all start with $n = 1$, we can write the resultant series in the form

$$\overline{E}_s(\overline{R}) = \frac{-k\omega\mu_0 c}{4\pi}\sum_{n=1}^{\infty}\left\{\frac{c_n}{n(n+1)}\overline{M}^{(1)}_{o1n}(k) + \left[\frac{c_{n-1}}{n(2n-1)} - \frac{c_{n+1}}{(n+1)(2n+3)} + d_n\right]\overline{N}^{(1)}_{e1n}\right\}. \tag{10.74}$$

Comparing (10.74) with the scattered part of (10.67), the following relations must hold true:

$$c_n = (2n+1)h^{(1)}_n(\rho_b)a_n \tag{10.75}$$

$$\begin{aligned} d_n &= \frac{(2n+1)}{n(n+2)}\frac{\left[\rho_b h^{(1)}_n(\rho_b)\right]'}{\rho_b}b_n + \frac{c_{n+1}}{(n+1)(2n+3)} - \frac{c_{n-1}}{n(2n-1)} \\ &= \frac{(2n+1)}{n(n+1)}\frac{\left[\rho_b h^{(1)}_n(\rho_b)\right]'}{\rho_b}b_n + \frac{h^{(1)}_{n+1}(\rho_b)}{n+1}a_{n+1} + \frac{h^{(1)}_{n-1}(\rho_b)}{n-1}a_{n-1}. \end{aligned} \tag{10.76}$$

By substituting the expressions for a_n and b_n given by (10.56) and (10.57) into (10.75) and (10.76), and applying the recurrence relations of the spherical Bessel functions of the type (10.39) and (10.40) with j_n replaced by $h^{(1)}_n$, we find that the coefficients c_n and d_n so obtained are indeed the same as the ones derived by Nomura, as they should be. The original verification of the identity between (10.67) and (10.70) was carried out by this author [Tai, 1952] without the aid of the dyadic Green functions technique. It was done by transforming the incident field as represented by (10.69) into two series involving both the $\overline{M}^{(1)}_{01n}$ and $\overline{N}^{(1)}_{e1n}$ functions. It seems obvious from this discussion that the use of the x-type of functions for the sphere problem makes the formulation considerably more complicated. The interpretation of the result based on Nomura's formula is also not as simple as the one offered by (10.67) from the eigenfunction point of view.

We shall now rederive Mie's series solution for the diffraction of a plane wave by a sphere from (10.67) by removing the dipole far away from the sphere. When kb is large, the spherical Hankel function has the asymptotic form

$$h_n^{(1)}(kb) \simeq (-i)^{n+1}\frac{e^{ikb}}{kb}$$

and

$$\frac{\left[kbh_n^{(1)}(kb)\right]'}{kb} \simeq (-i)^n \frac{e^{ikb}}{kb}\,.$$

Substituting these values into (10.67) for $R < b$, and identifying the amplitude of the plane wave as being given by

$$E_0 = \frac{i\omega\mu_0 c e^{ikb}}{kb} \quad (kb \gg 1),$$

we obtain

$$\begin{aligned}\overline{E}(\overline{R}) = E_0 \sum_{n=1}^{\infty}(-i)^n \frac{2n+1}{n(n+1)} \cdot \Big\{&\left[\overline{M}_{01n}(k) + a_n \overline{M}_{01n}^{(1)}(k)\right] \\ &+ i\left[\overline{N}_{e1n}(k) + b_n \overline{N}_{e1n}^{(1)}(k)\right]\Big\}\,. \end{aligned} \tag{10.77}$$

This is Mie's series as presented by Stratton [1941, p. 563] starting with the eigenfunction expansion of a plane wave using the spherical vector wave functions as the constituents; we note, however, that the direction of propagation of the plane wave created here is opposite to the one considered by Stratton.

Before we conclude this section, two additional formulas are given below. Their derivation is similar to the one that results in (10.67):

Case 1. Vertical Electric Dipole of Current Moment $\overline{c} = c\hat{z}$ Placed at $R' = b, \theta' = 0$

$$\begin{aligned}\overline{E}(\overline{R}) =& \frac{-k\omega\mu_0 c}{4\pi kb}\sum_{n=1}^{\infty} \\ &\cdot \begin{cases} \left[m_n(kb) + b_n h_n^{(1)}(kb)\right] \overline{N}_{e0n}^{(1)}(k), & R > b \\ h_n^{(1)}(kb)\left[\overline{N}_{e0n}(k) + b_n \overline{N}_{e0n}^{(1)}(k)\right], & R < b \end{cases} \\ b_n =& \frac{-\left[kaj_n(ka)\right]'}{\left[kah_n^{(1)}(kb)\right]'}.\end{aligned} \tag{10.78}$$

Case 2. A Small Circular Aperture Located at the Top of a Conducting Sphere (Fig. 10-1) Excited by a Constant Field $\overline{E}_a' = E_0\hat{x}$, $\theta' \le \theta_0$

$$\overline{E}(\overline{R}) = \frac{E_0\theta_0^2}{4}\sum_{n=1}^{\infty}\frac{2n+1}{n(n+1)}$$

$$\left\{\frac{1}{h_n^{(1)}(ka)}\overline{M}_{01n}^{(1)} + \frac{ka}{\left[kah_n^{(1)}(ka)\right]'}\overline{N}_{e1n}^{(1)}\right\}. \tag{10.79}$$

In deriving (10.79), terms of the order θ_0^4 and higher have been neglected. Equation (10.79), with an appropriate change of a constant of proportionality, also represents the field produced by a horizontal magnetic dipole placed at the top of the sphere and pointed in the y-direction.

10-4 SPHERICAL CAVITY

The eigenfunctions to be used in the expansions of the dyadic Green functions of a spherical cavity are the spherical vector wave functions with discrete eigenvalues. Four sets of solenoidal functions are needed. They are defined by

$$\overline{M}_{\substack{e\\o}mn}(\kappa_p) = \nabla \times \left[j_n(\kappa_p)P_n^m(\cos\theta)\begin{matrix}\cos\\ \sin\end{matrix} m\phi\overline{R}\right] \tag{10.80}$$

$$\overline{M}_{\substack{e\\o}mn}(\kappa_q) = \nabla \times \left[j_n(\kappa_q)P_n^m(\cos\theta)\begin{matrix}\cos\\ \sin\end{matrix} m\phi\overline{R}\right] \tag{10.81}$$

$$\overline{N}_{\substack{e\\o}mn}(\kappa_p) = \frac{1}{\kappa_p}\nabla \times \overline{M}_{\substack{e\\o}mn}(\kappa_p) \tag{10.82}$$

$$\overline{N}_{\substack{e\\o}mn}(\kappa_q) = \frac{1}{\kappa_q}\nabla \times \overline{M}_{\substack{e\\o}mn}(\kappa_q). \tag{10.83}$$

The expressions for these functions are identical to the functions defined by (10.9) and (10.10) with κ therein replaced by either κ_p and κ_q. The eigenvalues κ_p and κ_q, after being multiplied by a, the radius of a spherical cavity with center at the origin of the spherical coordinate system, are the roots of the characteristic

$$j_n(\kappa_p a) = 0 \tag{10.84}$$

$$[\kappa_q a j_n(\kappa_q a)]' = 0, \tag{10.85}$$

where the prime in (10.85) denotes the derivatives of the function inside the brackets with respect to $\kappa_q a$.

For convenience, a condensed notation for these functions in the form of

$$\overline{M}_p = \overline{M}_{\substack{e\\o}mn}(\kappa_p)$$

$$\overline{M}_q = \overline{M}_{\substack{e\\o}mn}(\kappa_q)$$

$$\overline{N}_p = \frac{1}{\kappa_p}\nabla \times \overline{M}_p$$

$$\overline{N}_q = \frac{1}{\kappa_q}\nabla \times \overline{M}_q$$

will be used to describe the orthogonal relations and the normlization factors of these functions. They are

$$\iiint \overline{M}_p \cdot \overline{N}_q dV = 0 \tag{10.86}$$

$$\iiint \overline{M}_q \cdot \overline{N}_p dV = 0 \tag{10.87}$$

$$\begin{aligned} \iiint \overline{M}_p \cdot \overline{M}'_p dV &= \iiint \overline{N}_p \cdot \overline{N}'_p dV \\ &= \begin{cases} 0, & m \neq m', n \neq n', \kappa_p \neq \kappa_{p'} \\ \frac{4\pi}{C_p} I_p, & m = m', n = n', \kappa_p = \kappa_{p'}, \end{cases} \end{aligned} \tag{10.88}$$

where

$$\begin{aligned} C_p &= (2 - \delta_0) \frac{2n+1}{n(n+1)} \frac{(n-m)!}{(n+m)!} \\ I_p &= \int_0^a R^2 j_n^2(\kappa_p R) dR \\ &= \frac{a^3}{3} \left[\frac{\partial j_n(\kappa_p a)}{\partial(\kappa_p a)} \right]^2 \end{aligned}$$

$$\begin{aligned} \iiint \overline{M}_q \cdot \overline{M}_{q'} dV &= \iiint \overline{N}_q \cdot \overline{N}_{q'} dV \\ &= \begin{cases} 0, & m \neq m', n \neq n', \kappa_q \neq \kappa_{q'} \\ \frac{4\pi}{C_{mn}} I_q, & m = m', n = n', \kappa_q = \kappa_{q'}, \end{cases} \end{aligned} \tag{10.89}$$

where

$$\begin{aligned} I_q &= \int_0^a \frac{R^2}{2n+1} \left[(n+1) j_{n-1}^2(\kappa_q R) + n j_{n+1}^2(\kappa_q R) \right] dR \\ &= \frac{a^3}{2} \left[1 - \frac{n(n+1)}{\kappa_q^2 a^2} \right] j_n^2(\kappa_q a). \end{aligned} \tag{10.90}$$

The normalization factors in (10.89) and (10.90) apply to functions of the same species, either both even or both odd. Functions of different species, one even and another odd, are orthogonal. The proofs of these relations are very similar to the ones discussed in Chapter 6 for the vector wave functions encountered in the theory of cylindrical waveguide, but the proofs for the spherical case would take more time. They are assigned as exercises for the readers which are quite challenging. Once the orthogonal relations are known we can find the representations for $\nabla \times \left[\overline{\overline{I}} \delta(\overline{R} - \overline{R}') \right]$ and $\overline{\overline{G}}_{m2}$. They are

$$\nabla \times \left[\overline{\overline{I}} \delta(\overline{R} - \overline{R}') \right] = \sum_{\ell,m,n} \left(\frac{\kappa_p}{I_p} \overline{N}_p \overline{M}'_p + \frac{\kappa_q}{I_q} \overline{M}_q \overline{N}'_q \right) \tag{10.91}$$

$$\overline{\overline{G}}_{m2}(\overline{R},\overline{R}') = \sum_{\ell,m,n}\left[\frac{\kappa_p}{(\kappa_p^2-k^2)I_p}\overline{N}_p\overline{N}'_p + \frac{\kappa_q}{(\kappa_q^2-\kappa^2)I_q}\overline{M}_q\overline{N}'_q\right]. \tag{10.92}$$

The primed functions are defined with respect to (R', θ', ϕ') and the ordinal number ℓ represents the numerical number for the discrete eigenvalues κ_p and κ_q. It is recalled that for the function $\overline{\overline{G}}_{m0}$ we had a Hankel transform representation for that function and the discontinuous behavior of $\overline{\overline{G}}_{m0}$ at $R = R'$ can be spelled out in the form of $\overline{\overline{G}}^{\pm}_{m0}$. The discontinuous behavior of $\overline{\overline{G}}_{m2}$ in the present case, however, is not available. The explicit form of $\overline{\overline{G}}_{e1}$, therefore, can only be presented in the form

$$\begin{aligned}\overline{\overline{G}}_{e1}(\overline{R},\overline{R}') &= \frac{1}{k^2}\left[\nabla\times\overline{\overline{G}}_{m2}(\overline{R},\overline{R}') - \overline{\overline{I}}\delta(\overline{R}-\overline{R}')\right]\\ &= -\frac{1}{k^2}\overline{\overline{I}}\delta(\overline{R}-\overline{R}') + \frac{1}{k^2}\sum_{\ell,m,n}\\ &\left[\frac{\kappa_p^2}{\kappa_p^2-k^2}\overline{M}_p\overline{M}'_p + \frac{\kappa_q^2}{\kappa_q^2-k^2}\overline{N}_q\overline{N}'_q\right].\end{aligned} \tag{10.93}$$

In a work by Rozenfeld [1974], the method of $\overline{\overline{G}}_e$ was used to derive the expression for $\overline{\overline{G}}_{e1}$ which requires not only the solenoidal functions $\overline{M}_p$ and $\overline{N}_q$ but also the longitudinal function $\overline{L}_p$. After the elimination of the longitudinal function with the aid of the eigenfunction expansion for $\overline{\overline{I}}\delta(\overline{R}-\overline{R}')$, his expression for $\overline{\overline{G}}_{e1}$ is identical to (10.93). The simplicity of the method of $\overline{\overline{G}}_m$ is again demonstrated convincingly in this problem. The general method of $\overline{\overline{G}}_m$ as presented now in this book was not yet fully developed when Rozenfeld wrote his dissertation on this subject.

10-5 PERFECTLY CONDUCTING CONICAL STRUCTURES

There are two distinct types of perfectly conducting conical structures of which the electric dyadic Green functions of the first kind and the second kind can be found. The first type has the form of a single cone, and the second type is a bicone formed by two single cones. These two structures are shown in Fig. 10-2. They are both infinitely long in the radial direction. For both cases, it is assumed that the axis of the cone is aligned with the z-axis. We define the conical vector wave functions in exactly the same form as the spherical vector wave functions except that the eigenvalue n is now, in general, fractional. There are four kinds of these functions which are defined by

$$\overline{M}_{\substack{e\\o}m\mu}(k) = \nabla\times\left(\psi_{\substack{e\\o}m\mu}\overline{R}\right) \tag{10.94}$$

$$\overline{N}_{\substack{e\\o}m\mu}(k) = \frac{1}{k}\nabla\times\nabla\times\left(\psi_{\substack{e\\o}m\mu}\overline{R}\right) \tag{10.95}$$

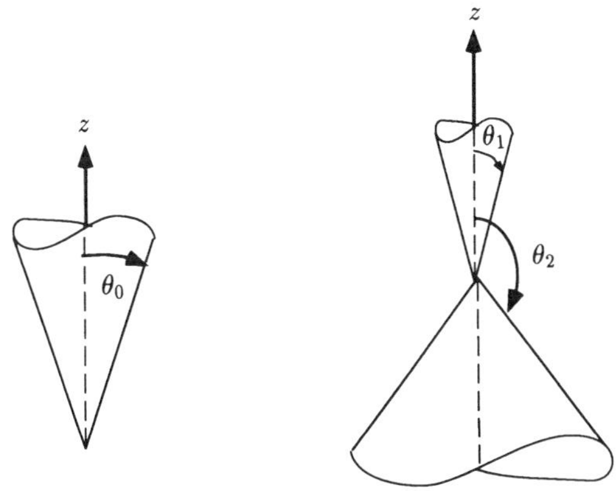

Fig. 10-2 (a) A single cone (b) A bicone

$$\overline{M}_{\substack{e\\o}m\lambda}(k) = \nabla \times \left(\psi_{\substack{e\\o}m\lambda}\overline{R}\right) \tag{10.96}$$

$$\overline{N}_{\substack{e\\o}m\lambda}(k) = \frac{1}{\kappa}\nabla \times \nabla \times (\psi_{\substack{e\\o}m\lambda}\overline{R}), \tag{10.97}$$

where

$$\psi_{\substack{e\\o}m\mu} = j_\mu(\kappa R)P_\mu^m(\cos\theta)\,{\cos \atop \sin}\, m\phi$$

$$\psi_{\substack{e\\o}m\lambda} = j_\lambda(\kappa R)P_\lambda^m(\cos\theta)\,{\cos \atop \sin}\, m\phi.$$

The eigenvalues μ and λ are determined from the characteristic equations that for a single cone

$$P_\mu^m(\cos\theta_0) = 0 \tag{10.98}$$

$$\frac{dP_\lambda^m(\cos\theta_0)}{d\theta_0} = 0, \tag{10.99}$$

and for a bicone they are

$$P_\mu^m(\cos\theta_1) = P_\mu^m(\cos\theta_2) = 0 \tag{10.100}$$

$$\frac{dP_\lambda^m(\cos\theta_1)}{d\theta_1} = \frac{dP_\lambda^m(\cos\theta_2)}{d\theta_2} = 0. \tag{10.101}$$

From now on we will not distinguish these two cases, as there is no difference in the formulation of the problems for the two types of cones except that the numerical values of μ and λ are different. Although we do not have a complete knowledge of these values, we are not prevented from formulating this class of problems pending further numerical computations.

The conical vector wave functions defined by (10.94) to (10.97) satisfy the boundary condition that at the surface(s) of the cone

$$\hat{\theta} \times \overline{M}_{\substack{e \\ o}n\lambda} = 0, \; \hat{\theta} \times \overline{N}_{\substack{e \\ o}m\mu} = 0$$
$$\hat{\theta} \times \nabla \times \overline{M}_{\substack{e \\ o}m\mu} = 0, \; \hat{\theta} \times \nabla \times \overline{N}_{\substack{e \\ o}m\lambda} = 0.$$

The functions which satisfy the Dirichlet boundary condition will be used for the construction of the electric dyadic Green function of the first kind, while the ones satisfying the Neumann boundary condition are involved in the function of the second kind. To determine the normalization factor for these functions, we must first give a brief review of the orthogonal relations of the associated Legendre functions.

Let us consider the case of a single cone. The associated Legendre function P_λ^m and P_μ^m satisfy, respectively, the differential equations

$$\frac{1}{\sin\theta}\frac{d}{d\theta}\left(\sin\theta\frac{dP_\lambda^m}{d\theta}\right) + \left[\lambda(\lambda+1) - \frac{m^2}{\sin^2\theta}\right]P_\lambda^m = 0 \tag{10.102}$$

$$\frac{1}{\sin\theta}\frac{d}{d\theta}\left(\sin\theta\frac{dP_\mu^m}{d\theta}\right) + \left[\mu(\mu+1) - \frac{m^2}{\sin^2\theta}\right]P_\mu^m = 0. \tag{10.103}$$

By multiplying (10.102) by $\sin\theta P_\mu^m$ and (10.103) by $\sin\theta P_\lambda^m$ and integrating the difference of the two resultant equations from $\theta = \theta_0$ to π, we obtain

$$[\lambda(\lambda+1) - \mu(\mu+1)]\int_{\theta_0}^{\pi} P_\lambda^m P_\mu^m \sin\,\theta d\theta =$$
$$\left[\sin\theta P_\lambda^m \frac{dP_\mu^m}{d\theta} - \sin\theta P_\mu^m \frac{dP_\lambda^m}{d\theta}\right] = 0;$$

hence

$$\int_{\theta_0}^{\pi} P_\lambda^m P_\mu^m \sin\,\theta\, d\theta = 0, \tag{10.104}$$

since μ and λ are distinct. For two functions of the same species, we have

$$\int_{\theta_0}^{\pi} P_\lambda^m P_{\lambda'}^m \sin\,\theta\, d\theta = \begin{cases} 0, & \lambda \neq \lambda' \\ I_{m\lambda}, & \lambda = \lambda' \end{cases}$$
$$\int_{\theta_0}^{\pi} P_\mu^m P_{\mu'}^m \sin\,\theta\, d\theta = \begin{cases} 0, & \mu \neq \mu' \\ I_{m\mu}, & \mu = \mu', \end{cases}$$

where $I_{m\lambda}$ and $I_{m\mu}$ are two normalization constants. It is not difficult to show by integration by parts that

$$\int_{\theta_0}^{\pi}\left[\left(\frac{dP_\lambda^m}{d\theta}\right)^2 + \left(\frac{mP_\lambda^m}{\sin\,\theta}\right)^2\right]\sin\theta\, d\theta = \lambda(\lambda+1)I_{m\lambda}$$
$$\int_{\theta_0}^{\pi}\left[\left(\frac{dP_\mu^m}{d\theta}\right)^2 + \left(\frac{mP_\mu^m}{\sin\,\theta}\right)^2\right]\sin\theta\, d\theta = \mu(\mu+1)I_\mu.$$

With the aid of these orthogonal relations, we can determine the orthogonal relations of the conical vector wave functions. They are

$$\iiint \overline{M}_{\substack{e\\o} m\lambda}(\kappa)\cdot \overline{N}_{\substack{e\\o} m'\mu'}(\kappa')\, dV = 0$$

$$\iiint \overline{M}_{\substack{e\\o} m\mu}(\kappa)\cdot \overline{N}_{\substack{e\\o} m'\mu'}(\kappa')\, dV = 0$$

$$\iiint \left\{ \begin{matrix} \overline{M}_{\substack{e\\o} m\lambda}(\kappa)\cdot \overline{M}_{\substack{e\\o} m'\lambda'}(\kappa') \\ \overline{N}_{\substack{e\\o} m\lambda}(\kappa)\cdot \overline{N}_{\substack{e\\o} n'\lambda'}(\kappa) \end{matrix} \right\} dV$$

$$= \begin{cases} 0, m \neq m', \lambda \neq \lambda' \\ \dfrac{(1+\delta_0)\pi^2\lambda(\lambda+1)I_{m\lambda}}{2k^2}\delta(\kappa-\kappa') \end{cases}$$

$$\iiint \left\{ \begin{matrix} \overline{M}_{\substack{e\\o} m\mu}(\kappa)\cdot \overline{M}_{\substack{e\\o} m'\mu'}(\kappa') \\ \overline{N}_{\substack{e\\o} m\mu}(\kappa)\cdot \overline{N}_{\substack{e\\o} m'\mu'}(\kappa') \end{matrix} \right\} dV$$

$$= \begin{cases} 0, m \neq m', \mu \neq \mu' \\ \dfrac{(1+\delta_0)\pi^2\mu(\mu+1)I_{m\mu}}{2k^2}\delta(\kappa-\kappa'). \end{cases}$$

Knowing these relations for the conical vector wave functions, we can easily derive the functions of the first and the second kind by the method of $\overline{\overline{G}}_m$ Omitting the details, we find

$$\overline{\overline{G}}_{e1}(\overline{R},\overline{R}') = \frac{ik}{2\pi}\sum_m (2-\delta_0)\left[\sum_\lambda \frac{1}{\lambda(\lambda+1)I_{m\lambda}} \left\{ \begin{matrix} \overline{M}^{(1)}_{\substack{e\\o} m\lambda}(k)\overline{M}'_{\substack{e\\o} m\lambda}(k) \\ \overline{M}_{\substack{e\\o} m\lambda}(k)\overline{M}'^{(1)}_{\substack{e\\o} m\lambda}(k) \end{matrix} \right\}\right.$$

$$\left. + \sum_\mu \frac{1}{\mu(\mu+1)I_{m\mu}} \left\{ \begin{matrix} \overline{N}^{(1)}_{\substack{e\\o} m\mu}(k)\overline{N}'_{\substack{e\\o} m\mu}(k) \\ \overline{N}_{\substack{e\\o} m\mu}(k)\overline{N}'^{(1)}_{\substack{e\\o} m\mu}(k) \end{matrix} \right\}\right], \begin{matrix} R > R' \\ R < R'. \end{matrix} \tag{10.105}$$

In the function of the second kind, we replace $\overline{M}_{\substack{e\\o} m\lambda}$ and $\overline{N}_{\substack{e\\o} m\mu}$ in (10.105) respectively, by $\overline{N}_{\substack{e\\o} m\lambda}$ and $\overline{M}_{\substack{e\\o} m\mu}$.

10-6 CONE WITH A SPHERICAL SECTOR

When a spherical sector is attached to a cone (Fig. 10-3), the dyadic Green functions pertaining to such a composite body can also be derived.

Thus for a perfectly conducting sphere, we obtain, by the method of scattering superposition, for $R > R'$,

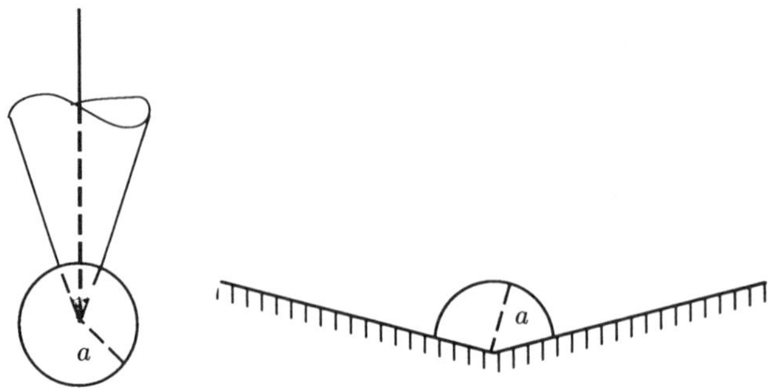

Fig. 10-3 A composite body made of a cone and a spherical sector

$$
\begin{aligned}
\overline{\overline{G}}_{e1}(\overline{R},\overline{R}') =& \frac{ik}{2\pi}\sum_m (2-\delta_0) \\
&\cdot \left\{ \sum_\lambda \frac{1}{\lambda(\lambda+1)I_{m\lambda}} \overline{M}^{(1)}_{\substack{e\\o}m\lambda}(k) \left[\overline{M}'_{\substack{e\\o}m\lambda}(k) + \alpha_{\substack{e\\o}m\lambda} \overline{M}'^{(1)}_{\substack{e\\o}m\lambda}(k) \right] \right. \\
&\left. + \sum_\mu \frac{1}{\mu(\mu+1)I_{m\mu}} \overline{N}^{(1)}_{\substack{e\\o}m\mu}(k) \left[\overline{N}'_{\substack{e\\o}m\mu}(k) + \beta_{\substack{e\\o}m\mu} \overline{N}'^{(1)}_{\substack{e\\o}m\mu}(k) \right] \right\},
\end{aligned}
$$

where

$$
\begin{aligned}
\alpha_{\substack{e\\o}m\lambda} &= \frac{-j_\lambda(ka)}{h^{(1)}_\lambda(ka)} \\
\beta_{\substack{e\\o}m\mu} &= \frac{-\left[kaj_\mu(ka)\right]'}{\left[kah^{(1)}_\mu(ka)\right]'}.
\end{aligned}
$$

The formulations provided here can be used to investigate many technical problems involving these structures.

11

Planar Stratified Media

The cylindrical vector wave functions introduced previously can also be used in the eigenfunction expansion of the dyadic Green functions associated with plane stratified media. The free-space Green function will be transformed into an integral form suitable for the construction of the function of the third kind. For flat earth the present formulation will be compared with Sommerfeld's classical work, and certain unique features of the dyadic Green function technique will be pointed out. Finally, other stratified problems will also be considered in this chapter.

11-1 FLAT EARTH

When a space is partitioned into two halves, one of which is filled with air and the other half with a homogeneous lossy dielectric as shown in Fig. 11-1, the geometry corresponds to that of a flat earth. We assume the earth is characterized by the constitutive constants ϵ, μ_0, and σ. For convenience, we will designate the propagation constants in the two media, respectively, by

$$k_1 = \omega(\mu_0\epsilon_0)^{\frac{1}{2}}, \; k_2 = \omega\left[\mu_0\epsilon\left(1+\frac{i\sigma}{\omega\epsilon}\right)\right]^{\frac{1}{2}}.$$

The very nature of the composition implies that the pertinent dyadic Green functions under consideration are functions of the third kind. To find these functions, we shall first transform the free-space function into an integral form that enables us to construct the functions of the third kind by the method of scattering superposition. For this reason, we would like to find a Fourier-Bessel or a Hankel transform representation of $\overline{\overline{G}}_{eo}$ first.

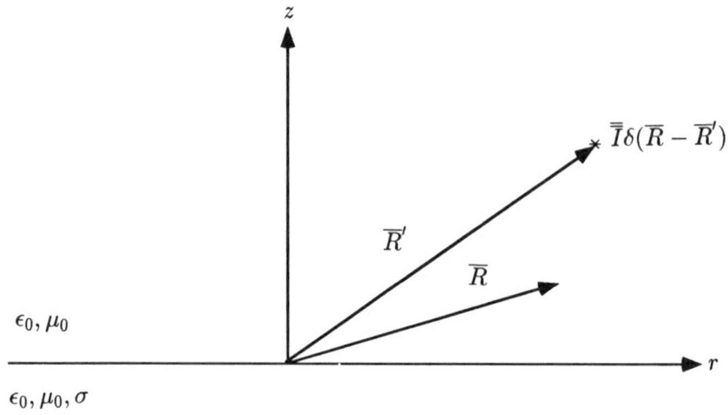

Fig. 11-1 Flat earth

According to (7.14), the double integral representation of $\overline{\overline{G}}_{m0}$ is given by

$$\overline{\overline{G}}_{m0}(\overline{R},\overline{R}') = \int_0^\infty d\lambda \int_{-\infty}^\infty dh \sum_n \left[(2-\delta_0)\kappa/4\pi^2\lambda(\kappa^2-k_1^2)\right] \cdot \left[\overline{N}_\lambda(h)\overline{M}'_\lambda(-h)+\overline{M}_\lambda(h)\overline{N}'_\lambda(-h)\right], \tag{11.1}$$

where the wave number k in (7.14) is now replaced by k_1 and $\kappa^2 = h^2+\lambda^2$. A condensed notation has been used for the vector wave functions as had been explained in the paragraph immediately after (7.14). The Fourier integral in (11.1) can be evaluated with the aid of the residue theorem in the h- plane. The poles of the integrand are located at $h = \pm h_1$, where $h_1 = \left(k_1^2-\lambda\right)^{\frac{1}{2}}$. The result yields

$$\overline{\overline{G}}^{\pm}_{m0}(\overline{R},\overline{R}') = k\int_0^\infty d\lambda \sum_n C^{(1)}$$

$$\cdot \left[\overline{N}_\lambda\left(\pm h_1\right)\overline{M}'_\lambda\left(\mp h_1\right)+\overline{M}_\lambda\left(\pm h_1\right)\overline{N}'_\lambda\left(\mp h_1\right)\right], \; z \gtrless z', \tag{11.2}$$

where

$$C^{(1)} = \frac{(2-\delta_0)}{4\pi\lambda h_1}.$$

The plane of discontinuity for the free-space magnetic dyadic Green function is now located at $z = z'$. It is recalled that for cylindrical problems we get rid of the λ-integration and have preserved the Fourier integral, but now we have retained the Fourier-Bessel integral.

By means of the method of $\overline{\overline{G}}_m$, the expression for $\overline{\overline{G}}_{eo}$ can now be written in the form

$$\begin{aligned}\overline{\overline{G}}_{eo}^{(1)}(\overline{R},\overline{R}') =& \frac{1}{k_1^2}\left[-\hat{z}\hat{z}\delta(\overline{R}-\overline{R}') + \left(\nabla\times\overline{\overline{G}}_{mo}^{+}\right)U(z-z')\right.\\ &\left.-\left(\nabla\times\overline{\overline{G}}_{mo}^{-}\right)U(z'-z)\right]\\ =& -\frac{1}{k_1^2}\hat{z}\hat{z}\delta(\overline{R}-\overline{R}')\\ &+\int_0^\infty d\lambda\sum_n C^{(1)}\left[\overline{M}(\pm h_1)\overline{M}'(\mp h_1)\right.\\ &\left.+\overline{N}(\pm h_1)\overline{N}'(\mp h_1)\right],\ z\gtrless z'.\end{aligned} \tag{11.3}$$

We have deleted the superscript "λ" attached to the vector wave functions. The superscript in $\overline{\overline{G}}_{eo}^{(1)}$ is to indicate that the function is defined with respect to k_1. For a flat earth with its surface located at $z=0$ as shown in Fig. 11-1, we identify region 1 to be above the earth and region 2 inside the earth, and the source is assumed to be located in region 1, then the pertinent functions involved are $\overline{\overline{G}}_e^{(11)}$ and $\overline{\overline{G}}_e^{(21)}$. By the method of scattering superposition, we let

$$\overline{\overline{G}}_e^{(11)}(\overline{R},\overline{R}') = \overline{\overline{G}}_{eo}^{(1)}(\overline{R},\overline{R}') + \overline{\overline{G}}_{es}^{(11)}(\overline{R},\overline{R}') \tag{11.4}$$

and

$$\overline{\overline{G}}_e^{(21)}(\overline{R},\overline{R}'). = \overline{\overline{G}}_{es}^{(21)}(\overline{R},\overline{R}'). \tag{11.5}$$

As had been done many times before, the scattered terms must have the form

$$\overline{\overline{G}}_{es}^{(11)}(\overline{R},\overline{R}') = \int_0^\infty d\lambda\sum_n C^{(1)}\left[a\overline{M}(h_1)\overline{M}'(h_1) + b\overline{N}(h_1)\overline{N}'(h_1)\right] \tag{11.6}$$

$$\overline{\overline{G}}_{es}^{(21)}(\overline{R},\overline{R}') = \int_0^\infty d\lambda\sum_n C^{(1)}\left[c\overline{M}(-h_2)\overline{M}'(h_1) + d\overline{N}(-h_2)\overline{N}'(h_1)\right], \tag{11.7}$$

where $h_2 = \left(k_2^2-\lambda^2\right)^{\frac{1}{2}}$. The functions $\overline{M}(-h_2)$ and $\overline{N}(-h_2)$ are wave functions which are solutions of the wave equation in region 2 with wave number k_2. At the interface, $z=0$, the boundary conditions are

$$\hat{z}\times\left[\overline{\overline{G}}_e^{(11)}(\overline{R},\overline{R}') - \overline{\overline{G}}_e^{(21)}(\overline{R},\overline{R}')\right] = 0 \tag{11.8}$$

and

$$\hat{z}\times\left[\nabla\times\overline{\overline{G}}_e^{(11)}(\overline{R},\overline{R}') - \nabla\times\overline{\overline{G}}_e^{(21)}(\overline{R},\overline{R}')\right] = 0, \tag{11.9}$$

where we have assumed $\mu_1=\mu_2=\mu_0$. The coefficients a, b, c, and d thus can be determined; they are given by

$$a = \frac{(h_1 - h_2)}{(h_1 + h_2)}$$
$$b = \frac{(k_2^2 h_1 - k_1^2 h_2)}{(k_2^2 h_1 + k_1^2 h_2)}$$
$$= \frac{(n^2 h_1 - h^2)}{(n^2 h_1 + h^2)}$$
$$c = \frac{2h_1}{(h_1 + h_2)}$$
$$d = \frac{2nh_1}{(n^2 h_1 + h^2)}, \quad n = \frac{k_2}{k_1},$$

where n denotes the complex index of refraction of the earth medium. Knowing $\overline{\overline{G}}_e^{(11)}$ and $\overline{\overline{G}}_e^{(21)}$, we can find the electric field in the two regions due to a current distribution in region 1 by the formulas

$$\overline{E}_1(\overline{R}) = i\omega\mu_0 \iiint \overline{\overline{G}}_e^{(11)}(\overline{R}, \overline{R}') \cdot \overline{J}_1(\overline{R}')\, dV' \tag{11.10}$$
$$\overline{E}_2(\overline{R}) = i\omega\mu_0 \iiint \overline{\overline{G}}_e^{(21)}(\overline{R}, \overline{R}') \cdot \overline{J}_1(\overline{R}')\, dV'. \tag{11.11}$$

11-2 RADIATION FROM ELECTRIC DIPOLES IN THE PRESENCE OF A FLAT EARTH AND SOMMERFELD'S THEORY

For an infinitesimal vertical electric dipole with current moment $c\hat{z}$ located at $(0, 0, z_0)$ in air, we write

$$\overline{J}_1(\overline{R}') = c\hat{z}\delta\,(x' - 0)\,\delta\,(y' - 0)\,\delta\,(z' - z_0)\,.$$

In order to compare with Sommerfeld's theory later, we choose c to be numerically equal to $4\pi k_1^2 / i\omega\mu_0$. Then according to (11.10) we obtain

$$\overline{E}_1(\overline{R}) = 4\pi k_1^2 \overline{\overline{G}}_e^{(11)}(\overline{R}, \overline{R}') \cdot \hat{z}\Big|_{\overline{R}'=(0,0,z_0)}. \tag{11.12}$$

Using the expression for $\overline{\overline{G}}_e^{(11)}$ found in the previous section and the expressions for $\overline{M}'_{\substack{e\\o}n\lambda}$ and $\overline{N}'_{\substack{e\\o}n\lambda}$ defined by (B.13) and (B.14) of Appendix B, but in the primed variables, we find that only $\overline{N}'_{e0\lambda} \cdot \hat{z}$ survives when we let $\overline{R}' = (0, 0, z_0)$. For simplicity, let us consider just the case corresponding to $z > z_0$; then

$$\overline{E}_1(\overline{R}) = ik_1 \int_0^\infty \left(\frac{\lambda}{h_1}\right) \overline{N}_{e0\lambda}(h_1) \left[e^{-ih_1 z_0} + be^{ih_1 z_0}\right] d\lambda, \tag{11.13}$$

where

$$b = \frac{(n^2 h_1 - h_2)}{(n^2 h_1 + h_2)}.$$

For an infinitesimal horizontal electric dipole with the same current moment pointed in the x-direction, we have

$$\overline{E}_1(\overline{R}) = 4\pi k_1^2 \overline{\overline{G}}_e^{(11)}(\overline{R}, \overline{R}') \cdot \hat{x} \Big|_{\overline{R}'=(0,0,z_0)}. \tag{11.14}$$

In this case, only $\overline{M}_{01\lambda}$ and $\overline{N}_{e1\lambda}$ survive, and we obtain

$$\begin{aligned}\overline{E}_1(\overline{R}) = ik_1 \int_0^\infty & \{(h_1/h_1)\overline{M}_{01\lambda}(h_1)\left[e^{-ih_1 z_0} + ae^{ih_1 z_0}\right] \\ & - i\overline{N}_{e1\lambda}(h_1)\left[e^{-ih_1 z_0} - be^{ih_1 z_0}\right]\} \, d\lambda,\end{aligned} \tag{11.15}$$

where b is the same as the one defined previously and

$$a = \frac{(h_1 - h_2)}{(h_1 + h_2)}.$$

We shall show now that the expressions for the electric field as given by (11.13) and (11.14) are equivalent to the ones obtained by Sommerfeld in his famous work and treated very completely in Chapter 6 of his book on partial differential equations [1949, pp. 236–261].

Sommerfeld formulated these problems by the method of potentials. He used an electric Hertzian potential alone, so that

$$\overline{E}_1(\overline{R}) = k_1^2 \overline{\pi} + \nabla\nabla \cdot \overline{\pi}. \tag{11.16}$$

For $\overline{R} \neq \overline{R}'$, (11.16) is equivalent to

$$\overline{E}_1(\overline{R}) = \nabla \times \nabla \times \overline{\pi}. \tag{11.17}$$

In the case of a vertical dipole he showed that a z- component of $\overline{\pi}$ alone is sufficient to formulate and to solve the problem. The resultant expression for π_z, including both the primary and the scattered field, for $z > z_0$, is given by

$$\pi_z = \int_0^x \left(\frac{\lambda}{h_1}\right) J_0(\lambda r) \, e^{ih_1 z} \left(e^{-ih_1 z_0} + be^{ih_1 z_0}\right) d\lambda. \tag{11.18}$$

We have written all his parameters in our notation. Substituting (10.18) into (11.17) we obtain

$$\overline{E}_1(\overline{R}) = ik_1 \int_0^\infty \left(\frac{\lambda}{h_1}\right) \overline{N}_{e0\lambda}(h_1) \left(e^{-ih_1 z_0} + be^{ih_1 z_0}\right) d\lambda, \tag{11.19}$$

which is identical to (11.13).

In the case of a horizontal dipole, he showed that two components of $\overline{\pi}$, corresponding to π_x and π_z are required. For $z > z_0$, they are

$$\pi_x = i \int_0^\infty \left(\frac{\lambda}{h_1}\right) J_0(\lambda r) e^{ih_1 z} \left(e^{-ih_1 z_0} + ae^{ih_1 z_0}\right) d\lambda \tag{11.20}$$

and

$$\pi_z = -\,2\cos\phi \int_0^\infty \frac{\lambda^2(h_1 - h_2)}{(k_2^2 h_1 + k_1^2 h_2)} \cdot J_1(\lambda r) e^{ih_1(z+z_0)}\, d\lambda. \tag{11.21}$$

It is obvious that π_z and the part of π_x containing the coefficient a represent the scattered field. Substituting (11.20) and (11.21) into (11.17) we obtain

$$\begin{aligned}\overline{E}_1 \overline{R} = {} & ik_1 \int_0^\infty \left(\frac{\lambda}{h_1}\right) \overline{N}^{(x)}_{e0\lambda}(h_1)\left(e^{ih_1 z_0} + a e^{ih_1 z_0}\right) d\lambda \\ & - 2k_1 \int_0^\infty \left[\frac{\lambda^2(h_1 - h_2)}{(k_2^2 h_1 + k_1^2 h_2)}\right] \overline{N}_{e1\lambda}(h_1) e^{ih_1 z_0}\, d\lambda,\end{aligned} \tag{11.22}$$

where $\overline{N}^{(x)}_{eo\lambda}$ represents a cylindrical vector wave function of the x- type defined by

$$\overline{N}^{(x)}_{eo\lambda}(h_1) = \left(\frac{1}{k_1}\right) \nabla \times \nabla \times \left[J_0(\lambda r) e^{ih_1 z} \hat{x}\right].$$

As shown in Appendix B, this type of function is related to the functions of the z-type which we have used exclusively in the construction of the dyadic Green function. In particular, when $n = 0$, Eq. (B.18) of Appendix B becomes

$$\overline{N}^{(x)}_{e0\lambda}(h_1) = \frac{k_1 \overline{M}_{01\lambda}(h_1) - ih_1 \overline{N}_{e1\lambda}(h_1)}{\lambda}.$$

Substituting it into (11.22) and making use of the identity that

$$\frac{2\lambda^2\,(h_1 - h_2)}{(k_2^2 h_1 + k_1^2 h_2)} = a + b,$$

where a and b are the same coefficients defined in (11.13) and (11.14), we thus find that (11.22), indeed, is identical to (11.15).

In retrospect, we see that Sommerfeld's original formulation used only the electric Hertzian potential. The formulation discussed here, from the point of view of potential theory, uses both the electric Hertzian potential and the magnetic Hertzian potential. The former generates the $\overline{N}$ functions, and the latter generates the $\overline{M}$ functions. Whereas Sommerfeld's treatment is very ingenious, ours is more or less methodical. One unique feature of the dyadic Green function formulation is that for any other current distributions the same technique applies. In doing so we have bypassed the need of determining specifically the potential functions for each problem.

Finally, we would like to mention the fine book written by Baños [1966] on dipole radiation in the presence of a flat earth. He has systematically reduced the various integrals into some basic ones for the convenience of numerical computation as well as for correlating different problems. Undoubtedly the integrals resulting from the present formulation can all be expressed in terms of Baños basic functions. The books by Wait [1962] and King, Wu, and Owen [1991] also contain much useful information. It is beyond the scope of this book to discuss the numerical aspect of these problems. Some simple asymptotic expressions of the integrals, however, will be considered here.

Under certain conditions to be specified later, we can find the asymptotic expression for the dyadic Green function of the third kind by the method of saddle-point integration. To derive this expression we shall first transform the integral representation of $\bar{\bar{G}}_e^{(11)}$ as given by (11.3) and (11.6) from a semi-infinite path to an infinite path in the λ-plane.

We consider an integral of the type

$$F = \int_0^\infty f_n(\lambda) J_n(\lambda r)\, d\lambda, \tag{11.23}$$

which can be written in the form

$$F = \frac{1}{2}\int_0^\infty f_n(\lambda)\left[H_n^{(1)}(\lambda r) + H_n^{(2)}(\lambda r)\right] d\lambda, \tag{11.24}$$

where $H_n^{(1)}$ and $H_n^{(2)}$ denote, respectively, the Hankel functions of the first and the second kind. The integral involving the second kind in (10.24) can be transformed into the first kind with a different path as follows:

$$\begin{aligned}
&\frac{1}{2}\int_0^\infty f_n(\lambda) H_n^{(2)}(\lambda r)\, d\lambda \\
&= \frac{1}{2}\int_{-\infty}^0 f_n(-\lambda)\, H_n^{(2)}(-\lambda r)\, d\lambda = -\frac{1}{2}\int_{-\infty}^0 f_n(-\lambda)\, e^{in\pi} H_n^{(1)}(\lambda r)\, d\lambda,
\end{aligned}$$

where we have made use of the half-circuit relation between the two kinds of Hankel functions [Sommerfeld, 1949, p. 315]. If $f_n(\lambda)$ satisfies the relation that

$$f_n(\lambda) = -e^{in\pi} f_n(-\lambda),$$

then the original integral becomes

$$F = \int_{-\infty}^\infty f_n(\lambda) H_n^{(1)}(\lambda r)\, dr. \tag{11.25}$$

The transform from (11.23) to (11.25) can be extended to a dyadic function by an operational method. Thus, if we let

$$\bar{\bar{F}} = \int_0^\infty \bar{\bar{f}}_n(\lambda)\, [J_n(\lambda r)]\, d\lambda, \tag{11.26}$$

where $\bar{\bar{f}}_n$ denotes a dyadic operator which satisfies the relation that

$$\bar{\bar{f}}_n(\lambda) = -e^{in\pi}\bar{\bar{f}}_n(-\lambda), \tag{11.27}$$

then

$$\bar{\bar{F}} = \int_{-\infty}^{\infty} \bar{\bar{f}}_n(\lambda)\left[H_n^{(1)}(\lambda r)\right] d\lambda. \tag{11.28}$$

Now we consider a typical term of $\bar{\bar{G}}_{eo}^{(1)}$ described by (11.3) for $z \geq z_0$,

$$\bar{\bar{g}} = \int_0^{\infty} C^{(1)}\overline{M}(h_1)\overline{M}'(-h_1)\, d\lambda. \tag{11.29}$$

Since $\overline{M}(h_1)$ contains $J_n(\lambda r)$, (11.29) can be written in operational form as (11.26) and the operator does satisfy the relation described by (11.27). Consequently, in view of (11.28), (11.29) can be transformed to

$$\bar{\bar{g}} = \int_{-\infty}^{\infty} C^{(1)}\overline{M}^{(1)}(h_1)\overline{M}'(-h_1)\, d\lambda, \tag{11.30}$$

where $\overline{M}^{(1)}$ is defined with respect to the Hankel function of the first kind. The operational method avoids the long and tedious task of applying (11.25), a scalar relation, repeatedly to each individual term contained in $\overline{M}(h_1)\overline{M}'(-h_1)$. In view of (11.30) and similar transformations for the other terms, $\bar{\bar{G}}_e^{(11)}$ for $z > z'$ can be written in the form

$$\begin{aligned}\bar{\bar{G}}_e^{(11)}(\overline{R}, \overline{R}') = \int_{-\infty}^{\infty} \sum_n C^{(1)} \\ \cdot \left\{\overline{M}^{(1)}(h_1)\left[\overline{M}'(-h_1) + a\overline{M}'(h_1)\right]\right. \\ \left. + \overline{N}^{(1)}(h_1)\left[\overline{N}'(-h_1) + b\overline{N}'(h_1)\right]\right\}.\end{aligned} \tag{11.31}$$

Having changed the contour of integration into an infinite path, we can apply the method of saddle-point integration to (11.31) provided that $kR \gg 1$. Following the same procedure as described in Sec. 7-4, we found that the first-order solution for the far-zone field is given by the following expression:

$$\begin{aligned}\bar{\bar{G}}_e^{(11)}(\overline{R}, \overline{R}') = \left(\frac{e^{ikR}}{4\pi kR\sin\theta}\right)\sum_{n=0}^{\infty}(2-\delta_0)\,(-i)^{n+1}\,{}^{\cos}_{\sin}\, n\phi \\ \cdot \left\{\left[\overline{M}'(-k_1\cos\theta) + a(\theta)\overline{M}'(k_1\cos\theta)\right]\hat{\phi}\right. \\ \left. - i\left[\overline{N}'(-k_1\cos\theta) + b(\theta)\overline{N}'(k_1\cos\theta)\right]\hat{\theta}\right\},\end{aligned} \tag{11.32}$$

where

$$a(\theta) = \frac{\left[\cos\theta - \left(n^2 - \sin^2\theta\right)^{\frac{1}{2}}\right]}{\left[\cos\theta + \left(n^2 - \sin^2\theta\right)^{\frac{1}{2}}\right]}$$

$$b(\theta) = \frac{\left[n\cos\theta - \left(n^2 - \sin^2\theta\right)^{\frac{1}{2}}\right]}{\left[n^2\cos\theta + \left(n^2 - \sin^2\theta\right)^{\frac{1}{2}}\right]}$$

$$\cos\theta = \left(\hat{R}\cdot\hat{z}\right).$$

It is recognized that the coefficients $a(\theta)$ and $b(\theta)$ play the same role as the plane-wave reflection coefficients for an incident E-field either perpendicular or parallel to the plane of incidence. Being the first-order solution, it does not take into consideration the proximity effect when the saddle point is very close to the poles of the coefficients a and b lying in the λ-plane. In practice, (11.32) only represents the contribution due to the so-called *space wave*. It is a valid approximation when the point of observation is not near the ground. For a detailed discussion of the high- order solution, readers are referred to Baños's book quoted previously and the detailed discussion by Feynberg [1961] on the field near the ground, or the so-called *ground wave*. The high-order solution for this kind of problem has also been discussed very thoroughly by Felsen and Marcuvitz [1973] and in the recent book by King, Wu, and Owens [1991].

11-3 A DIELECTRIC LAYER ON A CONDUCTING PLANE

The structure under consideration is shown in Fig. 11-2. The site of the interface is denoted by S, the site of the conducting plane by S_0. If the conducting plane has an aperture, that portion of the plane will be denoted by S_A. The wave number in region 1 (air) and that in region 2 (dielectric) will be denoted, respectively, by k_1 and k_2. We assume $\mu_1 = \mu_2 = \mu_0$, but the dielectric constant of the layer can be complex. Two cases will be treated in detail depending on the location of the electric current source. The electric dyadic Green functions to be used are $\overline{\overline{G}}_{e1}^{(11)}, \overline{\overline{G}}_{e1}^{(22)}, \overline{\overline{G}}_{e1}^{(12)}$, and $\overline{\overline{G}}_{e1}^{(21)}$. All these functions are of the third kind as implied by the superscripts. They are also functions of the first kind as indicated by the subscript "$e1$" because of the presence of a conducting plane for the composite structure. We will first derive the integral expressions for the electric field in the two regions and then find the eigenfunction expansions of the relevant dyadic Green functions.

Case 1. An Electric Current Source in Region 1

In this case, the differential equations for the electric fields in the two regions are

$$\nabla\times\nabla\times\overline{E}_1(\overline{R}) - k_1^2\overline{E}_1(\overline{R}) = i\omega\mu_0\overline{J}_1(\overline{R}) \tag{11.33}$$

$$\nabla\times\nabla\times\overline{E}_2(\overline{R}) - k_2^2\overline{E}_2(\overline{R}) = 0. \tag{11.34}$$

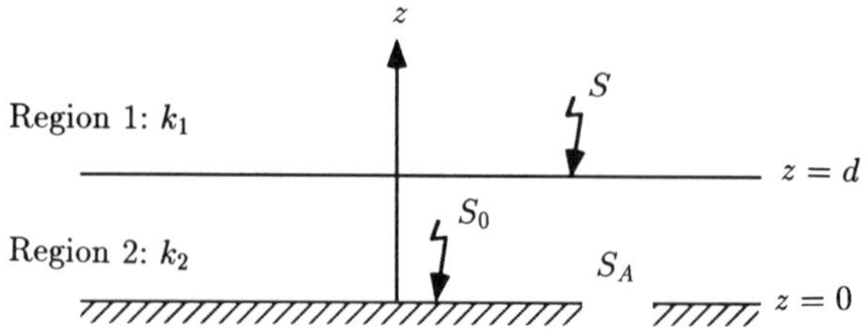

Fig. 11-2 A dielectric layer on a conducting ground plane with an aperture

To find the integral expressions for $\overline{E}_1$ and $\overline{E}_2$, we need the electric dyadic Green functions $\overline{\overline{G}}_{e1}^{(11)}, \overline{\overline{G}}_{e1}^{(22)}, \overline{\overline{G}}_{e1}^{(12)}$, and $\overline{\overline{G}}_{e1}^{(21)}$ which satisfy the differential equations

$$\nabla \times \nabla \times \overline{\overline{G}}_{e1}^{(11)}(\overline{R}, \overline{R}') - k_1^2 \overline{\overline{G}}_{e1}^{(11)}(\overline{R}, \overline{R}') = \overline{\overline{I}}\delta(\overline{R} - \overline{R}') \tag{11.35}$$

$$\nabla \times \nabla \times \overline{\overline{G}}_{e1}^{(22)}(\overline{R}, \overline{R}') - k_2^2 \overline{\overline{G}}_{e1}^{(22)}(\overline{R}, \overline{R}') = \overline{\overline{I}}\delta(\overline{R} - \overline{R}') \tag{11.36}$$

$$\nabla \times \nabla \times \overline{\overline{G}}_{e1}^{(12)}(\overline{R}, \overline{R}') - k_1^2 \overline{\overline{G}}_{e1}^{(12)}(\overline{R}, \overline{R}') = 0 \tag{11.37}$$

$$\nabla \times \nabla \times \overline{\overline{G}}_{e1}^{(21)}(\overline{R}, \overline{R}') - k_2^2 \overline{\overline{G}}_{e1}^{(21)}(\overline{R}, \overline{R}') = 0. \tag{11.38}$$

The boundary conditions imposed on these functions at S are

$$\hat{z} \times \left[\overline{\overline{G}}_{e1}^{(11)} - \overline{\overline{G}}_{e1}^{(21)}\right] = 0 \tag{11.39}$$

$$\hat{z} \times \left[\nabla \times \overline{\overline{G}}_{e1}^{(11)} - \nabla \times \overline{\overline{G}}_{e1}^{(21)}\right] = 0 \tag{11.40}$$

$$\hat{z} \times \left[\overline{\overline{G}}_{e1}^{(22)} - \overline{\overline{G}}_{e1}^{(12)}\right] = 0 \tag{11.41}$$

$$\hat{z} \times \left[\nabla \times \overline{\overline{G}}_{e1}^{(22)} - \nabla \times \overline{\overline{G}}_{e1}^{(12)}\right] = 0 \tag{11.42}$$

and on S_0, the boundary conditions imposed on $\overline{\overline{G}}_{e1}^{(21)}$ and $\overline{\overline{G}}_{e1}^{(22)}$ are

$$\hat{z} \times \overline{\overline{G}}_{e1}^{(21)} = 0 \tag{11.43}$$

$$\hat{z} \times \overline{\overline{G}}_{e1}^{(22)} = 0. \tag{11.44}$$

Now we apply the vector-dyadic Green's theorem

$$\begin{aligned} &\iiint \left[\overline{P} \cdot \nabla \times \nabla \times \overline{\overline{Q}} - (\nabla \times \nabla \times \overline{P}) \cdot \overline{\overline{Q}}\right] dV \\ &= -\oint\!\!\!\oint \left[(\hat{n} \times \nabla \times \overline{P}) \cdot \overline{\overline{Q}} + (\hat{n} \times \overline{P}) \cdot \nabla \times \overline{\overline{Q}}\right] dS \end{aligned} \tag{11.45}$$

to region 1 with $\bar{P} = \bar{E}_1, \bar{\bar{Q}} = \bar{\bar{G}}_{e1}^{(11)}$. In view of (11.33) and (11.35) we obtain

$$\begin{aligned}\bar{E}_1(\bar{R}') = & i\omega\mu_0 \iiint_{V_1} \bar{J}_1(\bar{R}) \cdot \bar{\bar{G}}_{e1}^{(11)}(\bar{R}, \bar{R}')dV \\ & + \iint_S \left\{ \left[\hat{z} \times \nabla \times \bar{E}_1(\bar{R})\right] \cdot \bar{\bar{G}}_{e1}^{(11)}(\bar{R}, \bar{R}') \right. \\ & \left. + \left[\hat{z} \times \bar{E}_1(\bar{R})\right] \cdot \nabla \times \bar{\bar{G}}_{e1}^{(11)}(\bar{R}, \bar{R}') \right\} dS. \end{aligned} \tag{11.46}$$

We have already deleted the surface integral at infinity in that region as a result of the radiation condition. By applying (11.45) to region 2 with $\bar{P} = \bar{E}_2$ and $\bar{\bar{Q}} = \bar{\bar{G}}_{e1}^{(21)}$ we find that the volume integral vanishes because of (11.34) and (11.38); thus

$$\begin{aligned} & - \iint_S \left\{ \left[\hat{z} \times \nabla \times \bar{E}_2(\bar{R})\right] \cdot \bar{\bar{G}}_{e1}^{(21)}(\bar{R}, \bar{R}') \right. \\ & \left. + \left[\hat{z} \times \bar{E}_2(\bar{R})\right] \cdot \nabla \times \bar{\bar{G}}_{e1}^{(21)}(\bar{R}, \bar{R}') \right\} dS \\ & + \iint_{S_0} \left\{ \left[\hat{z} \times \nabla \times \bar{E}_2(\bar{R})\right] \cdot \bar{\bar{G}}_{e1}^{(21)}(\bar{R}, \bar{R}') \right. \\ & \left. + \left[\hat{z} \times \bar{E}_2(\bar{R})\right] \cdot \nabla \times \bar{\bar{G}}_{e1}^{(21)}(\bar{R}, \bar{R}') \right\} dS = 0. \end{aligned} \tag{11.47}$$

On $S_0, \bar{\bar{G}}_{e1}^{(21)}$ satisfies the dyadic Dirichlet condition stated by (11.43) and $\hat{z} \times \bar{E}_2(\bar{R})$ vanishes elsewhere except at S_A; (11.47), therefore, reduces to

$$\begin{aligned} & \iint_S \left\{ \left[\hat{z} \times \nabla \times \bar{E}_2(\bar{R})\right] \cdot \bar{\bar{G}}_{e1}^{(21)}(\bar{R}, \bar{R}') \right. \\ & \left. + \left[\hat{z} \times \bar{E}_2(\bar{R})\right] \cdot \nabla \times \bar{\bar{G}}_{e1}^{(21)}(\bar{R}, \bar{R}') \right\} dS \\ & = \iint_{S_A} \left[\hat{z} \times \bar{E}_2(\bar{R})\right] \cdot \nabla \times \bar{\bar{G}}_{e1}^{(21)}(\bar{R}, \bar{R}')\, dS. \end{aligned} \tag{11.48}$$

Because of the boundary conditions stated by (11.39) and (11.40) and two similar vector conditions for $\bar{E}_1$ and $\bar{E}_2$ and $\nabla \times \bar{E}_1$ and $\nabla \times \bar{E}_2$, the two surface integrals on S in (11.46) and (11.48) are equal to each other. By eliminating the two surface integrals from these two equations, we obtain

$$\begin{aligned}\bar{E}_1(\bar{R}') = & i\omega\mu_0 \iiint_{V_1} \bar{J}_1(\bar{R}) \cdot \bar{\bar{G}}_{e1}^{(11)}(\bar{R}, \bar{R}')\, dV \\ & + \iiint_{S_A} \left[\hat{z} \times \bar{E}_2(\bar{R})\right] \cdot \nabla \times \bar{\bar{G}}_{e1}^{(21)}(\bar{R}, \bar{R}')\, dS. \end{aligned} \tag{11.49}$$

By interchanging $\bar{R}'$ with $\bar{R}$ and making use of the transposed property of the scalar product of a dyadic function with a vector function we can transform

(11.49) into the form

$$\overline{E}_1(\overline{R}) = i\omega\mu_0 \iiint_V \left[\overline{\overline{G}}_{e1}^{(11)}(\overline{R}', \overline{R})\right]^T \cdot \overline{J}_1(\overline{R}')dV' + \oiint_{S_A} \left[\nabla' \times \overline{\overline{G}}_{e1}^{(21)}(\overline{R}', \overline{R})\right]^T \cdot \left[\hat{z} \times \overline{E}_2(\overline{R}')\right] dS'. \tag{11.50}$$

By definition,

$$\nabla \times \overline{\overline{G}}_{e1}^{(21)}(\overline{R}', \overline{R}) = \overline{\overline{G}}_{m2}^{(21)}(\overline{R}', \overline{R}), \tag{11.51}$$

where $\overline{\overline{G}}_{m2}^{(21)}(\overline{R}', \overline{R})$ denotes the magnetic dyadic Green function of the second kind as well as of the third kind; thus an alternative formula for (11.50) is

$$\overline{E}_1(\overline{R}) = i\omega\mu_0 \iiint_{V_1} \left[\overline{\overline{G}}_{e1}^{(11)}(\overline{R}', \overline{R})\right]^T \cdot \overline{J}_1(\overline{R}')\, dV' + \oiint_{S_A} \left[\overline{\overline{G}}_{m2}^{(21)}(\overline{R}', \overline{R})\right]^T \cdot \left[\hat{z} \times \overline{E}_2(\overline{R}')\right] dS'. \tag{11.52}$$

Before we discuss the symmetrical relationships of the transposed functions in (11.52), let us write down the expression for $\overline{E}_2(\overline{R})$, with the current source still in region 1. By applying (11.45) first to region 1 with $\overline{P} = \overline{E}_1, \overline{\overline{Q}} = \overline{\overline{G}}_{e1}^{(12)}$ and then to region 2 with $\overline{P} = \overline{E}_2, \overline{\overline{Q}} = \overline{\overline{G}}_{e1}^{(22)}$ we can readily derive the following expression for $\overline{E}_2(\overline{R})$:

$$\overline{E}_2(\overline{R}) = i\omega\mu_0 \iiint \left[\overline{\overline{G}}_{e1}^{(12)}(\overline{R}', \overline{R})\right]^T \cdot \overline{J}_1(\overline{R})\, dV' + \oiint_{S_A} \left[\overline{\overline{G}}_{m2}^{(22)}(\overline{R}', \overline{R})\right]^T \cdot \left[\hat{z} \times \overline{E}_2(\overline{R}')\right] dS'. \tag{11.53}$$

Case 2. An Electric Current Source in Region 2

The differential equations for the two electric fields in this case are

$$\nabla \times \nabla \times \overline{E}_1(\overline{R}) - k_1^2 \overline{E}_1(\overline{R}) = 0 \tag{11.54}$$

$$\nabla \times \nabla \times \overline{E}_2(\overline{R}) - k_2^2 \overline{E}_2(\overline{R}) = i\omega\mu_0 \overline{J}_2(\overline{R}). \tag{11.55}$$

The four electric dyadic Green functions introduced in the previous case are again needed. Without repeating much of the same procedure we merely give the results as follows:

$$\overline{E}_1(\overline{R}) = i\omega\mu_0 \iiint_{V_2} \left[\overline{\overline{G}}_{e1}^{(21)}(\overline{R}', \overline{R})\right]^T \cdot \overline{J}_2(\overline{R}')\, dV' + \oiint_{S_A} \left[\overline{\overline{G}}_{m2}^{(21)}(\overline{R}', \overline{R})\right]^T \cdot \left[\hat{z} \times \overline{E}_2(\overline{R}')\right] dS' \tag{11.56}$$

$$\overline{E}_2(\overline{R}) = i\omega\mu_0 \iiint_{V_2} \left[\overline{\overline{G}}_{e1}^{(22)}(\overline{R}', \overline{R})\right]^T \cdot \overline{J}_2(\overline{R}')\, dV' + \oiint_{S_A} \left[\overline{\overline{G}}_{m2}^{(22)}(\overline{R}', \overline{R})\right]^T \cdot \left[\hat{z} \times \overline{E}_2 \overline{R}\right] dS'. \tag{11.57}$$

Equations (11.52)–(11.53) and (11.56)–(11.57) involve six distinct transposed functions. To find the symmetrical relationships of these functions, we will make use of the relevant reciprocity theorems for the composite structure under consideration; they are the Rayleigh-Carson theorem and the complementary $\bar{J} \cdot \bar{H}$ reciprocity theorem. Some simple versions of these two theorems have already been introduced in Chap. 4.

11-4 RECIPROCITY THEOREMS FOR STRATIFIED MEDIA

The Rayleigh-Carson reciprocity theorem is applicable to the structure shown in Fig. 11-3, where S_e is the site of an electrically perfect conducting ground plane and S is the site of the interface of two isotropic media. By applying Stratton's vector Green's theorem to the two different regions, several formulas can be derived. The procedure is very similar to the treatment found in Sec. 4-5. There are three distinct cases.

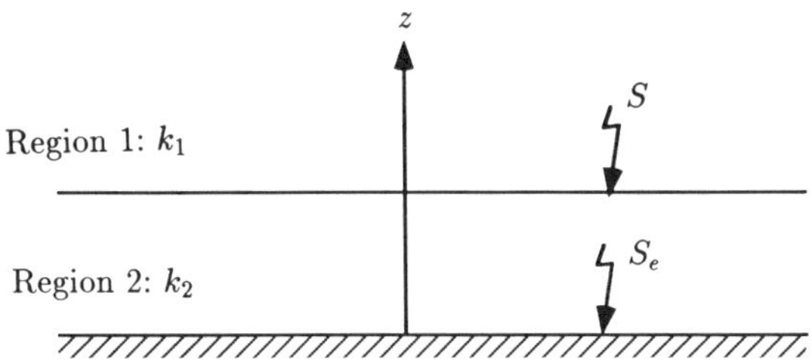

Fig. 11-3 A dielectric layer on a perfectly conducting plane

Case 1. Two sets of fields due to two distinct electric current sources located *both* in region 1. The Rayleigh-Carson theorem in this case states

$$\iiint_{V_a} \bar{J}_{1a} \cdot \bar{E}_{1b}\, dV = \iiint_{V_b} \bar{J}_{1b} \cdot \bar{E}_{1a}\, dV, \tag{11.58}$$

where $\bar{E}_{1a}$ denotes the electric field produced by $\bar{J}_{1a}$ and $\bar{E}_{1b}$ the electric field produced by $\bar{J}_{1b}$. All these quantities are defined in region 1. V_a and V_b denote, respectively, the volume occupied by the two current sources. The two fields, of course, are the fields produced by the current sources in the presence of the layered medium on an electrically perfect conducting plane.

Case 2. Two sets of fields due to two distinct electric current sources located both in region 2.

In this case we have

$$\iiint_{V_a} \bar{J}_{2a} \cdot \bar{E}_{2b}\, dV = \iiint_{V_b} \bar{J}_{2b} \cdot \bar{E}_{2a} dV. \tag{11.59}$$

Case 3. One current source, $\bar{J}_{1a}$ located in region 1 and another current source,

$\overline{J}_{2b}$, located in region 2.

In this case, the reciprocity theorem has the form

$$\iiint_{V_a} \overline{J}_{1a} \cdot \overline{E}_{1b} \, dV = \iiint_{V_b} \overline{J}_{2b} \cdot \overline{E}_{2a} \, dV, \tag{11.60}$$

where $\overline{E}_{1b}$ is the electric field in region 1 due to a current source $\overline{J}_{2b}$ placed in region 2. From these formulas, the symmetrical relationships of the electric dyadic Green functions can be readily derived.

Let the current source $\overline{J}_{1a}$ be an infinitesimal electric dipole located in region 1 at $\overline{R}_a$ and pointed in the $\hat{x}_i$-direction, and let

$$\overline{J}_{1a} = \frac{\hat{x}_i \delta(\overline{R} - \overline{R}_a)}{i\omega\mu_1}. \tag{11.61}$$

Similarly, we let

$$\overline{J}_{1b} = \frac{\hat{x}_j \delta(\overline{R} - \overline{R}_b)}{i\omega\mu_1} \tag{11.62}$$

representing a current source located at $\overline{R}_b$, pointing in the $\hat{x}_j$-direction. By substituting (11.61) and (11.62) into (11.58), we obtain

$$\hat{x}_i \cdot \overline{E}_{1b} \left(\overline{R}_a\right) = \hat{x}_j \cdot \overline{E}_{1a} \left(\overline{R}_b\right). \tag{11.63}$$

By definition of the electric dyadic Green functions

$$\overline{E}_{1a}(\overline{R}_b) = \overline{\overline{G}}_{e1}^{(11)}(\overline{R}_b, \overline{R}_a) \cdot \hat{x}_j \tag{11.64}$$

$$\overline{E}_{1b}(\overline{R}_a) = \overline{\overline{G}}_{e1}^{(11)}(\overline{R}_a, \overline{R}_b) \cdot \hat{x}_i; \tag{11.65}$$

thus

$$\begin{aligned} &\hat{x}_i \cdot \overline{\overline{G}}_{e1}^{(11)}(\overline{R}_b, \overline{R}_a) \cdot \hat{x}_j \\ &\quad = \hat{x}_j \cdot \overline{\overline{G}}_{e1}^{(11)}(\overline{R}_a, \overline{R}_b) \cdot \hat{x}_i. \end{aligned} \tag{11.66}$$

In the language of dyadic analysis, (11.66), with $\overline{R}_a$ and $\overline{R}_b$ replaced by $\overline{R}'$ and $\overline{R}$, means

$$\left[\overline{\overline{G}}_{e1}^{(11)}(\overline{R}', \overline{R})\right]^T = \overline{\overline{G}}_{e1}^{(11)}(\overline{R}, \overline{R}'). \tag{11.67}$$

Similarly, by means of (11.59) we can derive

$$\left[\overline{\overline{G}}_{e1}^{(22)}(\overline{R}', \overline{R})\right]^T = \overline{\overline{G}}_{e1}^{(22)}(\overline{R}, \overline{R}'). \tag{11.68}$$

The two electric fields in (11.60) are produced by currents located in different regions. Thus, if we have

$$\overline{J}_{1a} = \frac{\hat{x}_i \delta(\overline{R} - \overline{R}_a)}{i\omega\mu_1}$$
$$\overline{J}_{2b} = \frac{\hat{x}_j \delta(\overline{R} - \overline{R}_b)}{i\omega\mu_2},$$

(11.60) yields

$$\frac{1}{\mu_1}\hat{x}_i \cdot \overline{E}_{1b}(\overline{R}_a) = \frac{1}{\mu_2}\hat{x}_j \cdot \overline{E}_{2a}(\overline{R}_b). \tag{11.69}$$

In this case,

$$\overline{E}_{1b}(\overline{R}_a) = \overline{\overline{G}}_{e1}^{(12)}(\overline{R}_a, \overline{R}_b) \cdot \hat{x}_j$$
$$\overline{E}_{2a}(\overline{R}_b) = \overline{\overline{G}}_{e1}^{(21)}(\overline{R}_b, \overline{R}_a) \cdot \hat{x}_i;$$

therefore, the symmetrical relationship between the two functions is

$$\frac{1}{\mu_2}\left[\overline{\overline{G}}_{e1}^{(21)}(\overline{R}', \overline{R})\right]^T = \frac{1}{\mu_1}\overline{\overline{G}}_{e1}^{(12)}(\overline{R}, \overline{R}'). \tag{11.70}$$

Equations (11.67), (11.68), and (11.70) are the three symmetrical relations which we are seeking. They can be condensed into one formula in the form

$$\frac{1}{\mu_i}\left[\overline{\overline{G}}_{e1}^{(ij)}(\overline{R}', \overline{R})\right]^T = \frac{1}{\mu_j}\overline{\overline{G}}_{e1}^{(ji)}(\overline{R}, \overline{R}'), \tag{11.71}$$

with $i, j = 1, 2$ where i and j can be equal or different. Equation (11.71) is an extension of (4.206). The latter does not involve an electrically conducting surface or a conducting body.

To find the symmetrical relations for the magnetic dyadic Green functions based on reciprocity theorems we have to derive a complementary $\overline{J} \cdot \overline{H}$ theorem for the problem under consideration. The derivation is much more involved but the result is comparable to the Rayleigh-Carson theorem or the $\overline{J} \cdot \overline{E}$ theorem in appearance.

Figure 11-4 shows two models. In model A, the medium constants are μ_1, ϵ_1 and μ_2, ϵ_2 in the two regions with wave number k_1 and k_2. These constants are assumed to be known. The ground plane in region 2 is an electrically perfect conducting plane. In model B, the medium constants are μ_1', ϵ_1' with wave number k_1' in region 1 and the same constants in region 2 as in model A. The ground plane in model B is a *magnetically* perfect conducting plane. The constants μ_1' and ϵ_1' are so far unspecified; they will be determined later.

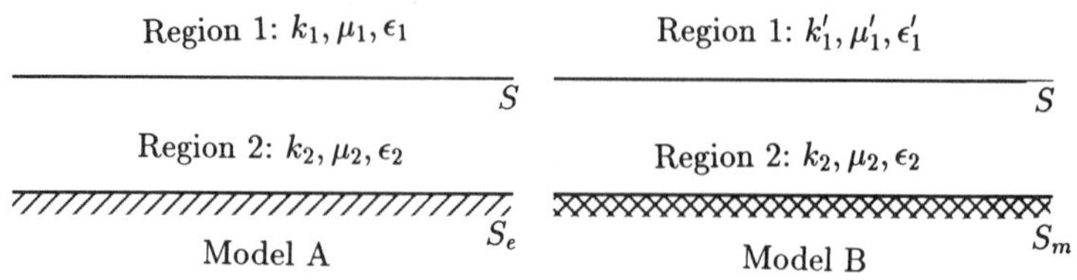

Fig. 11-4 Model A: Two plane stratified media in contact with an electrically perfect conducting wall; model B: two plane stratified media with a magnetically perfect conducting wall

The differential equations and the boundary conditions for the fields excited by electric current sources are

$$\nabla \times \nabla \times \overline{E}_{1A} - k_1^2 \overline{E}_{1A} = i\omega\mu_1 \overline{J}_{1A} \tag{11.72}$$

$$\nabla \times \nabla \times \overline{E}_{2A} - k_2^2 \overline{E}_{2A} = i\omega\mu_2 \overline{J}_{2A} \tag{11.73}$$

$$\nabla \times \nabla \times \overline{E}_{1B} - k_1'^2 \overline{E}_{1B} = i\omega\mu_1' \overline{J}_{1B} \tag{11.74}$$

$$\nabla \times \nabla \times \overline{E}_{2B} - k_2^2 \overline{E}_{2B} = i\omega\mu_2' \overline{J}_{2B} \tag{11.75}$$

on S,

$$\hat{z} \times \left(\overline{E}_{1A} - \overline{E}_{2A}\right) = 0 \tag{11.76}$$

$$\hat{z} \times \left(\overline{H}_{1A} - \overline{H}_{2A}\right) = 0 \tag{11.77}$$

$$\hat{z} \times \left(\overline{E}_{1B} - \overline{E}_{2B}\right) = 0 \tag{11.78}$$

$$\hat{z} \times \left(\overline{H}_{1B} - \overline{H}_{2B}\right) = 0 \tag{11.79}$$

on S_e,

$$\hat{z} \times \overline{E}_{2A} = 0 \tag{11.80}$$

on S_m,

$$\hat{z} \times \overline{H}_{2B} = 0. \tag{11.81}$$

Several different cases will be considered.

Case 1. Currents $\overline{J}_{1A}$ and $\overline{J}_{1B}$ present in region 1, no currents in region 2:

Now we apply the vector Green's theorem (4.212) to region 1 with $\overline{P} = \overline{E}_{1A}$ and $\overline{Q} = \overline{H}_{1B}$ where $\overline{H}_{1B}$ satisfies the differential equation

$$\nabla \times \nabla \times \overline{H}_{1B} - k_1'^2 \overline{H}_{1B} = \nabla \times \overline{J}_{1B}, \tag{11.82}$$

which yields

$$\iiint_{V_1} \left[\overline{E}_{1A} \cdot \left(k_1'^2 \overline{H}_{1B} + \nabla \times \overline{J}_{1B}\right) - \overline{H}_{1B} \cdot \left(k_1^2 \overline{E}_{1A} + i\omega\mu_1 \overline{J}_{1A}\right)\right] dV$$
$$= \iint_S \hat{z} \cdot \left[i\omega\mu_1 \overline{H}_{1A} \times \overline{H}_{1B} + \overline{E}_{1A} \times \left(\overline{J}_{1B} - i\omega\epsilon_1' \overline{E}_{1B}\right)\right] dS, \tag{11.83}$$

where we have already deleted the surface integral at infinity due to the radiation condition. One of the surface integrals in (11.83) can be decomposed into two parts, namely,

$$\iint_S \hat{z}\cdot(\overline{E}_{1A}\times\overline{J}_{1B})\,dS = \iiint_{V_1}\nabla\cdot(\overline{E}_{1A}\times\overline{J}_{1B})\,dV$$
$$= \iiint_{V_1}(\overline{J}_{1B}\cdot\nabla\times\overline{E}_{1A}-\overline{E}_{1A}\cdot\nabla\times\overline{J}_{1B})\,dV$$
$$= \iiint_{V_1}(i\omega\mu_1\overline{J}_{1B}\cdot\overline{H}_{1A}-\overline{E}_{1A}\cdot\nabla\times\overline{J}_{1B})\,dV. \tag{11.84}$$

By substituting (11.84) into (11.83) and rearranging the terms, we obtain

$$\iiint_{V_1}\left[\left(k_1'^2-k_1^2\right)\overline{E}_{1A}\cdot\overline{H}_{1B}+i\omega\mu_1\left(\overline{J}_{1B}\cdot\overline{H}_{1A}-\overline{J}_{1A}\cdot\overline{H}_{1B}\right)\right]dV$$
$$= \iint_S \hat{z}\cdot\left(i\omega\mu_1\overline{H}_{1A}\times\overline{H}_{1B}-i\omega\epsilon_1'\overline{E}_{1A}\times\overline{E}_{1B}\right)dS. \tag{11.85}$$

We now impose the relationship such that

$$k_1' = k_1$$

or

$$\mu_1'\epsilon_1' = \mu_1\epsilon_1. \tag{11.86}$$

Since μ_1 and ϵ_1 are given constants, (11.86) puts a constraint on the product of μ_1' and ϵ_1' but not individually. Equation (11.86) will be referred to as the wave number matching condition. Under this condition (11.85) can be written in the form

$$\iiint_{V_1}(\overline{J}_{1B}\cdot\overline{H}_{1A}-\overline{J}_{1A}\cdot\overline{H}_{1B})\,dV$$
$$= \iint_S \hat{z}\cdot\left[\overline{H}_{1A}\times\overline{H}_{1B}-\left(\frac{\epsilon_1'}{\mu_1}\right)\overline{E}_{1A}\times\overline{E}_{1B}\right]dS. \tag{11.87}$$

The vector Green's theorem is now applied to region 2 with $\overline{P} = \overline{E}_{2A}$ and $\overline{Q} = \overline{H}_{2B}$. The volume integral vanishes because both $\overline{E}_{2A}$ and $\overline{H}_{2B}$ satisfy the homogeneous wave equation with the same wave number k_2. And on S_e, $\hat{z}\times\overline{E}_{2A} = 0$ and on $\S_m$, $\hat{z}\times\overline{H}_{2B} = 0$, the result yields

$$\iint_S \hat{z}\cdot\left[\overline{H}_{2A}\times\overline{H}_{2B}-\left(\frac{\epsilon_2}{\mu_2}\right)\overline{E}_{2A}\times\overline{E}_{2B}\right]dS = 0. \tag{11.88}$$

Now we impose a condition on ϵ_1' such that

$$\epsilon_1' = \left(\frac{\mu_1}{\mu_2}\right)\epsilon_2. \tag{11.89}$$

Under this condition the surface integral in (11.87) is equal to the surface integral in (11.88) because of the continuity condition of the tangential components of the $\bar{E}$ and $\bar{H}$ fields on S. Thus (11.87) becomes

$$\iiint_{V_1} \left(\bar{J}_{1A} \cdot \bar{H}_{1B} - \bar{J}_{1B} \cdot \bar{H}_{1A}\right) dV = 0. \tag{11.90}$$

This is the complementary reciprocity theorem for the two sets of magnetic field in model A and model B, designated as the $\bar{J} \cdot \bar{H}$ theorem. By combining (11.86) with (11.89), it can be shown readily that

$$\left(\frac{\mu_1 \mu_1'}{\epsilon_1 \epsilon_1'}\right)^{\frac{1}{2}} = \frac{\mu_2}{\epsilon_2}$$

or

$$z_1 z_1' = z_2^2, \tag{11.91}$$

which is the complementary impedance condition for the wave impedances in the three media of the two models. When $\mu_1 = \mu_2 = \mu_0, \epsilon_1 = \epsilon_0$, and $\epsilon_2 = \epsilon$ we have $\epsilon_1' = \epsilon$ and $\mu_1' = (\epsilon_0/\epsilon)\,\mu_0$. Unlike in network synthesis the physical realizability of model B is not an issue in this theory. The model is introduced mainly to derive a mathematical theorem to be used to find the symmetrical relations of the magnetic dyadic Green functions. The situation is similar to the use of vector potential function in electromagnetic theory which is not a physically measurable quantity. It is introduced simply as a mathematical tool. Following a similar procedure the $\bar{J} \cdot \bar{H}$ theorem for the other cases can be derived accordingly. They are stated below.

Case 2. Currents $\bar{J}_{2A}$ and $\bar{J}_{2B}$ in region 2, no currents in region 1:

$$\iiint_{V_2} \left(\bar{J}_{2A} \cdot \bar{H}_{2B} - \bar{J}_{2B} \cdot \bar{H}_{2A}\right) dV = 0. \tag{11.92}$$

Case 3. Current $\bar{J}_{1A}$ in region 1 and current $\bar{J}_{2B}$ in region 2, $\bar{J}_{2A} = 0, \bar{J}_{1B} = 0$:

$$\iiint_{V_1} \bar{J}_{1A} \cdot \bar{H}_{1B}\, dV = \iiint_{V_2} \bar{J}_{2B} \cdot \bar{H}_{2A}\, dV. \tag{11.93}$$

Case 4. Current $\bar{J}_{1B}$ in region 1 and current $\bar{J}_{2A}$ in region 2, $\bar{J}_{1A} = 0, \bar{J}_{2B} = 0$:

$$\iiint_{V_1} \bar{J}_{1B} \cdot \bar{H}_{1A}\, dV = \iiint_{V_2} \bar{J}_{2A} \cdot \bar{H}_{2B}\, dV. \tag{11.94}$$

It should be emphasized that although we have used the plane stratified structure to derive the $\bar{J} \cdot \bar{E}$ and $\bar{J} \cdot \bar{H}$ theorems, they are valid for similar stratified structures such as a conducting cylinder or a sphere coated with a layer of dielectric material. The theorem can also be extended to multiple layers of isotropic media placed above an electrically perfect conducting plane (model A).

In general, for $n = 1, 2, \ldots, N$, where the uppermost region ($n = 1$) may either extend to infinity or be terminated by an electric wall in model A and a magnetic wall in model B, the general theorem is

$$\iiint_{V_i} \overline{J}_{iA} \cdot \overline{H}_{iB}\, dV = \iiint_{V_j} \overline{J}_{jB} \cdot \overline{H}_{jA}\, dV \tag{11.95}$$

where $i, j = 1, 2, \cdots N$, derivable under the condition

$$\begin{aligned} k_n &= k_n', & n &= 1, 2, \ldots, N \\ z_n z_n' &= z_N^2, & n &= 1, 2, \ldots, N, \end{aligned}$$

where the last layer ($n = N$) is the one in contact with either an electric wall or a magnetic wall.

With the complementary $\overline{J} \cdot \overline{H}$ reciprocal theorem at our disposal the symmetrical relationships of the magnetic dyadic Green functions can readily be found.

For example, we consider Case 3 with two localized currents

$$\overline{J}_{1A} = \frac{\hat{x}_i \delta(\overline{R} - \overline{R}_a)}{i\omega\mu_1} \tag{11.96}$$

$$\overline{J}_{2B} = \frac{\hat{x}_j \delta(\overline{R} - \overline{R}_b)}{i\omega\mu_2}; \tag{11.97}$$

then, by definition,

$$\overline{H}_{2A} = \overline{\overline{G}}_{m2}^{(21)}(\overline{R}, \overline{R}_a) \cdot \frac{\hat{x}_i}{i\omega\mu_2} \tag{11.98}$$

$$\overline{H}_{1B} = \overline{\overline{G}}_{m1}^{(12)}(\overline{R}, \overline{R}_b) \cdot \frac{\hat{x}_j}{i\omega\mu_1'}. \tag{11.99}$$

The magnetic dyadic Green function of the second kind is involved in model A because at S_d, $\overline{H}_{2A}$ satisfies the vector Neumann boundary condition, while in model B, $\overline{H}_{1B}$ satisfies the vector Dirichlet boundary condition at S_m; hence the corresponding magnetic dyadic Green function must be of the first kind. Of course, the superscript implies that both are of functions of the third kind too. By substituting (11.96)–(11.99) into (11.93) we obtain

$$\begin{aligned} &\hat{x}_i \cdot \overline{\overline{G}}_{m1}^{(12)}(\overline{R}_a, \overline{R}_b) \cdot \frac{\hat{x}_j}{\mu_1 \mu_1'} \\ &= \hat{x}_j \cdot \overline{\overline{G}}_{m2}^{(21)}(\overline{R}_b, \overline{R}_a) \cdot \frac{\hat{x}_i}{\mu_2^2}, \end{aligned}$$

which is equivalent to

$$\frac{1}{k_1^2}\left[\overline{\overline{G}}_{m1}^{(12)}(\overline{R}, \overline{R}')\right] = \frac{1}{k_2^2}\left[\overline{\overline{G}}_{m2}^{(21)}(\overline{R}', \overline{R})\right]^T \tag{11.100}$$

because from (11.86) and (11.89) one finds $\mu_1\mu_1'/\mu_2^2 = k_1^2/k_2^2$.

Likewise, we can derive

$$\overline{\overline{G}}_{m1}^{(22)}(\overline{R},\overline{R}') = \left[\overline{\overline{G}}_{m2}^{(22)}(\overline{R}',\overline{R})\right]^T. \tag{11.101}$$

The above two transposed functions are the ones that appeared in (11.52), (11.53), (11.56), and (11.57). With the aid these symmetrical relationships for the dyadic Green functions, the expressions for the electric fields in different cases can now be written in the form

$$\begin{aligned}\overline{E}_1(\overline{R}) &= i\omega\mu_0 \iiint_{V_1} \overline{\overline{G}}_{e1}^{(11)}(\overline{R},\overline{R}') \cdot \overline{J}_1(\overline{R}')\, dV' \\ &+ \left(\frac{k_2}{k_1}\right)^2 \iint_{S_A} \overline{\overline{G}}_{m1}^{(12)}(\overline{R},\overline{R}') \cdot \left[\hat{z} \times \overline{E}_2(\overline{R}')\right]\, dS' \end{aligned} \tag{11.102}$$

$$\begin{aligned}\overline{E}_2(\overline{R}) &= i\omega\mu_0 \iiint \overline{\overline{G}}_{e1}^{(21)}(\overline{R},\overline{R}') \cdot \overline{J}_1(\overline{R})\, dV' \\ &+ \iint_{S_A} \overline{\overline{G}}_{m1}^{(22)}(\overline{R},\overline{R}') \cdot \left[\hat{z} \times \overline{E}_2(\overline{R}')\right] dS' \end{aligned} \tag{11.103}$$

$$\begin{aligned}\overline{E}_1(\overline{R}) &= i\omega\mu_0 \iiint_{V_2} \overline{\overline{G}}_{e1}^{(12)}(\overline{R},\overline{R}') \cdot \overline{J}_2(\overline{R}')\, dV' \\ &+ \left(\frac{k_2}{k_1}\right)^2 \iint_{S_A} \left[\overline{\overline{G}}_{m1}^{(12)}(\overline{R},\overline{R}')\right] \cdot \left[\hat{z} \times \overline{E}_2(\overline{R}')\right] dS' \end{aligned} \tag{11.104}$$

$$\begin{aligned}\overline{E}_2(\overline{R}) &= i\omega\mu_0 \iiint_{V_2} \overline{\overline{G}}_{e1}^{(22)}(\overline{R},\overline{R}') \cdot \overline{J}_2(\overline{R}')\, dV' \\ &+ \iint_{S_A} \left[\overline{\overline{G}}_{m1}^{(22)}(\overline{R},\overline{R}')\right] \cdot \left[\hat{z} \times \overline{E}_2(\overline{R}')\right] dS'. \end{aligned} \tag{11.105}$$

This completes our long and tedious derivations of these formulas. In retrospect, model B was introduced merely to derive the $\overline{J} \cdot \overline{H}$ theorem and to find the function $\overline{\overline{G}}_{m1}^{(12)}$. In particular, no physical significance should be attached to the constitutive constants in that model.

11-5 EIGENFUNCTION EXPANSIONS

The eigenfunction expansions of the dyadic Green functions for a dielectric layer on a conducting plane which appeared in (11.102)–(11.105) will be derived in this section. The derivation of one of the functions will be treated in detail. The formulas for the others will be listed.

We consider the function $\overline{\overline{G}}_{e1}^{(11)}$. By the method of scattering superposition we let

$$\overline{\overline{G}}_{e1}^{(11)}(\overline{R},\overline{R}') = \overline{\overline{G}}_{eo}^{(1)}(\overline{R},\overline{R}') + \overline{\overline{G}}_{es}^{(11)}(\overline{R},\overline{R}') \tag{11.106}$$

$$\overline{\overline{G}}_{e1}^{(21)}(\overline{R},\overline{R}') = \overline{\overline{G}}_{es}^{(21)}, \tag{11.107}$$

where the function $\overline{\overline{G}}_{eo}^{(1)}$ denotes the free- space electrical dyadic Green function defined in a medium of the same constitutive constants as that of region 1 (air). The single superscript is used for this identification. The expression for $\overline{\overline{G}}_{eo}^{(1)}$ is given by (11.3) with k replaced by k_1; that is,

$$\begin{aligned}\overline{\overline{G}}_{eo}^{(1)}(\overline{R},\overline{R}') = &-\frac{1}{k_1^2}\hat{z}\hat{z}\delta(\overline{R}-\overline{R}') \\ &+\int_0^\infty d\lambda \sum_n C^{(1)}\left[\overline{M}\left(\pm h_1\right)\overline{M}'\left(\mp h_1\right)\right. \\ &\left.+\overline{N}\left(\pm h_1\right)\overline{N}'\left(\mp h_1\right)\right], \quad z \gtrless z', \end{aligned} \tag{11.108}$$

where

$$\begin{aligned} C^{(1)} &= \frac{i\left(2-\delta_0\right)}{4\pi\lambda h_1} \\ h_1 &= \left(k_1^2-\lambda^2\right)^{\frac{1}{2}}. \end{aligned}$$

A condensed notation for the terms in $\overline{\overline{G}}_{eo}^{(1)}$ has been used; namely,

$$\begin{aligned} \overline{M}(\pm h_1) &= \overline{M}_{\substack{e\\o}n\lambda}(\pm h_1) \\ \overline{N}(\pm h_1) &= \left(\frac{1}{k_1}\right)\nabla\times\overline{M}(\pm h_1). \end{aligned}$$

The wavelets in the scattered terms are excited by the downward-going wavelets of $\overline{\overline{G}}_{eo}^{(1)}$ with excitation coefficients $\overline{M}'(h_1)$ and $\overline{N}'(h_1)$ and the field functions in $\overline{\overline{G}}_{es}^{(11)}$ must consist of upward-going wavelets $\overline{M}(h_1)$ and $\overline{N}(h_1)$. In $\overline{\overline{G}}_{es}^{(21)}$, the field functions must consist of both upward- and downward-going wavelets. With these considerations the scattered functions must have the form

$$\begin{aligned} \overline{\overline{G}}_{es}^{(11)}(\overline{R},\overline{R}') = &\int_0^\infty d\lambda \sum_n C^{(1)} \\ &\left[a_1\overline{M}(h_1)\overline{M}'(h_1)+b_1\overline{N}(h_1)\overline{N}'(h_1)\right] \end{aligned} \tag{11.109}$$

$$\begin{aligned} \overline{\overline{G}}_{es}^{(21)}(\overline{R},\overline{R}') = &\int_0^\infty d\lambda \sum_n C^{(1)} \\ &\left\{\left[a_2^+\overline{M}(h_2)+a_2^-\overline{M}(-h_2)\right]\overline{M}'(h_1)\right. \\ &\left.\left[b_2^+\overline{N}(h_2)+b_2^-\overline{N}(-h_2)\right]\overline{N}'(h_1)\right\}, \end{aligned} \tag{11.110}$$

where $h_2 = \left(k_2^2-\lambda^2\right)^{\frac{1}{2}}$. The boundary conditions to be satisfied are

$$\text{at } z=0, \quad \hat{z}\times\overline{\overline{G}}_{e1}^{(21)} = 0$$

$$\text{at} \quad z = d, \quad \hat{z} \times \left[\overline{\overline{G}}_{e1}^{(11)} - \overline{\overline{G}}_{e1}^{(21)}\right] = 0$$

$$\hat{z} \times \left[\nabla \times \overline{\overline{G}}_{e1}^{(11)} - \nabla \times \overline{\overline{G}}_{e1}^{(21)}\right] = 0,$$

where we have assumed $\mu_1 = \mu_2 = \mu_0$. Based on these conditions we find

$$\begin{aligned}
a_1 &= -e^{i2(\Delta_2 - \Delta_1)} \frac{\left(1 + \rho e^{-i2\Delta_2}\right)}{D} \\
b_1 &= e^{i2(\Delta_2 - \Delta_1)} \frac{\left(1 - \rho' e^{-i2\Delta_2}\right)}{D'} \\
a_2^+ &= -a_2^- = -e^{i(\Delta_2 - \Delta_1)} \frac{(1 - \rho)}{D} \\
b_2^+ &= b_2^- = \left(\frac{k_1}{k_2}\right) e^{i(\Delta_2 - \Delta_1)} \frac{(1 - \rho')}{D'},
\end{aligned}$$

where

$$\begin{aligned}
\Delta_1 &= h_1 d, \ \Delta_2 = h_2 d \\
h_1 &= \left(k_1^2 - \lambda^2\right)^{\frac{1}{2}}, \ h_2 = \left(k_2^2 - \lambda^2\right)^{\frac{1}{2}} \\
k_1 &= \omega \left(\mu_0 \epsilon_1\right)^{\frac{1}{2}}, \ k_2 = \omega \left(\mu_0 \epsilon_2\right)^{\frac{1}{2}} \\
\rho &= \frac{(h_2 - h_1)}{(h_2 + h_1)} \\
\rho' &= \frac{\left(k_1^2 h_2 - k_2^2 h_1\right)}{\left(k_1^2 h_2 - k_2^2 h_1\right)} \\
D &= 1 + \rho^{i2\Delta_2}, \ D' = 1 - \rho' e^{i2\Delta_2}.
\end{aligned}$$

The formulas for $\overline{\overline{G}}_{e1}^{(22)}$ and $\overline{\overline{G}}_{e1}^{(12)}$ can be derived in a similar fashion. The compositions of these two functions and the coefficients attached to various terms are listed below.

$$\overline{\overline{G}}_{e1}^{(22)}(\overline{R}, \overline{R}') = \overline{\overline{G}}_{eo}^{(2)}(\overline{R}, \overline{R}') + \overline{\overline{G}}_{es}^{(22)}(\overline{R}, \overline{R}') \tag{11.111}$$

$$\overline{\overline{G}}_{e1}^{(12)}(\overline{R}, \overline{R}') = \overline{\overline{G}}_{es}^{(12)}(\overline{R}, \overline{R}') \tag{11.112}$$

$$\begin{aligned}
\overline{\overline{G}}_{eo}^{(2)}(\overline{R}, \overline{R}') = &- \frac{1}{k_2^2} \hat{z}\hat{z} \delta(\overline{R} - \overline{R}') \\
&+ \int_0^\infty d\lambda \sum_n C^{(2)} \left[\overline{M}\left(\pm h_2\right) \overline{M}'\left(\mp h_2\right)\right. \\
&\left. + \overline{N}\left(\pm h_2\right) \overline{N}'\left(\mp h_2\right)\right], \ z \gtrless z',
\end{aligned} \tag{11.113}$$

where

$$C^{(2)} = \frac{i\,(2-\delta_0)}{4\pi\lambda h_2}$$
$$h_2 = \left(k_2^2 - \lambda^2\right)^{\frac{1}{2}}$$

and

$$\begin{aligned}\overline{\overline{G}}{}_{es}^{(22)}(\overline{R},\overline{R}') = \int_0^\infty d\lambda \sum_n C^{(2)} \\ \cdot\left\{\overline{M}(h_2)\left[A_2^+\overline{M}'(h_2) + A_2^-\overline{M}'(-h_2)\right]\right. \\ +\overline{M}(-h_2)\left[B_2^+\overline{M}'(h_2) + B_2^-\overline{M}'(-h_2)\right] \\ +\overline{N}(h_2)\left[C_2^+\overline{N}'(h_2) + C_2^-\overline{N}'(-h_2)\right] \\ \left.+\overline{N}(-h_2)\left[D_2^+\overline{N}'(h_2) + D_2^-\overline{N}'(-h_2)\right]\right\}\end{aligned} \quad (11.114)$$

$$\begin{aligned}\overline{\overline{G}}{}_{es}^{(12)}(\overline{R},\overline{R}') = \int_0^\infty d\lambda \sum_n C^{(2)} \cdot \\ \cdot\left\{\overline{M}(h_1)\left[A_1^+\overline{M}'(h_2) + A_1^-\overline{M}'(-h_2)\right]\right. \\ \left.+\overline{N}(h_1)\left[C_1^+\overline{N}'(h_2) + C_1^-\overline{N}'(-h_2)\right]\right\},\end{aligned} \quad (11.115)$$

where

$$\begin{aligned}A_1^+ &= -A_1^- - e^{i(\Delta_2-\Delta_1)}\frac{(1+\rho)}{D} \\ C_1^+ &= C_1^- = \left(\frac{k_2}{k_1}\right)e^{i(\Delta_2-\Delta_1)}\frac{(1+\rho')}{D'} \\ A_2^+ &= \frac{-1}{D} \\ A_2^- &= B_2^+ = -B_2^- = \frac{-\rho e^{i2\Delta_2}}{D} \\ C_2^+ &= \frac{1}{D'} \\ C_2^- &= D_2^+ = D_2^- = \frac{\rho' e^{i2\Delta_2}}{D'}.\end{aligned}$$

The coefficients ρ, ρ', D, and D' are the same as the ones defined previously following (11.109) and (11.110).

The formulas for the functions $\overline{\overline{G}}{}_{m1}^{(22)}(\overline{R},\overline{R}')$ and $\overline{\overline{G}}{}_{m1}^{(12)}(\overline{R},\overline{R}')$ contained in (11.102) and (11.105) can be obtained by taking advantage of the symmetrical relations:

$$\begin{aligned}\overline{\overline{G}}{}_{m1}^{(22)}(\overline{R},\overline{R}') &= \left[\overline{\overline{G}}{}_{m2}^{(22)}(\overline{R}',\overline{R})\right]^T \\ &= \left[\nabla' \times \overline{\overline{G}}{}_{e1}^{(22)}(\overline{R}',\overline{R})\right]^T\end{aligned} \quad (11.116)$$

$$\overline{\overline{G}}_{m1}^{(12)}(\overline{R},\overline{R}') = \left(\frac{k_1}{k_2}\right)^2 \left[\overline{\overline{G}}_{m2}^{(21)}(\overline{R}',\overline{R})\right]^T$$
$$= \left(\frac{k_1}{k_2}\right)^2 \left[\nabla' \times \overline{\overline{G}}_{e1}^{(21)}(\overline{R}',\overline{R})\right]^T. \quad (11.117)$$

Since we have already found $\overline{\overline{G}}_{e1}^{(22)}(\overline{R},\overline{R}')$ and $\overline{\overline{G}}_{e1}^{(21)}(\overline{R},\overline{R}')$, it is a matter of interchanging the roles of $\overline{R}'$ and $\overline{R}$ and then taking the transpose of $\nabla' \times \overline{\overline{G}}_{e1}^{(22)}(\overline{R}',\overline{R})$ and $\nabla' \times \overline{\overline{G}}_{e1}^{(21)}(\overline{R}',\overline{R})$ to obtain $\overline{\overline{G}}_{m1}^{(22)}(\overline{R},\overline{R}')$ and $\overline{\overline{G}}_{m1}^{(12)}(\overline{R},\overline{R}')$. The results are

$$\begin{aligned}\overline{\overline{G}}_{m1}^{(22)}(\overline{R},\overline{R}') = \int_0^\infty d\lambda \sum_n \frac{i(2-\delta_0)k_2}{4\pi\lambda h_2} \cdot \\ \cdot \big\{ \left[\overline{M}(\pm h_2)\overline{N}'(\mp h_2) + \overline{N}(\pm h_2)\overline{M}'(\mp h_2)\right] \\ + \left[A_2^+ \overline{M}(h_2) + A_2^- \overline{M}(-h_2)\right] \overline{N}'(h_2) \\ + \left[B_2^+ \overline{M}(h_2) + B_2^- \overline{M}(-h_2)\right] \overline{N}'(h_2) \\ + \left[C_2^+ \overline{N}(h_2) + C_2^- \overline{N}(-h_2)\right] \overline{M}'(h_2) \\ + \left[D_2^+ \overline{N}(h_2) + D_2^- \overline{N}(-h_2)\right] \overline{M}'(h_2)\big\},\ z \gtrless z' \end{aligned} \quad (11.118)$$

and

$$\begin{aligned}\overline{\overline{G}}_{m1}^{(12)}(\overline{R},\overline{R}') = \left(\frac{k_1}{k_2}\right)^2 \int_0^\infty d\lambda \sum_n \frac{i(2-\delta_0)k_2}{4\pi\lambda h_2}. \\ \cdot \big\{ \overline{M}(h_1)\left[a_2^+ \overline{N}'(h_2) + a_2^- \overline{N}'(-h_2)\right] \\ + \overline{N}(h_1)\left[b_2^+ \overline{M}'(h_2) + b_2^- \overline{M}'(-h_2)\right]\big\}, \end{aligned} \quad (11.119)$$

where the coefficients are defined in the expressions of $\overline{\overline{G}}_{e1}^{(22)}$ and $\overline{\overline{G}}_{e1}^{(21)}$. It can be proved that at $z = 0$, the site of S_m in model B

$$\hat{z} \times \overline{\overline{G}}_{m1}^{(22)}(\overline{R},\overline{R}') = 0, \quad (11.120)$$

which shows that the magnetic dyadic Green function of the first kind indeed satisfies the Dirichlet boundary condition. It should be emphasized that the functions $\overline{\overline{G}}_{m1}^{(22)}(\overline{R},\overline{R}')$ and $\overline{\overline{G}}_{m1}^{(12)}(\overline{R},\overline{R}')$ are defined in model B with constitutive constants μ_1', ϵ_1' in region 1 and μ_2, ϵ_2 in region 2. Under the condition $\mu_1 = \mu_2 = \mu_0$ for model A, we have

$$\epsilon_1' = \epsilon_2$$

and

$$\mu_1' = \mu_0 \left(\frac{\epsilon_1}{\epsilon_2}\right) = \mu_0 \left(\frac{k_1^2}{k_2^2}\right).$$

An exercise has been assigned to find $\overline{\overline{G}}_{m1}^{(22)}$ and $\overline{\overline{G}}_{m1}^{(12)}$ directly without using the known expressions of $\overline{\overline{G}}_{e1}^{(22)}$ and $\overline{\overline{G}}_{e1}^{(21)}$ of model A. That exercise would demonstrate very clearly the significance of the complementary reciprocity theorems.

11-6 A DIELECTRIC SLAB IN AIR

The structure of the problem under consideration is shown in Fig. 11-5. There are three regions; hence nine electric dyadic Green functions of the third kind are involved, namely, $\overline{\overline{G}}_{e}^{(ij)}$ with $i, j = 1, 2, 3$. The wave numbers of the three regions are k_1 (air), k_2 (dielectric), and k_3 (air). We will consider the functions $\overline{\overline{G}}_{e}^{(i2)}$ only with $i = 1, 2, 3$. The formulation is very similar to that of a dielectric layer placed on a conducting plane except that there is an additional region. Only the final result of various expressions will be given here. They are listed as follows:

$$\overline{\overline{G}}_{e}^{(22)}(\overline{R}, \overline{R}') = \overline{\overline{G}}_{eo}^{(2)}(\overline{R}, \overline{R}') + \overline{\overline{G}}_{es}^{(22)}(\overline{R}, \overline{R}') \tag{11.121}$$

$$\overline{\overline{G}}_{e}^{(12)}(\overline{R}, \overline{R}') = \overline{\overline{G}}_{es}^{(12)} \tag{11.122}$$

$$\overline{\overline{G}}_{e}^{(32)}(\overline{R}, \overline{R}') = \overline{\overline{G}}_{es}^{(32)}. \tag{11.123}$$

In (11.121) the free-space electric dyadic Green function is the same as (11.113). $\overline{\overline{G}}_{es}^{(22)}$ and $\overline{\overline{G}}_{es}^{(12)}$ have the same forms as (11.114) and (11.115), except that the coefficients attached to the wave functions would have different values. However, the same notations for these coefficients will be retained. $\overline{\overline{G}}_{es}^{(32)}$ must have the form

$$\overline{\overline{G}}_{es}^{(32)}(\overline{R}, \overline{R}') = \int_0^\infty d\lambda \sum_n C^{(2)} \left\{ \overline{M}(-h_1) \left[b_3^{\pm} \overline{M}'(\pm h_2) \right] \right.$$
$$\left. + \overline{N}(-h_1) \left[d_3^{\pm} \overline{N}'(\pm h_2) \right] \right\}. \tag{11.124}$$

It is understood that $k_1 = k_3$ and $h_1 = h_3$. By applying the boundary conditions for $\overline{\overline{G}}_{e}^{(i2)}$ and $\nabla \times \overline{\overline{G}}_{e}^{(i2)}$ at the interfaces S_{12} and S_{23}, the 16 unknown coefficients can be determined. They are

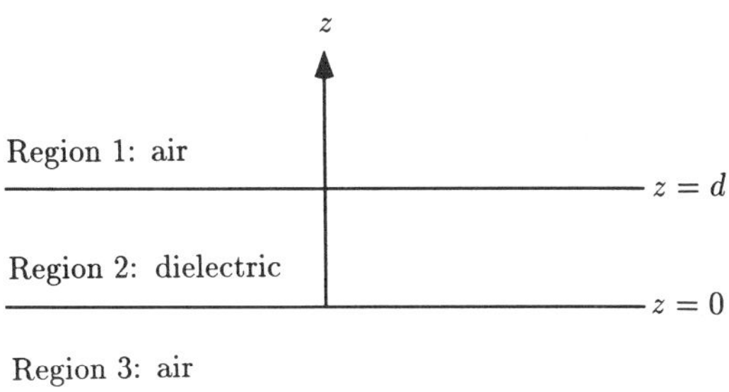

Fig. 11-5 A dielectric slab in air

$$
\begin{aligned}
a_2^+ &= \frac{\rho}{\Gamma}, \ a_2^- = \frac{\rho^2 e^{i2\Delta_2}}{\Gamma} \\
b_2^+ &= \frac{\rho^2 e^{i2\Delta_2}}{\Gamma}, \ b_2^- = \frac{\rho e^{i2\Delta_2}}{\Gamma} \\
c_2^+ &= \frac{\rho'}{\Gamma'}, \ c_2^- = \frac{\rho'^2 e^{i2\Delta_2}}{\Gamma'} \\
d_2^+ &= \frac{\rho'^2 e^{i2\Delta_2}}{\Gamma'}, \ d_2^- = \frac{\rho' e^{i2\Delta_2}}{\Gamma'} \\
a_1^+ &= \frac{\rho(1+\rho)e^{i(\Delta_2-\Delta_1)}}{\Gamma} \\
a_1^- &= \frac{(1+\rho)e^{i(\Delta_2-\Delta_1)}}{\Gamma} \\
c_1^+ &= \frac{k_2\rho(1+\rho)e^{i(\Delta_2-\Delta_1)}}{k_1\Gamma'} \\
c_1^- &= \frac{k_2\rho(1+\rho)e^{i(\Delta_2-\Delta_1)}}{k_1\Gamma'} \\
b_3^+ &= \frac{(1+\rho)}{\Gamma}, \ b_3^- = \frac{\rho(1+\rho)e^{i2\Delta_2}}{\Gamma} \\
d_3^+ &= \frac{k_2(1+\rho')}{k_1\Gamma'} \\
d_3^- &= \frac{k_2\rho'(1+\rho')e^{i2\Delta_2}}{k_1\Gamma'},
\end{aligned}
$$

where Δ_1, Δ_2, ρ, and ρ' are the same parameters defined in the lines following (11.115). Γ and Γ' are defined by

$$\Gamma = 1 - \rho^2 e^{i2\Delta_2}, \ \Gamma' = 1 - \rho'^2 e^{i2\Delta_2}.$$

The physical interpretation of the wave functions contained in the three Green functions has been discussed by Cheng [1986] based on multiple wave reflections and refractions. The formulas for the other two sets of Green functions corresponding to sources located in region 1 or region 3 are also found in that reference.

The eigenfunction expansions which we have developed so far for the plane stratified media contain the Fourier-Bessel integrals and the associated Fourier series. When these formulas are applied to practical problems, we have to evaluate integrals commonly referred to as Sommerfeld integrals. The recent advance in the technique of fast Fourier transform (FTT) [Nussbaumer, 1982] suggests that an alternative representation of the Green functions for planar stratified media is to cast the eigenfunction expansion in the form of two-dimensional Fourier transform. We will use the free-space dyadic Green functions to illustrate this formulation.

11-7 TWO-DIMENSIONAL FOURIER TRANSFORM OF THE DYADIC GREEN FUNCTIONS

The desirable vector wave functions to be used to represent the free-space dyadic Green functions, both electric and magnetic, are defined by

$$\begin{aligned}\overline{M}(\overline{\kappa}) &= \nabla \times \left[e^{i\overline{\kappa}\cdot\overline{R}}\hat{z}\right] \\ &= i\left(\kappa_2\hat{x} - \kappa_1\hat{y}\right)e^{i\overline{\kappa}\cdot\overline{R}}\end{aligned} \tag{11.125}$$

$$\begin{aligned}\overline{N}(\overline{\kappa}) &= \frac{1}{\kappa}\nabla \times \overline{M}(\overline{\kappa}) \\ &= \frac{1}{\kappa}\left[-\kappa_1\kappa_3\hat{x} - \kappa_2\kappa_2\hat{y} + \left(\kappa_1^2 + \kappa_2^2\right)\hat{z}\right]e^{i\overline{\kappa}\cdot\overline{R}},\end{aligned} \tag{11.126}$$

where

$$\begin{aligned}&\overline{\kappa} = \kappa_1\hat{x} + \kappa_2\hat{y} + \kappa_3\hat{z}, \kappa = \left(\kappa_1^2 + \kappa_2^2 + \kappa_3^2\right)^{\frac{1}{2}} \\ &\overline{R} = x\hat{x} + y\hat{y} + z\hat{z}.\end{aligned}$$

It can be shown that $\overline{M}$ and $\overline{N}$ are orthogonal; that is,

$$\iiint \overline{M}(\overline{\kappa}) \cdot \overline{N}\left(-\kappa'\right) dV = 0. \tag{11.127}$$

The normalization factor of the vector wave function is given by

$$\begin{aligned}&\iiint \overline{M}(\overline{\kappa}) \cdot \overline{M}(-\overline{\kappa}')\, dV \\ &= \iiint \overline{N}(\overline{\kappa}) \cdot \overline{N}(-\overline{\kappa}')\, dV \\ &= (2\pi)^3\left(\kappa_1^2 + \kappa_2^2\right)\delta\left(\overline{\kappa} - \overline{\kappa}'\right).\end{aligned} \tag{11.128}$$

The volume of integration in the above integrals covers the entire space. By the method of $\overline{\overline{G}}_m$, we let

$$\begin{aligned}&\nabla \times \left[\overline{\overline{I}}\delta(\overline{R} - \overline{R}')\right] \\ &= \iiint_{-\infty}^{\infty} d\kappa_1\, d\kappa_2\, d\kappa_3 \left[\overline{M}(\overline{\kappa})\overline{A}(\overline{\kappa}) + \overline{N}(\overline{\kappa})\overline{B}(\overline{\kappa})\right].\end{aligned} \tag{11.129}$$

With the aid of the orthogonal property of $\overline{M}$ and $\overline{N}$ and their normalization, we find

$$\overline{A}(\overline{\kappa}) = \frac{\kappa\overline{N}'(-\overline{\kappa})}{(2\pi)^3}\left(\kappa_1^2 + \kappa_2^2\right) \tag{11.130}$$

$$\overline{B}(\overline{\kappa}) = \frac{\kappa\overline{M}'(-\overline{\kappa})}{(2\pi)^3}(\kappa_1^2 + \kappa_2^2), \tag{11.131}$$

where

$$\overline{M}'(-\overline{\kappa}) = \nabla' \times \left[e^{-i\overline{\kappa}\cdot\overline{R}'} \hat{z} \right] \tag{11.132}$$

$$\overline{N}'(-\overline{\kappa}) = \frac{1}{\kappa} \nabla' \times \overline{M}(-\overline{\kappa}). \tag{11.133}$$

Since $\overline{\overline{G}}_{m0}$ satisfies the equation

$$\nabla \times \nabla \times \overline{\overline{G}}_{m0} - \kappa^2 \overline{\overline{G}}_{m0} = \nabla \times \left[\overline{\overline{I}} \delta(\overline{R} - \overline{R}') \right],$$

it should have the following integral representation:

$$\begin{aligned} \overline{\overline{G}}_{m0}(\overline{R}, \overline{R}') = & \frac{1}{(2\pi)^3} \iiint_{-\infty}^{\infty} d\kappa_1\, d\kappa_2\, d\kappa_3\, \kappa \left[(\kappa_1^2 + \kappa_2^2)(\kappa^2 - k^2) \right]^{-1} \\ & \cdot \left[\overline{N}(\overline{\kappa}) \overline{M}'(-\overline{\kappa}) + \overline{M}(\overline{\kappa}) \overline{N}'(-\overline{\kappa}) \right] \end{aligned} \tag{11.134}$$

and

$$\begin{aligned} \nabla \times \overline{\overline{G}}_{m0}(\overline{R}, \overline{R}') = & \frac{1}{(2\pi)^3} \iiint_{-\infty}^{\infty} d\kappa_1\, d\kappa_2 d\, \kappa_3\, \kappa^2 \left[\left(\kappa_1^2 + \kappa_2^2 \right) \left(\kappa^2 - k^2 \right) \right]^{-1} \\ & \cdot \left[\overline{M}(\overline{\kappa}) \overline{M}'(-\overline{\kappa}) + \overline{N}(\overline{\kappa}) \overline{N}'(-\overline{\kappa}) \right]. \end{aligned} \tag{11.135}$$

By integrating (11.135) with respect to κ_3 with the aid of the residue theorem, we find

$$\nabla \times \overline{\overline{G}}{}_{m0}^{\pm}(\overline{R}, \overline{R}') = \frac{ik^2}{8\pi^2} \iint_{-\infty}^{\infty} d\kappa_1\, d\kappa_2 \left[h \left(\kappa_1^2 + \kappa_2^2 \right) \right]^{-1}$$

$$\cdot \left[\overline{M}(\pm h) \overline{M}'(\mp h) + \overline{N}(\pm h) \overline{N}'(\mp h) \right], \; z \gtrless z', \tag{11.136}$$

where

$$\overline{M}(\pm h) = \nabla \times \left[e^{i(\kappa_1 x + \kappa_2 y \pm hz)} \hat{z} \right] \tag{11.137}$$

$$\overline{N}(\pm h) = \frac{1}{\kappa} \nabla \times \overline{M}\,(\pm h) \tag{11.138}$$

$$h = (k^2 - \kappa_1^2 - \kappa_2^2)^{\frac{1}{2}}.$$

According to the method of $\overline{\overline{G}}_m$, the expression for $\overline{\overline{G}}_{e0}$ would be

$$\begin{aligned} \overline{\overline{G}}_{eo}(\overline{R}, \overline{R}) = & \frac{1}{k^2} \left[\nabla \times \overline{\overline{G}}_{m0} - \overline{\overline{I}} \delta(\overline{R} - \overline{R}') \right] \\ = & - \frac{1}{k^2} \hat{z}\hat{z} \delta(\overline{R} - \overline{R}') + \frac{1}{k^2} \left[\nabla \times \overline{\overline{G}}{}_{m0}^{\pm}(\overline{R}, \overline{R}') \right] \\ = & - \frac{1}{k^2} \hat{z}\hat{z} \delta(\overline{R} - \overline{R}') \\ & + \iint_{-\infty}^{\infty} d\kappa_1\, d\kappa_2\, C_\kappa \left[\overline{M}(\pm h) \overline{M}'(\mp h) \right. \\ & \left. + \overline{N}(\pm h) \overline{N}'(\mp h) \right], \; z \gtrless z', \end{aligned} \tag{11.139}$$

where

$$C_\kappa = \frac{i}{8\pi^2 h\left(\kappa_1^2 + \kappa_2^2\right)}.$$

By comparing (11.139) with (11.3), we see clearly the one-to-one correspondence between these two formulas. For clarity we also use k instead of k_1 to denote the wave number in (11.3), then the corresponding terms are listed below, where "CW" denotes the cylindrical wave formulation and 'PW' the plane wave formulation.

CW: $$\int_0^\infty d\lambda \sum_n C^{(1)}\left[\overline{M}(h_1), \ldots\right], \quad C^{(1)} = \frac{2-\delta_0}{4\pi\lambda h_1}$$

PW: $$\iint_{-\infty}^{\infty} C_\kappa\left[\overline{M}(h), \ldots\right], \quad C_\kappa = \frac{i}{8\pi^2 h(\kappa_1^2 + \kappa_2^2)}$$

CW: $$h_1 = (k^2 - \lambda^2)^{\frac{1}{2}}, \quad \overline{M}(h_1) = \nabla \times \left[J_n(\lambda r)\begin{matrix}\cos\\ \sin\end{matrix} n\phi e^{ih_1 z}\hat{z}\right]$$

PW: $$h = (k^2 - \kappa_1^2 - \kappa_2^2)^{\frac{1}{2}}, \quad \overline{M}(h) = \nabla \times \left[e^{i(\kappa_1 x + \kappa_2 y + hz)}\hat{z}\right]$$

CW: $$\overline{N}(h_1) = \frac{1}{k}\nabla \times \overline{M}(h_1)$$

PW: $$\overline{N}(h) = \frac{1}{k}\nabla \times \overline{M}(h).$$

Now if (11.139) is used to construct the scattered terms for problems involving planar layered media, the coefficients associated with the vector wave functions will have exactly the same forms as in the "CW" formulation. For example, the scattering function previously described by (11.115) is now replaced by

$$\overline{\overline{G}}_{es}^{(12)}(\overline{R}, \overline{R}') = \iint_{-\infty}^{\infty} d\kappa_1 d\kappa_2 C_\kappa^{(2)} \left\{\overline{M}(h_1)\left[a_1^{\mp}\overline{M}'(\pm h_2)\right]\right. \cdot$$
$$\left. + \overline{N}(h_1)\left[c_1^{\mp}\overline{N}'(\pm h_2)\right]\right\}, \qquad (11.140)$$

where

$$C_\kappa^{(2)} = \frac{i}{8\pi^2 h_2\left(\kappa_1^2 + \kappa_2^2\right)}$$
$$\overline{M}(h_1) = \nabla \times \left[e^{i(\kappa_1 x + \kappa_2 y + h_1 z)}\hat{z}\right]$$
$$\overline{N}(h_1) = \frac{1}{k_1}\nabla \times \overline{M}(h_1)$$
$$\overline{M}(h_2) = \nabla \times \left[e^{i(\kappa_1 x + \kappa_2 y + h_2 z)}\hat{z}\right]$$
$$\overline{N}(h_2) = \frac{1}{k_2}\nabla \times \overline{M}(h_2)$$

and the coefficients $a_1^{\mp}, c_1^{\mp}$ have the same expressions as the ones listed after

(11.115) except that the parameters h_1 and h_2 are replaced by

$$h_1 = (k_1^2 - \kappa_1^2 - \kappa_2^2)^{\frac{1}{2}}$$
$$h_2 = (k_2^2 - \kappa_1^2 - \kappa_2^2)^{\frac{1}{2}}.$$

That the coefficients $a_1^{\pm}$ and $c_1^{\pm}$ have the same forms in the two alternative representations is due to the fact that the boundary conditions are only dependent on the parameters k_1, k_2, h_1, and h_2 and the functions $e^{\pm ih_1 z}$ and $e^{\pm ih_2 z}$, and the dependence is identical in both formulations.

12

Inhomogeneous Media and Moving Medium

This chapter contains some generalizations of the dyadic Green function technique to more complex media, including particularly the inhomogeneous media and moving isotropic media. Vector wave functions for plane stratified and spherically stratified media are introduced. Several spherical lens functions are treated in detail, as they have not previously been covered in books either on electromagnetic theory or on differential equations. The remaining part of the chapter deals with the topics on moving isotropic media. Maxwell's equations with the constitutive relations based on Minkowski's relativistic formulation are solved for monochromatically oscillating excitation and for transient current source. In the latter case, the differential equations are first transformed into a spatial and pseudotime domain and then solved by the method of Fourier transform. This approach avoids the necessity of introducing four-dimensional space and time operator as done by Compton [1966] and several new mathematical theorems involving this operator. The chapter concludes with the derivation of the dyadic Green functions for waveguides filled with a moving medium and for a conducting cylinder placed in such a medium.

12-1 VECTOR WAVE FUNCTIONS FOR PLANE STRATIFIED MEDIA

When the permittivity and the permeability of a medium are functions of position, we refer to such a medium as inhomogeneous. For an inhomogeneous isotropic medium, Maxwell's equations for a monochromatically oscillating field read

$$\nabla \times \overline{E} = i\omega\mu_0\mu_r(\overline{R})\overline{H}$$
$$\nabla \times \overline{H} = \overline{J} - i\omega\epsilon_0\epsilon_r(\overline{R})\overline{E},$$

where $\mu_r(\overline{R})$ and $\epsilon_r(\overline{R})$ denote, respectively, the relative permeability and permittivity functions of the medium. In many practical cases of interest the inhomogeneity is normally due to variation of the permittivity only so that $\mu_r(\overline{R})$ is constant. In this section and that which follows, we shall restrict ourselves to this case only and let $\mu_r(\overline{R})$ be equal to unity. Under this condition the wave equations for $\overline{E}$ and $\overline{H}$ are given by

$$\nabla \times \nabla \times \overline{E} - k^2\epsilon_r(\overline{R})\overline{E} = i\omega\mu_0\overline{J} \tag{12.1}$$

$$\nabla \times \left[\frac{1}{\epsilon_r(\overline{R})}\nabla \times \overline{H}\right] - k^2\overline{H} = \nabla \times \left[\frac{\overline{J}}{\epsilon_r(\overline{R})}\right]. \tag{12.2}$$

We are seeking the dyadic Green functions pertaining to these two equations under various boundary conditions, particularly the eigenfunction expansions of these functions for bodies made of such an inhomogeneous medium but of different shapes. To that end, we must first discuss the vector wave functions which are solutions for the inhomogeneous vector wave equations of the form

$$\nabla \times \nabla \times \overline{E} - k^2\epsilon_r(\overline{R})\overline{E} = 0 \tag{12.3}$$

$$\nabla \times \left[\frac{1}{\epsilon_r(\overline{R})}\nabla \times \overline{H}\right] - k^2\overline{H} = 0. \tag{12.4}$$

If $\epsilon_r(\overline{R})$ is a function of all three variables in an orthogonal system, we have so far no general method of finding these vector wave functions. In fact, we have only two classes of inhomogeneous media for which we know how to find the appropriate vector wave functions. The first class consists of plane stratified media in which $\epsilon_r(\overline{R})$ is a function of one of the Cartesian coordinate variables only, such as z. In the second class, the medium is stratified in the radial direction in the spherical coordinate system so that the permittivity is a function of R only. Strangely enough, we have not been able to find the general vector wave functions for a cylindrically stratified medium, except under the condition that the problems under consideration are either rotationally symmetric or when the field is a two-dimensional one being independent of the longitudinal axis. For this reason we shall discuss only the vector wave functions for plane and spherically stratified media.

For plane stratified media, we assume that the stratification occurs along the z-direction so that $\epsilon_r(\overline{R}) = \epsilon_r(z)$. Under that condition, we have

$$\nabla \times \nabla \times \overline{E} - k^2\epsilon_r(z)\overline{E} = 0 \tag{12.5}$$

$$\nabla \times \left[\frac{1}{\epsilon_r(z)}\nabla \times \overline{H}\right] - k^2\overline{H} = 0. \tag{12.6}$$

It can easily be verified that the following two sets of vector wave functions are solutions of (12.5):

$$\overline{M}^{(m)} = \nabla \times (\Psi \hat{z}) \tag{12.7}$$

and

$$\overline{N}^{(e)} = \frac{1}{k\epsilon_r(z)} \nabla \times \nabla \times (\Phi \hat{z}), \tag{12.8}$$

where the two generating functions Ψ and Φ satisfy, respectively, the differential equations

$$\nabla^2 \Psi + k^2 \epsilon_r(z) \Psi = 0 \tag{12.9}$$

and

$$\nabla^2 \Phi - \frac{1}{\epsilon_r} \frac{d\epsilon_r}{dz} \frac{\partial \Phi}{\partial z} + k^2 \epsilon_r(z) \Phi = 0. \tag{12.10}$$

The superscript m attached to the M- function signifies that it is of the magnetic or transverse electric type with respect to the z-axis, while the superscript e represents the electric or transverse magnetic type. Similarly, the vector wave functions which are solutions of (12.6) can be represented by

$$\overline{M}^{(e)} = \nabla \times (\Phi \hat{z}) \tag{12.11}$$

and

$$\overline{N}^{(m)} = \frac{1}{k} \nabla \times \nabla \times (\Psi \hat{z}). \tag{12.12}$$

These four kinds of vector wave functions satisfy the symmetrical relations that

$$\begin{aligned}
\overline{N}^{(e)} &= \frac{1}{k\epsilon_r} \nabla \times \overline{M}^{(e)} \\
\overline{M}^{(e)} &= \frac{1}{k} \nabla \times \overline{N}^{(e)} \\
\overline{N}^{(m)} &= \frac{1}{k} \nabla \times \overline{M}^{(m)} \\
\overline{M}^{(m)} &= \frac{1}{k\epsilon_r} \nabla \times \overline{N}^{(m)}.
\end{aligned}$$

We repeat that $\overline{M}^{(m)}$ and $\overline{N}^{(e)}$ are solutions for the vector wave equation satisfied by the electric E-field; that is,

$$\nabla \times \nabla \times \left\{ \begin{matrix} \overline{M}^{(m)} \\ \overline{N}^{(e)} \end{matrix} \right\} - k^2 \epsilon_r(z) \left\{ \begin{matrix} \overline{M}^{(m)} \\ \overline{N}^{(e)} \end{matrix} \right\} = 0$$

and $\overline{M}^{(e)}$ and $\overline{N}^{(m)}$ are solutions for the vector wave equation satisfied by the magnetic H-field:

$$\nabla \times \frac{1}{\epsilon_r(z)} \nabla \times \left\{ \begin{matrix} \overline{M}^{(e)} \\ \overline{N}^{(m)} \end{matrix} \right\} - k^2 \left\{ \begin{matrix} \overline{M}^{(e)} \\ \overline{N}^{(m)} \end{matrix} \right\} = 0.$$

We cannot find a better notation for these functions without causing some kind of confusion. To find the eigenfunctions to be used in the dyadic Green function expansion, we must specify not only the functional representation for $\epsilon_r(z)$ but also the geometrical shape of the structure to be considered. For example, if we have a rectangular waveguide filled with a medium stratified in the z-direction, corresponding to the longitudinal axis of the guide, then (12.9) and (12.10) must be solved in the rectangular coordinate system. The appropriate solutions for Ψ and Φ are

$$\Psi = \Psi_{emn} = \cos\frac{m\pi x}{a}\cos\frac{n\pi y}{b}F_1(z)$$

and

$$\Phi = \Phi_{omn} = \sin\frac{m\pi x}{a}\sin\frac{n\pi y}{b}F_2(z),$$

where $F_1(z)$ and $F_2(z)$ satisfy, respectively, the equations

$$\frac{d^2F_1(z)}{dz^2} + \left[k^2\epsilon_r(z) - \left(\frac{m\pi}{a}\right)^2 - \left(\frac{n\pi}{b}\right)^2\right]F_1(z) = 0 \tag{12.13}$$

$$\frac{d^2F_2(z)}{dz^2} - \frac{1}{\epsilon_r(z)}\frac{d\epsilon_r(z)}{dz}\frac{dF_2(z)}{dz} + \left[k^2\epsilon_r(z) - \left(\frac{m\pi}{a}\right)^2 - \left(\frac{n\pi}{b}\right)^2\right]F_2(z) = 0. \tag{12.14}$$

The two complementary solutions of F_1 or F_2 play the same role as the experimental functions $e^{\pm ihz}$ for a homogeneous medium. Only a limited number of stratified profiles have so far been investigated whereby the solutions can be expressed in terms of well-known functions. A few of them are found in Wait [1962, Ch. 3]. No attempt is made here to cover this large class of problems in this book. It suffices to point out that once we have at our disposal the available solutions for $F_1(z)$ and $F_2(z)$, we can follow the same procedure as in the homogeneous case for the construction of the dyadic Green functions. The above remark also applies to problems involving a cylindrical waveguide with a longitudinal stratification or for an inhomogeneous flat earth. For a plane stratified flat earth the appropriate solutions for Ψ and Φ would be

$$\Psi_{\substack{e\\o}n\lambda} = J_n(\lambda r)\,{\cos \atop \sin}\, n\phi F_1(z)$$

$$\Phi_{\substack{e\\o}n\lambda} = J_n(\lambda_r)\,{\cos \atop \sin}\, n\phi F_2(z),$$

where F_1 and F_2 are again solutions for (12.13) and (12.14). We, of course, use the ones which would satisfy the radiation condition at $z = -\infty$ in the construction of $\overline{\overline{G}}_e^{(22)}$ and $\overline{\overline{G}}_e^{(12)}$. This concludes our brief discussion for the general method of finding the dyadic Green functions for plane stratified media.

12-2 VECTOR WAVE FUNCTIONS FOR SPHERICALLY STRATIFIED MEDIA

When the permittivity is a function of the spherical variable R only, the wave equations for $\overline{E}$ and $\overline{H}$ according to (12.3) and (12.4) become

$$\nabla \times \nabla \times \overline{E} - k^2 \epsilon_r(R) \overline{E} = 0 \tag{12.15}$$

$$\nabla \times \left[\frac{1}{\epsilon_r(R)} \nabla \times \overline{H} \right] - k^2 \overline{H} = 0. \tag{12.16}$$

We now define four inhomogeneous, spherical vector wave functions [Tai, 1958a] which are generalizations of the functions defined for homogeneous media. They are

$$\overline{M}^{(m)} = \nabla \times \left(\Psi \overline{R} \right) \tag{12.17}$$

$$\overline{N}^{(m)} = \frac{1}{k} \nabla \times \nabla \times \left(\Psi \overline{R} \right) \tag{12.18}$$

$$\overline{M}^{(e)} = \nabla \times \left(\Phi \overline{R} \right) \tag{12.19}$$

$$\overline{N}^{(e)} = \frac{1}{k \epsilon_r(R)} \nabla \times \nabla \times \left(\Phi \overline{R} \right). \tag{12.20}$$

It can be verified that (12.17) and (12.20) are solutions to (12.1) and (12.4) and (12.5) are solutions to (12.2), provided that Ψ and Φ satisfy the following scalar equations:

$$\nabla^2 \Psi + k^2 \epsilon_r(R) \Psi = 0 \tag{12.21}$$

$$\nabla^2 \Phi - \frac{1}{\epsilon_r(R)} \frac{d\epsilon_r(R)}{dR} \frac{\partial \Phi}{\partial R} + k^2 \epsilon_r(R) \Phi = 0. \tag{12.22}$$

The four sets of vector wave functions have certain symmetrical relations. They are

$$\overline{N}^{(e)} = \frac{1}{k \epsilon_r(R)} \nabla \times \overline{M}^{(e)} \tag{12.23}$$

$$\overline{M}^{(e)} = \frac{1}{k} \nabla \times \overline{N}^{(e)} \tag{12.24}$$

$$\overline{N}^{(m)} = \frac{1}{k} \nabla \times \overline{M}^{(m)} \tag{12.25}$$

$$\overline{M}^{(m)} = \frac{1}{k \epsilon_r(R)} \nabla \times \overline{N}^{(m)}. \tag{12.26}$$

By applying the method of separation of variables to (12.21) and (12.22) in spherical coodinate systems, we find that the eigenfunctions can be written in the form

$$\Psi_{\substack{e \\ o} n\lambda} = \frac{1}{R} S_n(kR) P_n^m(\cos\theta) \begin{matrix}\cos \\ \sin\end{matrix} m\phi \tag{12.27}$$

$$\Phi_{\substack{e \\ o} n\lambda} = \frac{1}{R} T_n(kR) P_n^m(\cos\theta) \begin{matrix}\cos \\ \sin\end{matrix} m\phi, \tag{12.28}$$

where $S_n(kR)$ and $T_n(kR)$ satisfy, respectively, the differential equations

$$\frac{d^2 S_n}{dR^2} + \left[k^2 \epsilon_r(R) - \frac{n(n+1)}{R^2}\right] S_n = 0 \tag{12.29}$$

$$\epsilon_R(R)\frac{d}{dR}\left[\frac{1}{\epsilon_r(R)}\frac{dT_n}{dR}\right] + \left[k^2\epsilon_r(R) - \frac{n(n+1)}{R^2}\right] T_n = 0. \tag{12.30}$$

The functions S_n and T_n play the same role as $Rj_n(kR)$ in the homogeneous case. It is obvious that when $\epsilon_r(R)$ is equal to unity both (12.29) and (12.30) reduce to the equation satisfied by $Rj_n(kR)$ as discussed in Sec. 10-1. By using (12.27) and (12.28) as the generating functions the complete expressions for the four vector wave functions can be found. They are given by

$$\overline{M}^{(m)}_{\substack{e\\o}mn} = \frac{S_n(kR)}{R}\overline{m}_{\substack{e\\o}mn} \tag{12.31}$$

$$\overline{N}^{(m)}_{\substack{e\\o}mn} = \frac{1}{k}\left[\frac{S_n(kR)}{R^2}\overline{\ell}_{\substack{e\\o}mn} + \frac{1}{R}\frac{\partial S_n(kR)}{\partial R}\hat{R}\times\overline{m}_{\substack{e\\o}mn}\right] \tag{12.32}$$

$$\overline{M}^{(e)}_{\substack{e\\o}mn} = \frac{T_n(kR)}{R}\overline{m}_{\substack{e\\o}mn} \tag{12.33}$$

$$\overline{N}^{(e)}_{\substack{e\\o}mn} = \frac{1}{k\epsilon_r(R)}\left[\frac{T_n(kR)}{R^2}\overline{\ell}_{\substack{e\\o}mn} + \frac{1}{R}\frac{\partial T_n(kR)}{\partial R}\hat{R}\times\overline{m}_{\substack{e\\o}mn}\right], \tag{12.34}$$

where

$$\overline{\ell}_{\substack{e\\o}mn} = n(n+1)P_n^m(\cos\theta)\,{}^{\cos}_{\sin}\, m\phi\hat{R}$$

$$\overline{m}_{\substack{e\\o}mn} = \mp\frac{mP_n^m}{\sin\theta}\,{}^{\sin}_{\cos}\, m\phi\hat{\theta} - \frac{\partial P_n^m}{\partial\theta}\,{}^{\cos}_{\sin}\, m\phi\hat{\phi}.$$

These vector zonal harmonic functions are analogous to the $\overline{P}$-, $\overline{C}$-, and $\overline{B}$-functions used by Morse and Feshbach [1953, pp. 1898–1899] in their presentation of the spherical vector wave functions, except that our normalization constants are different from theirs. These vector wave functions will now be used to construct the eigenfunction expansion of the dyadic Green function for the inhomogeneous spherical lenses.

12-3 INHOMOGENEOUS SPHERICAL LENSES

The various spherical lenses to be discussed are characterized by certain profiles of the permittivity function. Most of these profiles were originally discovered by the investigators whose names are associated with them and are based on Fermat's principle in the geometrical theory of optics. The Nomura-Takaku distribution, however, was used in a wave propagation study. Table 12-1 lists some of the profiles. The wave theory of these distributions will be discussed in this section.

TABLE 12-1 Profiles of Some Permittivity Functions

Maxwell fish-eyes (1860)	$\frac{4}{[1+(R/a)^2]}, a \geq R \geq 0$
Luneburg lens (1944)	$2 - \left(\frac{R}{a}\right)^2, a \geq R \geq 0$
Conical lens of Luneburg (1944)	$K[1 + \left(\frac{R_1}{R}\right)]$
Eaton lens (1952)	$\left(\frac{R}{a}\right)^2, a \geq R \geq 0$
Nomura-Takaku distribution (1955)	$\left(\frac{R}{a}\right)^{2q}$

We shall treat one profile in great detail showing how to find the solutions for S_n and T_n associated with a specific profile. The work [Tai, 1958a] contains some analyses which are not found in most books on the theory of differential equations or on electromagnetic theory. The case to be considered is the spherical Luneburg lens with the permittivity profile given by

$$\epsilon_r(R) = 2 - \left(\frac{R}{a}\right)^2, \quad a \geq R \geq 0, \tag{12.35}$$

where a denotes the radius of the inhomogeneous lens. According to Luneburg's original theory [1944], this lens will focus the rays, originated from a point source placed at the rim, to form a collimated beam when passing through the lens. Since the original theory is based on a scalar formulation, the polarization status of the source does not enter into the formulation. From the electromagnetic theory point of view, it is an important factor to be considered. A complete theory would also remove some of the uncertain characteristics obtained from the geometrical theory of diffraction.

By substituting (12.35) into (12.29), and making the following changes of variables

$$kR = \rho, \quad ka = \rho_a$$
$$S_n(kR) = \rho^{n+1} e^{-\rho^2/2\rho_a} U_n(\rho), \; n = 1, 2, \ldots,$$

we find that the function $U_n(\rho)$ must satisfy the following differential equation

$$\frac{d^2 U_n}{d\rho^2} + 2\left(\frac{n+1}{\rho} - \frac{\rho}{\rho_0}\right)\frac{dU_n}{d\rho} + \left(2 - \frac{2n+3}{\rho_a}\right) U_n = 0. \tag{12.36}$$

By letting

$$\frac{\rho^2}{\rho_a} = z$$
$$\nu = n + \frac{3}{2}$$
$$\alpha = \frac{1}{2}\left(n + \frac{3}{2} - \rho_a\right),$$

the above equation can be transformed into the standard form of the confluent hypergeometric equation [Copson, 1948, Ch. 10]; that is

$$z\frac{d^2U_n(z)}{dz^2} + (\nu - z)\frac{dU_n(z)}{dz} - \alpha U_n(z) = 0. \tag{12.37}$$

For the analysis of the Luneburg lens with the source placed at the rim we need the function which is regular at $R = 0$ that is represented by Kummer's functions,

$$U_n(z) = {}_1F_1(\alpha, \nu, z) = 1 + \frac{\alpha}{\nu}z + \frac{\alpha(\alpha+1)}{\nu(\nu+1)}\frac{z^2}{2!} + \cdots. \tag{12.38}$$

The other independent solution represented by

$$z^{1-\nu}\ {}_1F_1(\alpha - \nu + 1, 2 - \nu, z)$$

is not needed. Thus we identify the $S_n(\rho)$ function to be

$$S_n(\rho) = \rho^{n+1}e^{-\rho^2/2\rho_a}\ {}_1F_1\left(\alpha, \nu, \frac{\rho^2}{\rho_a}\right). \tag{12.39}$$

For the $T_n(kR)$ function the analysis is much more complicated. In the first place, (12.30) with $\epsilon_r(R)$ given by (12.35) cannot be reduced to a differential equation of the standard type. If we let

$$\begin{aligned} &kR = \rho, ka = \rho_a \\ &T_n(\rho) = \left(2\rho_a^2 - \rho^2\right)\rho^{n+1}e^{-\rho^2 2\rho_a}V_n(\rho), \end{aligned} \tag{12.40}$$

then the function $V_n(\rho)$ satisfies the following equation

$$\begin{aligned} &\frac{d^2V_n}{d\rho^2} + 2\left(\frac{n+1}{\rho} - \frac{\rho}{\rho_a}\right)\frac{dV_n}{d\rho} \\ &\quad + \left[2 - \frac{2n+3}{\rho_a} - \frac{1}{2\rho_a^2 - \rho^2} - \frac{3\rho^2}{\left(2\rho_a^2 - \rho^2\right)^2}\right]V_n = 0. \end{aligned}$$

A further transformation of the independent variable $z = \rho^2/\rho_a$ converts it into

$$\frac{d^2V_n}{dz^2} + \left(\frac{\nu}{z} - 1\right)\frac{dV_n}{dz} - \left[\frac{\alpha_1}{z} + \frac{2\alpha_2}{z - a_2} - \frac{2a_2\alpha_3}{(z - a_2)^2}\right]V_n = 0, \tag{12.41}$$

where the various constants are defined by

$$\begin{aligned} \nu &= n + \frac{3}{2} \\ \alpha_1 &= \frac{1}{2}\left(n + \frac{3}{2} - \rho_a + \frac{1}{4\rho_a}\right) \\ \alpha_2 &= -\frac{1}{16\rho_a},\ \alpha_3 = -\frac{3}{16\rho_a},\ a_2 = 2\rho_a. \end{aligned}$$

Equation (12.41) is similar to the confluent hypergeometric equation defined by (12.37), except that it has another regular singularity at $z = a_2$. The constants in the equation have purposely been so arranged in order to simplify the series solution discussed later. In order to appreciate more fully the nature of the solution to (12.41), it is profitable to digress to a discussion of the general characteristics of this equation. Like the hypergeometric equation, it has a regular singularity at $z = 0$ with exponents 0 and $1 - \nu$ [Copson, 1948, Ch. 10] and an irregular singularity at ∞. The exponents of the regular singularity at $z = a_2$ are equal to 3/2 and $-$ 1/2. From the theory of differential equations, it is known that the confluent hypergeometric equation can be obtained by the confluence of two regular singularities of the hypergeometric equation. The confluence is more easily seen by starting with the Papperitz equation of the Reimann P-equation [Morse and Feshbach, 1935, p. 539]. It is therefore anticipated that the new equation defined by (12.41) perhaps can be obtained by the suitable confluence of a number of regular singularities of a second-order differential equation with at least four or more regular singularities. It can be shown that unless the constants α_1, α_2, and α_3 are related in a certain special way, in general we need a second-order differential equation with five regular singularities to start with. The equation which fulfills this requirement is given by Whittaker and Watson [1943, p. 203] and can be written in the form

$$\frac{d^2u}{dz^2} + \sum_{r=1}^{4} \frac{1 - \lambda_r - \lambda_r'}{z - a_r} \frac{du}{dz}$$
$$+ \left[\frac{1}{\Pi_{r=1}^{4}} (z - a_r)\right] \left[Az^2 + Bz + C + \sum_{r=1}^{4} \frac{C_r}{(z - a_r)}\right] u = 0, \qquad (12.42)$$

where a_1, a_2, a_3, and a_4 denote the four regular singularities. Their exponents are λ_r and λ_r'. The point $z = \infty$ is the remaining regular singularity with exponents designated by μ and μ'. The constants A and the C_r's are given by

$$\alpha^2 - \left[3 - \sum_{r=1}^{4} (\mu_r + \lambda_r')\right] \alpha + A = 0, \quad \alpha = \mu \text{ or } \mu'$$
$$C_r = \frac{\lambda_r \lambda_r'}{\Pi_s (a_r - a_s)}, \; s = 1, 2, \ldots, s \neq r.$$

The constants B and C are arbitrary so far. The constants A, B, and C are not the same as those in the Whittaker and Watson equation. It is observed that because we have put (12.42) in a different form, the indicial equation for μ and μ' is also changed in appearance. Equation (12.42) is most conveniently presented by using the Reimann scheme

$$P \left\{ \begin{matrix} a_1 & a_2 & a_3 & a_4 & \infty & \\ \lambda_1 & \lambda_2 & \lambda_3 & \lambda_4 & \mu & z \\ \lambda_1' & \lambda_2' & \lambda_3' & \lambda_4' & \mu' & \end{matrix} \right\}.$$

In order to reduce (12.42) into (12.41) we first let

$$a_1 = 0,\ \lambda_1 = 0,\ \lambda_1' = 1 - \nu$$
$$a_2 = \alpha_2,\ \lambda_2 = \frac{3}{2},\ \lambda_2' = -\frac{1}{2},$$

which correspond to the two regular points and their exponents of (12.41). Without restricting ourselves to the exact procedure of the limiting process, if we assume then when a_3 and a_4 approach infinity

$$\frac{1-\lambda_3-\lambda_3'}{a_3} + \frac{1-\lambda_4-\lambda_4'}{a_4} \to -1,$$
$$\frac{A}{a_3a_4} \to 0,$$
$$\frac{B}{a_3a_a}, \frac{C}{a_3a_4}, \frac{\lambda_3\lambda_3'(a_3-a_4)}{a_4}, \frac{\lambda_4\lambda_4'(a_3-a_4)}{a_3} \to \text{finite},$$

then (12.42) reduces to the form

$$\frac{d^2u}{dz^2} + \left(\frac{\nu}{z} - 1\right)\frac{du}{dz} + \left[\frac{\beta}{z-a_2} + \frac{\delta}{z(z-a_2)} + \frac{\lambda_2\lambda_2'}{z(z-a_2)^2}\right]u = 0, \quad (12.43)$$

where β and δ are two arbitrary constants. By rearranging the terms in the last brackets and introducing two new arbitrary constants α_1 and α_2, we can write (12.43) in the form

$$\frac{d^2u}{dz^2} + \left(\frac{\nu}{z} - 1\right)\frac{du}{dz} - \left[\frac{\alpha_1}{z} + \frac{2\alpha_2}{z-a_2} - \frac{\lambda_2\lambda_2'}{(z-a_2)^2}\right]u = 0, \qquad (12.44)$$

which is identical in form to (12.41), observing that

$$\lambda_2\lambda_2' = 2a_2\alpha_3 = -\frac{3}{4}.$$

The foregoing discussion on the analytical nature of the function V_n is simply to show that while the confluent hypergeometric equation is obtained by the confluence of two regular singularities of the Papperitz equation, the "generalized" confluent hypergeometric equation as defined by (12.41) can be obtained by the confluence of three regular singularities (12.42). It may be remarked that the above procedure is not the only way to deduce (12.41) from a more general second- order linear differential equation. An alternative procedure is to start with an equation of five regular singularities and with the point at infinity considered to be an ordinary point. The same result can thus be achieved. If one uses the notation of Ince [1944, pp. 497–504], (12.41) can be classified as of the type $(0, 2, 1_2)$, which can be derived, by confluence of singularities, from the equations of the type $(8, 0, 0)$. It is therefore concluded that at least five regular singularities are needed to execute the proper confluence in obtaining (12.41). The foregoing discussion demonstrates the fact that analytically the differential equation for T_n is basically different from the differential equation for S_n and no simple connection exists between them. It also shows that the "generalized" confluent hypergeometric function is an entirely new function which is not related to other known functions.

For the Luneburg lens problem it is necessary to obtain a series expansion for V_n, hence T_n which is finite at $z = 0$. Recall that the two exponents at $z = 0$ are 0 and $1 - \gamma$, where γ is greater than one for this problem. The series solution which remains finite at the origin is therefore associated with the exponent equal to zero. On the other hand, the exponent $1 - \gamma$ leads to a solution which is singular at the origin. The desirable solution plays the same role as the Kummer function (12.38). The series solution of interest is of the form

$$V_n = \sum_{m=0}^{\infty} A_m z^m. \tag{12.45}$$

By substituting (12.45) into (12.41), one finds that

$$\begin{aligned}
\frac{A_1}{A_0} &= \frac{\alpha_1}{\gamma} \\
\frac{A_2}{A_0} &= \frac{1}{2}\frac{\alpha_1(\alpha_1+1)}{\gamma(\gamma+1)} - \frac{\alpha_2+\alpha_3}{(\gamma+1)a_2} \\
\frac{A_3}{A_0} &= \frac{1}{3!}\frac{\alpha_1(\alpha_1+1)(\alpha_1+2)}{\gamma(\gamma+1)(\gamma+2)} - \frac{1}{3(\gamma+2)} \\
&\quad \left[\frac{(\alpha_1+2)(\alpha_2+\alpha_3)}{(\gamma+1)a_2} + \frac{2\alpha_1(\alpha_2+\alpha_3)}{\gamma a_2} + \frac{2(\alpha_2+2\alpha_3)}{a_2^2}\right].
\end{aligned}$$

For $m \geq 3$, the coefficient can be obtained from the following four-term recurrence relationship

$$\begin{aligned}
&a_2^2(m+1)(m+\gamma)A_{m+1} - a_2\left[a_2(m+\alpha_1) + 2n(\gamma - m + 1)\right]A_m \\
&+ \left[2a_2(\alpha_1+\alpha_2+\alpha_3+m-1) + (n-1)(\gamma+m-2)\right]A_{m-1} \\
&- (\alpha_1 + 2\alpha_2 + m - 2)A_{m-2} = 0.
\end{aligned}$$

The series solution for V_n converges uniformly and absolutely for $z < a_2$ or $\rho < 2\rho_a$. Actually, in the Luneburg lens problem the values of ρ never exceed ρ_a. It is observed that for larger values of ρ_a, the leading term of A_m/A_0 is practically the same as the corresponding coefficient of the series expansion of the Kummer function. For example, when ρ_a is large,

$$\alpha_1 \simeq \alpha = \frac{1}{2}\left(n + \frac{3}{2} - \rho_a\right)$$

and

$$-\frac{\alpha_2+\alpha_3}{a_2}=\frac{5}{32\rho_a^2}.$$

The second term of A_2/A_0 is, therefore, much smaller than the first term. The T_n function, in view of (6), is then practically equal to the S_n function at $\rho=\rho_a$. The same is true for the derivative of T_n with respect to r when ρ_a is sufficiently large.

Knowing the two radial functions S_n and T_n, we can construct the vector wave functions according to (12.31) to (12.34), which can then be used to find the dyadic Green functions pertaining to the spherical Luneburg lens. According to the method of scattering superposition, we let

$$\bar{\bar{G}}_e^{(11)}(\bar{R},\bar{R}')=\bar{\bar{G}}_{eo}(\bar{R},\bar{R}')+\bar{\bar{G}}_{es}^{(11)}(\bar{R},\bar{R}') \tag{12.46}$$

$$\bar{\bar{G}}_e^{(21)}(\bar{R},\bar{R}')=\bar{\bar{G}}_{es}^{(21)}(\bar{R},\bar{R}'), \tag{12.47}$$

where $\bar{\bar{G}}_{eo}$ is represented by (10.24). The composition, or course, implies that the source is located outside of the lens. Following the same procedure as described in Sec. 10-3 for the dielectric sphere, we find that

$$\bar{\bar{G}}_{es}^{(11)}=\frac{ik}{4\pi}\sum_{n=1}^{\infty}\sum_{m=0}^{n}(2-\delta_0)\frac{2n+1}{n\,(n+1)}\frac{(n-m)!}{(n+m)!}\cdot\left[A_n\bar{M}^{(1)}(k)\bar{N}'^{(1)}(k)+B_n\bar{N}^{(1)}(k)\bar{N}'^{(1)}(k)\right] \tag{12.48}$$

$$\bar{\bar{G}}_{es}^{(21)}(\bar{R},\bar{R}')=\frac{ik}{4\pi}\sum_{n=1}^{\infty}\sum_{m=0}^{n}(2-\delta_0)\frac{2n+1}{n\,(n+1)}\frac{(n-m)!}{(n+m)!}\left[C_n\bar{M}^{(m)}(k)\bar{M}'^{(1)}(k)+D_n\bar{N}^{(e)}(k)\bar{N}'^{(1)}(k)\right], \tag{12.49}$$

where the coefficients A_n, B_n, C_n, and D_n are determined from the following system of equations resulting from the matching of boundary conditions for the function of the third kind:

$$P_a+A_nQ_a=C_nS_a$$
$$P_a'+A_nQ_a'=C_nS_a'$$
$$P_a+B_nQ_a=D_nT_a$$
$$P_n'+B_nQ_a'=D_nT_a',$$

where

$$P_a=\rho_a j_n(\rho_a),\quad P_a'=\frac{d}{d\rho_a}\left[\rho_a j_n(\rho_a)\right]$$

$$Q_a=\rho_a h_n^{(1)}(\rho_a)\,,\ Q_a'=\frac{d}{d\rho_a}\left[\rho_a h_n^{(1)}(\rho_a)\right]$$

$$S_a = S_n(\rho_a), \qquad S'_a = \frac{d}{d\rho_a} S_n(\rho_a)$$
$$T_a = T_n(\rho_a), \qquad T'_a = \frac{d}{d\rho_a} T_n(\rho_a).$$

We have deleted the subscript "${}_{\substack{e\\o}}mn$" for the wave functions in (12.48) and (12.49).

According to the excitation requirement of a Luneburg lens based on the geometrical theory of optics, the source should be located at the rim of the spherical lens. If we let it be represented by an electric dipole pointed in the x-direction with current moment equal to c so that

$$\bar{J}(\bar{R}') = c\,(x' - 0)\,\delta\,(y' - 0)\,\delta\,(z' + a)\,\hat{x}$$

as shown in Fig. 12-1, then the total electrical field outside of the lens is given by

$$\begin{aligned}\bar{E}(\bar{R}) =& i\omega\mu_0 \iiint \bar{\bar{G}}_e^{(11)}(\bar{R}, \bar{R}') \cdot \bar{J}(\bar{R}')\, dV' \\ =& \frac{-i\omega\mu_0 c}{4\pi a} \sum_{n=1}^{\infty} (-1)^{n+1} \frac{2n+1}{n(n+1)} \\ & \cdot \frac{1}{Q_a} \left(\left[\frac{1}{(Q'_a/Q_a - S'_a/S_a)} \right] \bar{M}_{01n}^{(1)} - \left\{ \frac{1}{[(T_a/T'_a)(Q'_a/Q_a) - 1]} \right\} \bar{N}_{e1n}^{(1)} \right).\end{aligned} \tag{12.50}$$

The far-zone expression for $\bar{E}(\bar{R})$, valid for large values of kR, is obtained by using the asymptotic expressions for $\bar{M}_{01n}^{(1)}$ and $\bar{N}_{e1n}^{(1)}$:

$$\bar{M}_{o1n}^{(1)} \simeq (-i)^{n+1} \frac{e^{ikR}}{kR} \bar{m}_{o1n}$$
$$\bar{N}_{e1n}^{(1)} \simeq (-i)^{n+1} \frac{e^{ikR}}{kR} \bar{n}_{e1n}.$$

These are the basic formulas which are needed for the evaluation of the radiation pattern of a spherical Luneburg lens when excited by an infinitesimal electric dipole. It is obvious that a similar expression can be derived if the exciting source is a magnetic dipole and the result will be slightly different.

As far as the wave theory of the spherical Luneburg lens is concerned, it is not necessary to arrive at this result by way of a dyadic Green formulation. By expanding the dipole field in terms of the spherical vector wave functions and then applying the scattering superposition theorem, we can derive the same result. In fact, this was the approach taken in the author's original paper quoted previously. The electromagnetic thoery of a spherical Luneburg lens was examined independently by Wilcox [1956]. There is, however, considerable difference in the two approaches. Wilcox found a series solution for T_n directly without examining in detail the analytical behavior of this function from the point of view of the theory of differential equations.

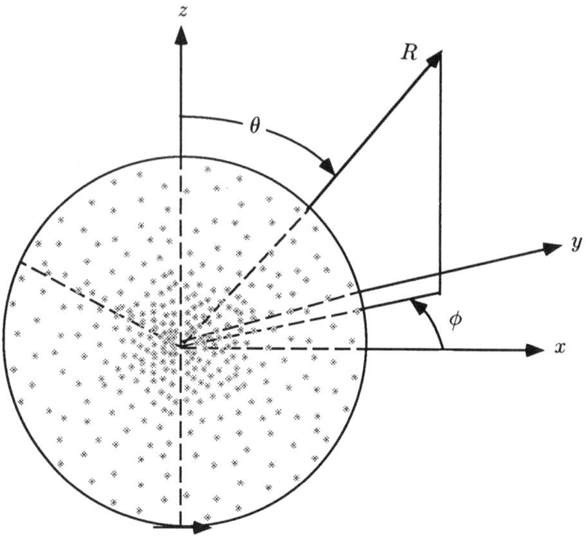

Fig. 12-1 Horizontal dipole placed at the surface of a spherical Luneberg lens

Although it is not yet feasible to construct the general dyadic Green functions for cylindrically stratified lenses, the electromagnetic theory for these lenses can be formulated if the field is a two- dimensional one, independent of the longitudinal variable or if it is a rotationally symmetrical field. The first electromagnetic theory of a Luneburg lens, in fact, was founded by Jasik [1954] using a cylindrically stratified lens with an electric line current as the source of excitation. The radial functions encountered in his work are expressed in terms of the confluent hypergeometric functions. A few calculations are found in Jasik's original work showing the radiation pattern of a moderately sized cylindrical Luneburg lens. The ideal characteristics of such a lens as deduced from the geometrical theory of optics are, of course, not preserved. Most important of all, the lens would emit a cylindrical wave in the far zone with a pattern function depending on size of the lens instead of a collimated beam as predicted from the geometrical theory of optics. A cylindrical Luneburg lens excited by a magnetic line source can be analyzed in a similar manner [Tai, 1956]. In that case we encountered again a radial function which is of the generalized confluent hypergeometric type.

For the conical lens of Luneburg, it can be shown that the S_n-function can be expressed in terms of the confluent hypergeometric function while the T_n-function is of the generalized confluent hypergeometric type. It is probably a coincidence that the same type of function is involved in Luneburg's two distinct lenses.

For the Maxwell fish-eyes, the permittivity function is described by

$$\epsilon_r(\bar{R}) = \frac{4}{\left[1 - (R/a)^2\right]^2}, \quad a \geq R \geq 0.$$

According to Maxwell's original theory, the lens will focus the rays emitted from a point source located at the surface of the lens to another focus at the opposite side of the lens, hence, the name "fish-eyes." If we let [Tai, 1958b]

$$\rho = kR, \; \rho_a = ka, \; \xi = -\left(\frac{\rho}{\rho_a}\right)^2$$

and

$$S_n(\xi) = \xi^{\frac{n+1}{2}} (\xi - 1)^{\mu} U_n(\xi) \tag{12.51}$$

$$T_n(\xi) = \xi^{\frac{n+1}{2}} (\xi - 1)^{\mu - 1} V_n(\xi), \tag{12.52}$$

where

$$\mu = \frac{1}{2}\left(1 + \sqrt{1 + 4\rho_a^2}\right),$$

then the functions $U_n(\xi)$ and $V_n(\xi)$ satisfy, respectively, the following two equations

$$\xi(\xi - 1)\frac{d^2U_n}{d\xi^2} + [(2\mu + \beta)\xi - \beta]\frac{dU_n}{d\xi} + \alpha U_n = 0 \tag{12.53}$$

$$\xi(\xi - 1)\frac{d^2V_n}{d\xi^2} + [(2\mu + \beta)\xi - \beta]\frac{dV_n}{d\xi} + \left(\alpha - \frac{1}{2}\right)V_n = 0, \tag{12.54}$$

where

$$\beta = n + \frac{3}{2}$$

$$\alpha = \beta\mu + \rho_a^2.$$

Equations (12.53) and (12.54) are identical in form to the hypergeometric equation normally written in the form [Copson, 1948, p. 247]

$$\xi(\xi - 1)\frac{d^2w}{d\xi^2} + [(a + b + 1)\xi - c]\frac{dw}{d\xi} + abw = 0. \tag{12.55}$$

It is relatively easy to determine the constants a, b, and c in terms of α, β, and μ for the two cases. For the analysis of the Maxwell fish-eyes, we only need the function $F(a, b, c, \xi)$ which is regular at $\xi = 0$. It is interesting to observe that the generalized confluent hypergeometric function is not encountered in the electromagnetic theory of Maxwell's fish-eyes.

In regard to the Nomura-Takaku distribution, as shown by these two authors, the differential equations for S_n and T_n are special cases of Malmsten's equation [Watson, 1922, p. 99]. These functions can be expressed in terms of the fractional order Bessel functions provided that $q \neq -1$ (see Table 12-1 for the permittivity function). The explicit relations are

$$S_n(\rho) = \rho^{\frac{1}{2}} J_\nu \left[\frac{\rho^{q+1}}{(q+1)\rho_a^q} \right] \tag{12.56}$$

and

$$T_n(\rho) = \rho^{q+\frac{1}{2}} J_{\nu'} \left[\frac{\rho^{q+1}}{(q+1)\rho_a^q} \right], \tag{12.57}$$

where

$$\rho = kR, \ \rho_a = ka$$
$$\nu = \frac{(n+1/2)}{(q+1)}$$
$$\nu' = \frac{\left[(n+1/2)^2 + q^2 + q\right]^{\frac{1}{2}}}{(q+1)}.$$

For $q = 1$, the profile reduces to that of Eaton's lens. For $q = -1$, the S_n- and T_n-functions are expressible in terms of elementary functions. Two typical solutions are

$$S_n = \rho^{\frac{1}{2}} \sin \left[\left(\rho_a^2 - n^2 - n - \frac{1}{4} \right)^{\frac{1}{2}} \ell n\rho \right] \tag{12.58}$$

$$T_n = \rho^{\frac{1}{2}} \left[\left(\rho_a^2 - n^2 - n - \frac{1}{4} \right)^{\frac{1}{2}} \ell n\rho \right]. \tag{12.59}$$

Because of that particular profile, which is physically unrealizable if we allow R to be approaching zero, the function of T_n has a singularity at $\rho = 0$.

Although we have not done any numerical work based on these formulas except for the Luneburg lens [Rozenfeld, 1976], the theoretical foundation provided here for various types of lenses should be useful to investigate not only the radiation pattern of these lenses but also the exact nature of a focal point, since the geometrical theory of optics fails to offer a valid description of the energy distribution at or in the neighborhood of such a point.

12-4 MONOCHROMATICALLY OSCILLATING FIELD IN A MOVING ISOTROPIC MEDIUM

Based on Minkowski's relativistic theory of electrodynamics [Minkowski, 1908; Sommerfeld, 1952], the constitutive relations between the field vectors in a moving isotropic medium can be described in a rather compact form [Tai, 1965a, 1967]. They are

$$\overline{B} = \mu\bar{\bar{\alpha}} \cdot \overline{H} - \overline{\Omega} \times \overline{E} \tag{12.60}$$

and

$$\overline{D} = \epsilon\bar{\bar{\alpha}} \cdot \overline{E} + \overline{\Omega} \times \overline{H} \tag{12.61}$$

where

$$\mu, \epsilon = \text{permeability and permittivity of the medium at rest}$$

$$\overline{\Omega} = \frac{\left(n^2 - 1\right)}{\left(1 - n^2\beta^2\right)c^2}$$

$$\bar{v} = v\hat{z} = \text{ velocity of the moving medium assuming to be constant and directed in the } z\text{-direction}$$

$$c = \left(\mu_0\epsilon_0\right)^{\frac{1}{2}} = \text{ velocity of light}$$

$$n = \left(\mu_0\epsilon_0\right)^{\frac{1}{2}} = \text{index of refraction}$$

$$\beta = \frac{v}{c}$$

$$\bar{\bar{\alpha}} = a\left(\hat{x}\hat{x} + \hat{y}\hat{y}\right) + \hat{z}\hat{z}$$

$$a = \frac{\left(1 - \beta^2\right)}{\left(1 - n^2\beta^2\right)}.$$

The dyadic coefficient $\bar{\bar{\alpha}}$ is a characteristic parameter of this theory. When $v = 0$ or $n = 1$, (12.60) and (12.61) reduce to the constitutive relations for a stationary medium or that of the vacuum (air) as it should be. For a monochromatically oscillating field with angular frequency ω, the definite form of Maxwell's equations are

$$\nabla \times \overline{E} = i\omega\left(\mu\bar{\bar{\alpha}} \cdot \overline{H} - \overline{\Omega} \times \overline{E}\right) \tag{12.62}$$

$$\nabla \times \overline{H} = \overline{J} - i\omega\left(\epsilon\bar{\bar{\alpha}} \cdot \overline{E} + \overline{\Omega} \times \overline{H}\right). \tag{12.63}$$

We now introduce two auxiliary field vectors $\overline{E}^{(b)}$ and $\overline{H}^{(b)}$ defined by

$$\overline{E} = e^{-i\omega\Omega z}\bar{\bar{b}} \cdot \overline{E}^{(b)} \tag{12.64}$$

$$\overline{H} = e^{-i\omega\Omega z}\bar{\bar{b}} \cdot \overline{H}^{(b)}, \tag{12.65}$$

where

$$\bar{\bar{b}} = \frac{1}{a}\left(\hat{x}\hat{x} + \hat{y}\hat{y}\right) + \hat{z}\hat{z},$$

which is reciprocal to $\bar{\bar{\alpha}}$ in the sense that

$$\bar{\bar{\alpha}} \cdot \bar{\bar{b}} = \bar{\bar{I}},$$

where $\bar{\bar{I}}$ denotes the idem factor. By substituting (12.64) and (12.65) into (12.62) and (12.63), we find that $\overline{E}^{(b)}$ and $\overline{H}^{(b)}$ satisfy the following pair of equations

$$\nabla \times \left(\bar{\bar{b}} \cdot \overline{E}^{(b)}\right) = i\omega\mu\overline{H}^{(b)} \tag{12.66}$$

$$\nabla \times \left(\bar{\bar{b}} \cdot \overline{H}^{(b)}\right) = \overline{J} e^{i\omega\Omega z} - i\omega\epsilon \overline{E}^{(b)}. \tag{12.67}$$

The wave equations for these two functions are

$$\nabla \times \left[\bar{\bar{b}} \cdot \nabla \times \left(\bar{\bar{b}} \cdot \overline{E}^{(b)}\right)\right] - k^2 \overline{E}^{(b)} = i\omega\mu e^{i\omega\Omega z} \overline{J} \tag{12.68}$$

$$\nabla \times \left[\bar{\bar{b}} \cdot \nabla \times \left(\bar{\bar{b}} \cdot \overline{H}^{(b)}\right)\right] - k^2 \overline{H}^{(b)} = \nabla \times \left(\bar{\bar{b}} \cdot \overline{J} e^{i\omega\Omega z}\right). \tag{12.69}$$

To find the integral solution for these two auxiliary field functions we introduce two dyadic Green functions $\overline{\overline{G}}_e^{(b)}$ and $\overline{\overline{G}}_m^{(b)}$ which satisfy the following system of equations:

$$\nabla \times \left(\bar{\bar{b}} \cdot \overline{\overline{G}}_e^{(b)}\right) = \overline{\overline{G}}_m^{(b)} \tag{12.70}$$

$$\nabla \times \left(\bar{\bar{b}} \cdot \overline{\overline{G}}_m^{(b)}\right) = \bar{\bar{I}} \delta(\overline{R} - \overline{R}') + k^2 \overline{\overline{G}}_e^{(b)} \tag{12.71}$$

$$\nabla \times \left[\bar{\bar{b}} \cdot \nabla \times \left(\bar{\bar{b}} \cdot \overline{\overline{G}}_e^{(b)}\right)\right] - k^2 \overline{\overline{G}}_e^{(b)} = \bar{\bar{I}} \delta(\overline{R} - \overline{R}') \tag{12.72}$$

$$\nabla \times \left[\bar{\bar{b}} \cdot \nabla \times \left(\bar{\bar{b}} \cdot \overline{\overline{G}}_m^{(b)}\right)\right] - k^2 \overline{\overline{G}}_m^{(b)} = \nabla \times \left[\bar{\bar{b}} \delta(\overline{R} - \overline{R}')\right]. \tag{12.73}$$

In order to integrate (11.68) and (11.69) with the aid of these two dyadic Green functions we need a vector-dyadic Green theorem more general than (1.51). For that purpose, we introduce a dyadic function $\bar{\bar{A}}$ defined by

$$\bar{A} = \left(\bar{\bar{b}} \cdot \overline{Q}\right) \times \left[\bar{\bar{b}} \cdot \nabla \times \left(\bar{\bar{b}} \cdot \overline{\overline{P}}\right)\right] + \left[\bar{\bar{b}} \cdot \nabla \times \left(\bar{\bar{b}} \cdot \overline{Q}\right)\right] \times \left(\bar{\bar{b}} \cdot \overline{\overline{P}}\right), \tag{12.74}$$

where $\overline{Q}$ is a vector function and $\overline{\overline{P}}$ is a dyadic function; then it can easily be verified that

$$\begin{aligned} \nabla \cdot \bar{A} = &\left\{\nabla \times \left[\bar{\bar{b}} \cdot \nabla \times \left(\bar{\bar{b}} \cdot Qv\right)\right] \cdot \left(\bar{\bar{b}} \cdot \overline{\overline{P}}\right)\right. \\ &\left. - \left(\bar{\bar{b}} \cdot \overline{Q}\right) \cdot \nabla \times \left[\bar{\bar{b}} \cdot \nabla \times \left(\bar{\bar{b}} \cdot \overline{\overline{P}}\right)\right]\right\}. \end{aligned} \tag{12.75}$$

Equation (12.75) is valid as long as $\bar{\bar{b}}$ is a symmetric dyadic with constant coefficients like the one in this formulation. By means of the divergence theorem,

$$\iiint \nabla \cdot \bar{A} dV = \oiint (\hat{n} \cdot \bar{A})\, dS. \tag{12.76}$$

We obtain the desired vector-dyadic theorem in the form

$$\begin{aligned} &\iiint \left(\left\{\nabla \times \left[\bar{\bar{b}} \cdot \nabla \times \left(\bar{\bar{b}} \cdot \overline{Q}\right)\right]\right\} \cdot \left(\bar{\bar{b}} \cdot \overline{\overline{P}}\right)\right. \\ &\left. - \left(\bar{\bar{b}} \cdot \overline{\overline{Q}}\right) \cdot \nabla \times \left[\bar{\bar{b}} \cdot \nabla \times \left(\bar{\bar{b}} \cdot \overline{\overline{P}}\right)\right]\right) dV \\ &= \oiint \left\{\left[\hat{n} \times \left(\bar{\bar{b}} \cdot \overline{Q}\right)\right] \cdot \bar{\bar{b}} \cdot \nabla \times \left(\bar{\bar{b}} \cdot \overline{\overline{P}}\right)\right. \\ &\left. + \left(\hat{n} \times \left[\bar{\bar{b}} \cdot \nabla \times \left(\bar{\bar{b}} \cdot \overline{Q}\right)\right]\right) \cdot \bar{\bar{b}} \cdot \overline{\overline{P}}\right\} dS. \end{aligned} \tag{12.77}$$

For an infinitely extended moving medium we let $\overline{Q} = \overline{E}^{(b)}(\overline{R})$ and $\overline{\overline{P}} = \overline{\overline{G}}_{eo}^{(b)}(\overline{R}, \overline{R}')$, the electric dyadic Green function in such an unbounded region. As a result of (12.67), (12.68), and (12.72), which now applies to $\overline{\overline{G}}_{eo}^{(b)}$, (12.77) yields

$$\begin{aligned}\bar{\bar{b}} \cdot \overline{E}^{(b)}(\overline{R}') =& i\omega\mu \iiint e^{i\omega\Omega z} \overline{J}(\overline{R}) \cdot \bar{\bar{b}} \cdot \overline{\overline{G}}_{eo}^{(b)}(\overline{R}, \overline{R}')\, dV \\ & - \oiint \left\{ \left(\hat{n} \times \left[\bar{\bar{b}} \cdot \overline{E}^{(b)}(\overline{R}) \right] \right) \cdot \bar{\bar{b}} \cdot \nabla \times \left[\bar{\bar{b}} \cdot \overline{\overline{G}}_{eo}^{(b)}(\overline{R}, \overline{R}') \right] \right. \\ & \left. + i\omega\mu \left(\hat{n} \times \left[\bar{\bar{b}} \cdot \overline{H}^{(b)}(\overline{R}) \right] \right) \cdot \bar{\bar{b}} \cdot \overline{\overline{G}}_{eo}^{(b)}(\overline{R}, \overline{R}') \right\} dS. \end{aligned} \tag{12.78}$$

By interchanging $\overline{R}$ and $\overline{R}'$ in (12.78) and using the symmetric relations

$$\left[\bar{\bar{b}} \cdot \overline{\overline{G}}_{eo}^{(b)}(\overline{R}', \overline{R}) \right]^T = \bar{\bar{b}} \cdot \overline{\overline{G}}_{eo}^{(b)}(\overline{R}, \overline{R}') \tag{12.79}$$

and

$$\bar{\bar{b}} \cdot \left[\nabla' \times \bar{\bar{b}} \cdot \overline{\overline{G}}_{eo}^{(b)}(\overline{R}', \overline{R}) \right]^T = \bar{\bar{b}} \cdot \nabla \times \left[\bar{\bar{b}} \cdot \overline{\overline{G}}_{eo}^{(b)}(\overline{R}, \overline{R}') \right], \tag{12.80}$$

(12.78) can be written in the form

$$\begin{aligned}\bar{\bar{b}} \cdot \overline{E}^{(b)}(\overline{R}) =& i\omega\mu \iiint e^{i\omega\Omega z'} \bar{\bar{b}} \cdot \overline{\overline{G}}_{eo}^{(b)}(\overline{R}, \overline{R}') \cdot \overline{J}(\overline{R}') dV' \\ & - \oiint \left\{ \bar{\bar{b}} \cdot \nabla \times \left[\bar{\bar{b}} \cdot \overline{\overline{G}}_{eo}^{(b)}(\overline{R}, \overline{R}') \right] \cdot \left(\hat{n} \times \left[\bar{\bar{b}} \cdot \overline{E}^{(b)}(\overline{R}') \right] \right) \right. \\ & \left. + i\omega\mu \bar{\bar{b}} \cdot \overline{\overline{G}}_{eo}^{(b)}(\overline{R}, \overline{R}') \cdot \left(\hat{n} \times \left[\bar{\bar{b}} \cdot \overline{H}^{(b)}(\overline{R}') \right] \right) \right\} dS'. \end{aligned} \tag{12.81}$$

When we let the surface of integration go to infinity the surface integral vanishes because of the radiation condition; then

$$\bar{\bar{b}} \cdot \overline{E}^{(b)}(\overline{R}) = i\omega\mu \iiint e^{i\omega\Omega z'} \bar{\bar{b}} \cdot \overline{\overline{G}}_{eo}^{(b)}(\overline{R}, \overline{R}') \cdot \overline{J}(\overline{R}')\, dV'.$$

Hence, in view of (12.64),

$$\overline{E}(\overline{R}) = i\omega\mu \iiint e^{-i\omega\Omega(z-z')} \bar{\bar{b}} \cdot \overline{\overline{G}}_{eo}^{(b)}(\overline{R}, \overline{R}') \cdot \overline{J}(\overline{R}')\, dV'. \tag{12.82}$$

For a region void of current source (12.81) becomes

$$\begin{aligned}\bar{\bar{b}} \cdot \overline{E}^{(b)}(\overline{R}) = & - \oiint_S \left\{ \bar{\bar{b}} \cdot \nabla \times \left[\bar{\bar{b}} \cdot \overline{\overline{G}}_{eo}^{(b)}(\overline{R}, \overline{R}') \right] \cdot \left(\hat{n} \times \left[\bar{\bar{b}} \cdot \overline{E}^{(b)}(\overline{R}') \right] \right) \right. \\ & \left. + i\omega\mu \bar{\bar{b}} \cdot \overline{\overline{G}}_{eo}^{(b)}(\overline{R}, \overline{R}') \cdot \left(\hat{n} \times \left[\bar{\bar{b}} \cdot \overline{H}^{(b)}(\overline{R}') \right] \right) \right\} dS'. \end{aligned} \tag{12.83}$$

In terms of the actual fields, $\overline{E}(\overline{R})$ and $\overline{H}(\overline{R})$, (12.83) leads to

$$\overline{E}(\overline{R}) = -\oiint_S e^{-i\omega\Omega(z-z')}\left\{\bar{\bar{b}}\cdot\nabla\times\left[\bar{\bar{b}}\cdot\overline{\overline{G}}_{eo}^{(b)}(\overline{R},\overline{R}')\right]\cdot\left[\hat{n}\times\overline{E}(\overline{R}')\right]\right.$$
$$\left. + i\omega\mu\bar{\bar{b}}\cdot\overline{\overline{G}}_{eo}^{(b)}(\overline{R},\overline{R}')\cdot\left[\hat{n}\times\overline{H}(\overline{R}')\right]\right\}dS', \tag{12.84}$$

which is a mathematical statement of Huygens' Principle in a moving isotropic medium. The remaining problem in this section is to find the explicit expression of $\overline{\overline{G}}_{eo}^{(b)}$. The most convenient approach is to adopt the operational method of Levine and Schwinger, reviewed in Chap. 4, originally applied to find $\overline{\overline{G}}_{eo}$ in a stationary medium.

We consider the equation for $\overline{\overline{G}}_{eo}^{(b)}$ given by (12.72) with $\overline{\overline{G}}_{e}^{(b)}$ replaced by $\overline{\overline{G}}_{eo}^{(b)}$; then

$$\nabla\times\overline{\overline{G}}_{eo}^{(b)}(\overline{R},\overline{R}') = -\frac{1}{k^2}\nabla\delta(\overline{R},\overline{R}') \tag{12.85}$$

It can be verified that

$$\nabla\times\left[\bar{\bar{b}}\cdot\nabla\times\left(\bar{\bar{b}}\cdot\overline{\overline{G}}_{eo}^{(b)}\right)\right] = \frac{1}{a}\left[\nabla_a\nabla\cdot\overline{\overline{G}}_{eo}^{(b)} - (\nabla_a\cdot\nabla)\,\overline{\overline{G}}_{eo}^{(b)}\right], \tag{12.86}$$

where the operators ∇_a and $(\nabla_a\cdot\nabla)$ are defined by

$$\nabla_a = \hat{x}\frac{\partial}{\partial x} + \hat{y}\frac{\partial}{\partial y} + \hat{z}\frac{\partial}{a\partial z} = \frac{1}{a}\bar{\bar{\alpha}}\cdot\nabla \tag{12.87}$$

and

$$\nabla_a\cdot\nabla = \frac{\partial^2}{\partial x^2} + \frac{\partial^2}{\partial y^2} + \frac{1}{a}\frac{\partial^2}{\partial z^2}. \tag{12.88}$$

The validity of (12.86) can most easily be demonstrated by replacing the dyadic function in that equation by a vector function.

As a result of (12.85) and (12.86), (12.72) can be written in the form

$$(\nabla_a\cdot\nabla)\,\overline{\overline{G}}_{eo}^{(b)}(\overline{R},\overline{R}') + k^2a\overline{\overline{G}}_{eo}^{(b)}(\overline{R},\overline{R}')$$
$$= -a\left(\bar{\bar{I}} + \frac{1}{k^2a^2}\bar{\bar{\alpha}}\cdot\nabla\nabla\right)\delta(\overline{R}-\overline{R}'). \tag{12.89}$$

If we relate $\overline{\overline{G}}_{eo}^{(b)}$ with a scalar function $G_o^{(b)}$ such that

$$\overline{\overline{G}}_{eo}^{(b)}(\overline{R},\overline{R}') = a\left(\bar{\bar{I}} + \frac{1}{k^2a^2}\bar{\bar{\alpha}}\cdot\nabla\nabla\right)\overline{G}_{o}^{(b)}(\overline{R},\overline{R}'). \tag{12.90}$$

Then (12.90) would be the solution to (12.72) if $G_o^{(b)}$ satisfies the differential equation

$$\left[(\nabla_a \cdot \nabla) + k^2 a\right] G_o^{(b)}(\bar{R}, \bar{R}') = -\delta(\bar{R} - \bar{R}')$$

or

$$\left(\frac{\partial^2}{\partial x^2} + \frac{\partial^2}{\partial y^2} + \frac{1}{a}\frac{\partial^2}{\partial z^2} + k^2 a\right) G_o^{(b)}(\bar{R}, \bar{R}') = -\delta(\bar{R} - \bar{R}'). \qquad (12.91)$$

The solution for $G_o^{(b)}$ depends on the value of "a", particularly its sign.

Case 1.

$$a = \frac{(1-\beta^2)}{(1-n^2\beta^2)} > 0, \text{ or } n\beta < 1.$$

$$G_o^{(b)}(\bar{R}, \bar{R}') = \frac{a^{\frac{1}{2}} e^{ika^{\frac{1}{2}} R_a}}{4\pi R_a} \qquad (12.92)$$

where

$$R_a = \left[(x-x')^2 + (y-y')^2 + a(z-z')^{\frac{1}{2}}\right]^{\frac{1}{2}}.$$

This solution is obtained by introducing a new variable $z_a = a^{\frac{1}{2}} z$ and treating (12.91) as an ordinary scalar wave equation in a pseudo-Cartesian coordinate system (x, y, z_a).

Case 2. $a < 0$ or $n\beta > 1$

In this case $a = - \mid a \mid$, (12.91) can be treated as a two-dimensional Klein-Gordon equation. Its solution is given by [Cohen, 1961]

$$G_0^{(b)}(\bar{R}, \bar{R}') = \begin{cases} 0, & \mid a \mid^{\frac{1}{2}} (z-z') < r \\ \dfrac{\mid a \mid^{\frac{1}{2}} \cos\left(k \mid a \mid^{\frac{1}{2}} R_a'\right)}{2\pi R_a'}, & \mid a \mid^{\frac{1}{2}} (z-z') > r, \end{cases} \qquad (12.93)$$

where

$$R_a' = \left[\mid a \mid^{\frac{1}{2}} (z-z')^2 - r^2\right]^{\frac{1}{2}}$$
$$r^2 = (x-x')^2 + (y-y')^2.$$

The discontinuous behavior of the function is a manifestation of the Cerenkov phenomenon, corresponding to the supersonic flow in hydrodynamics.

In the remaining part of this section we shall give the complete expressions of the electromagnetic field of a Hertzian dipole placed in a moving medium based on (12.82). For simplicity, we consider just the case corresponding to $n\beta < 1$. By combining (12.64) with (12.82) we obtain the formula

$$\bar{E}(\bar{R}) = i\omega\mu \iiint \bar{\bar{b}} \cdot \bar{\bar{G}}_{eo}^{(b)}(\bar{R}, \bar{R}') \cdot \bar{J}(\bar{R}') e^{-i\omega\Omega(z-z')}\, dV'.$$

It is sufficient for us to consider two distinct cases, corresponding to two different orientations of the dipole with respect to the direction of motion $\hat{z}$.

Case 1. Dipole Parallel to the Direction of Motion.

$$\bar{J}(\bar{R}') = c\delta\left(\bar{R}' - 0\right)\hat{z}.$$

In this case, we find that the expressions for the field are given by

$$\begin{aligned}
E_R =& \frac{\eta c k^2 a^2 e^{i(D-\omega\Omega z)}}{2\pi D^2}\left(1+\frac{i}{D}\right)\cos\theta \\
E_\theta =& \frac{-i\eta c k^2 a^2 e^i\left(D-\omega\Omega z\right)}{4\pi f^2 D} \\
& \cdot\left\{1+\frac{i}{D}\left(1+\frac{i}{D}\right)\left[1-2\left(a-1\right)\cos^2\theta\right]\right\}\sin\theta \\
H_\phi =& \frac{-ick^2 a^{3/2} e^{i(D-\omega\Omega z)}}{4\pi f^2 D}\left(1+\frac{i}{D}\right)\sin\theta,
\end{aligned}$$

where

$$\begin{aligned}
& D = ka^{\frac{1}{2}}Rf \\
& f = \left[1+(a-1)\cos^2\theta\right]^{\frac{1}{2}} \\
& R = \left(x^2+y^2+z^2\right)^{\frac{1}{2}} \\
& \eta = \left(\frac{\mu}{\epsilon}\right)^{\frac{1}{2}} \\
& c = \text{current moment.}
\end{aligned}$$

Case 2. Dipole Perpendicular to the Direction of Motion.

$$\bar{J}(\bar{R}') = c\delta\left(\bar{R}' - 0\right)\hat{x}.$$

$$\begin{aligned}
E_R =& \frac{\eta c k^2 a^2 e^{i(D-\omega\Omega z)}}{2\pi D^2}\left(1+\frac{i}{D}\right)\cos\theta\cos\phi \\
E_\theta =& \frac{i\eta c k^2 a^2 e^{i(D-\omega\Omega z)}}{4\pi f^2 D} \\
& \left\{1+\frac{i}{D}\left(1+\frac{i}{D}\right)\left[1+2\left(1-\frac{1}{a}\right)\sin^2\theta\right]\right\}\cos\theta\cos\phi \\
E_\phi =& \frac{-i\eta c k^2 a e^{i(D-\omega\Omega z)}}{4\pi D}\left(1+\frac{i}{D}\right)-\frac{i}{D^2}\sin\phi \\
\bar{H} =& \frac{ick^2 a^{3/2} e^{i(D-\omega\Omega z)}}{4\pi f D}\left(1+\frac{i}{D}\right)\left(\sin\phi\hat{\theta}+\cos\theta\cos\phi\hat{\phi}\right).
\end{aligned}$$

In a terrestrial environment, a is very near to unity and Ω is a very small quantity. The effect of the motion of the medium is, therefore, not significant. The theory presented here, however, does provide us with a complete understanding of the radiation process in such a medium.

The expressions derived here can also be obtained by a more meticulous method by the way of Fourier transform [Lee and Papas, 1964] at the expense of a rather elaborate analysis.

12-5 TIME-DEPENDENT FIELD IN A MOVING MEDIUM

When the current source is an arbitrary function of time, the fields in a moving isotropic nondispersive medium have to be determined by starting with Maxwell's equations in the space and time domain, namely,

$$\nabla \times \overline{E}(\overline{R}, t) = -\frac{\partial \overline{B}(\overline{R}, t)}{\partial t} \tag{12.94}$$

$$\nabla \times \overline{H}(\overline{R}, t) = \overline{J}(\overline{R}, t) + \frac{\partial \overline{D}(\overline{R}, t)}{\partial t}. \tag{12.95}$$

The constitutive relations of the field vectors have the same form as (12.60) and (12.61). However, the field vectors therein are all functions of space and time. They will be denoted with a dependence on both $\overline{R}$ and t in a parenthesis as in the above two equations. This problem has previously been investigated and solved by Compton [1966] based on a direct Fourier transform method which is quite complicated. Part of his analysis could be greatly simplified by making use of the result which was described in the previous section for the monochromatically oscillating fields.

In the present work, the formulation will be cast in an entirely new manner by introducing a new independent "time" variable. By means of this approach we can avoid the use of a space-time operator introduced by Compton. Conventional method is sufficient to solve the resultant equations. In order to coordinate the method of dyadic Green functions with the classical method, the solution of the new field equations will be obtained by means of the method of potentials.

The constitutive relations stated by (12.60) and (12.61) suggest that for a function such as $\overline{B}(\overline{R}, t)$, we can introduce a new *independent* variable τ defined by

$$\tau = t + \Omega z, \tag{12.96}$$

which is designated as the "pseudo-time variable," where

$$\Omega = \frac{(n^2 - 1)\, v}{(1 - n^2\beta^2)\, c^2}. \tag{12.97}$$

The constant Ω has the dimension of the reciprocal of velocity. Now by treating a scalar function $f(\overline{R}, t)$ as a function of $\overline{R}$ and τ, we have the following differential relations according to the calculus of implicit functional transformation

$$\frac{\partial f(\overline{R},t)}{\partial t}=\frac{\partial f(\overline{R},\tau)}{\partial \tau}\frac{\partial \tau}{\partial t}=\frac{\partial f(\overline{R},\tau)}{\partial \tau} \tag{12.98}$$

$$\begin{aligned}\frac{\partial f(\overline{R},t)}{\partial z}&=\frac{\partial f(\overline{R},\tau)}{\partial z}+\frac{\partial f(\overline{R},\tau)}{\partial \tau}\frac{\partial \tau}{\partial z}\\&=\frac{\partial f(\overline{R},\tau)}{\partial z}+\Omega\frac{\partial f(\overline{R},\tau)}{\partial \tau}.\end{aligned} \tag{12.99}$$

In the vector function $\nabla\times\overline{E}(\overline{R},t)$, the components containing the derivatives with respect to z are

$$\begin{aligned}&-\hat{x}\frac{\partial E_y(\overline{R},t)}{\partial z}+\hat{y}\frac{\partial E_x(\overline{R},t)}{\partial z}\\&=-\hat{x}\left[\frac{\partial E_y(\overline{R},\tau)}{\partial z}+\Omega\frac{\partial E_y(\overline{R},\tau)}{\partial \tau}\right]\\&\quad+\hat{y}\left[\frac{\partial E_x(\overline{R},\tau)}{\partial z}+\Omega\frac{\partial E_x(\overline{R},\tau)}{\partial \tau}\right]\\&=-\hat{x}\frac{\partial E_y(\overline{R},\tau)}{\partial z}+\hat{y}\frac{\partial E_x(\overline{R},\tau)}{\partial z}+\overline{\Omega}\times\frac{\partial \overline{E}(\overline{R},\tau)}{\partial \tau},\end{aligned} \tag{12.100}$$

where $\overline{\Omega}=\Omega\hat{z}$; hence

$$\nabla\times\overline{E}(\overline{R},t)=\nabla\times\overline{E}(\overline{R},\tau)+\overline{\Omega}\times\frac{\partial \overline{E}(\overline{R},\tau)}{\partial \tau}. \tag{12.101}$$

In view of (12.60) and (12.98) the time derivative of $\overline{B}(\overline{R},t)$ with respect to t in (12.94) can be written in the form

$$\begin{aligned}\frac{\partial \overline{B}(\overline{R},t)}{\partial t}&=\frac{\partial \overline{B}(\overline{R},\tau)}{\partial \tau}\\&=\mu\bar{\bar{\alpha}}\cdot\frac{\partial \overline{H}(\overline{R},\tau)}{\partial \tau}-\Omega\times\frac{\partial \overline{E}(\overline{R},\tau)}{\partial \tau}.\end{aligned} \tag{12.102}$$

By substituting (12.101) and (12.102) into (12.94) we obtain

$$\nabla\times\overline{E}(\overline{R},\tau)=-\mu\bar{\bar{\alpha}}\cdot\frac{\partial \overline{H}(\overline{R},\tau)}{\partial \tau}. \tag{12.103}$$

Similarly, (12.95) can be converted to

$$\nabla\times\overline{H}(\overline{R},\tau)=\overline{J}(\overline{R},\tau)+\epsilon\bar{\bar{\alpha}}\cdot\frac{\partial \overline{E}(\overline{R},\tau)}{\partial \tau}. \tag{12.104}$$

The important feature of this formulation is that we have changed the set of independent variables from $(x,y,z;t)$ to $(x,y,z;\tau)$ in contrast to the work of

Compton who invoked a space-time operator in his formulation. Our approach follows D'Alemberts' method of solving one-dimensional scalar wave equation [Sommerfeld, 1949, p. 41]. From now on we should forget about the original time variable "t" and treating the field vectors as functions of $(x, y, z; \tau)$. To solve (12.69) and (12.70), the method of Fourier transform will be used. We denote the Fourier transform of $\overline{E}(\overline{R}, \tau)$ simply by $\overline{E}(\overline{R})$, which is understood to be a function of ω; that is,

$$\overline{E}(\overline{R}) = \int_{-\infty}^{\infty} \overline{E}(\overline{R}, \tau) e^{i\omega\tau} d\tau \tag{12.105}$$

$$\overline{E}(\overline{R}, \tau) = \frac{1}{2\pi} \int_{-\infty}^{\infty} \overline{E}(\overline{R}) e^{-i\omega\tau} d\omega. \tag{12.106}$$

A Fourier transform of (12.103) and (12.104) yields

$$\nabla \times \overline{E}(\overline{R}) = i\omega\mu\bar{\bar{\alpha}} \cdot \overline{H}(\overline{R}) \tag{12.107}$$

$$\nabla \times \overline{H}(\overline{R}) = \overline{J}(\overline{R}) - i\omega\epsilon\bar{\bar{\alpha}} \cdot \overline{E}(R). \tag{12.108}$$

Since $\nabla \cdot \left[\bar{\bar{\alpha}} \cdot \overline{H}(\overline{R})\right] = 0$, the classical method of potentials with a slight modification could be used. Let

$$\mu\bar{\bar{\alpha}} \cdot \overline{H}(\overline{R}) = \nabla \times \left[\bar{\bar{b}} \cdot \overline{A}(\overline{R})\right], \tag{12.109}$$

where $\bar{\bar{b}} = [\bar{\bar{\alpha}}]^{-1}$ as introduced in the last section.

By substituting (12.109) into (12.107) we can relate the two vector functions $\overline{E}$ and $\bar{\bar{b}} \cdot \overline{A}$ with a dynamic scalar potential function φ such that

$$\overline{E}(\overline{R}) = i\omega\bar{\bar{b}} \cdot \overline{A}(\overline{R}) - \nabla\varphi(\overline{R}). \tag{12.110}$$

$\overline{A}(\overline{R})$ and $\varphi(\overline{R})$ are two potential functions introduced in this formulation. By substituting (12.109) and (12.110) into (12.108), we have

$$\begin{aligned} &\nabla \times \left\{\bar{\bar{b}} \cdot \nabla \times \left[\bar{\bar{b}} \cdot \overline{A}(\overline{R})\right]\right\} \\ &= \mu\left[\overline{J}(\overline{R}) + \omega^2\epsilon\overline{A}(\overline{R}) + i\omega\epsilon\bar{\bar{\alpha}} \cdot \nabla\varphi\right]. \end{aligned} \tag{12.111}$$

The first term in (12.111) can be split into two parts as shown by (12.86) with the dyadic function in that equation replaced by $\overline{A}(\overline{R})$ and the function $\bar{\bar{\alpha}} \cdot \nabla\varphi$ is equal to a $\nabla_a\varphi$ according to (12.87). The result is

$$\begin{aligned} &\nabla_a\nabla \cdot \overline{A}(\overline{R}) - (\nabla_a \cdot \nabla)\,\overline{A}(\overline{R}) \\ &= \mu a\left[\overline{J}(\overline{R}) + \omega^2\epsilon\overline{A}(\overline{R}) + i\omega\epsilon a\nabla_a\varphi\right]. \end{aligned} \tag{12.112}$$

A gauge condition is now imposed upon $\overline{A}$ and φ such that

$$\nabla \cdot \overline{A}(\overline{R}') = i\omega\mu\epsilon a^2\varphi. \tag{12.113}$$

Under this condition (12.112) reduces to

$$(\nabla_a \cdot \nabla)\, \bar{A}(\bar{R}) + k^2 a \bar{A}(\bar{R}) = -\mu a \bar{J}(\bar{R}), \tag{12.114}$$

where $k^2 = \omega^2 \mu\epsilon = (\omega n/c)^2$ and the operator $\nabla_a \cdot \nabla$ is defined by

$$\nabla_a \cdot \nabla = \frac{\partial^2}{\partial x^2} + \frac{\partial^2}{\partial y^2} + \frac{1}{a}\frac{\partial^2}{\partial z^2}. \tag{12.115}$$

To find the integral solution for $\bar{A}(\bar{R})$ we can use the same scalar Green function $G_0^{(b)}$ introduced in (12.91). Some mathematical theorems similar to the ordinary Green theorem of the second kind but involving the modified Laplacian operator $\nabla_a \cdot \nabla$ are needed.

We consider a vector function $\bar{f}$ defined by

$$\bar{f} = F\nabla_a G - G\nabla_a F,$$

where ∇_a is the modified "gradient" operator defined by (12.81), that is,

$$\nabla_a = \frac{1}{a}\bar{\bar{\alpha}} \cdot \nabla = \hat{x}\frac{\partial}{\partial x} + \hat{y}\frac{\partial}{\partial y} + \hat{z}\frac{1}{a}\frac{\partial}{\partial z};$$

then

$$\begin{aligned} \nabla \cdot \bar{f} &= F\nabla \cdot (\nabla_a G) - G\nabla \cdot (\nabla_a F) \\ &= F\,(\nabla \cdot \nabla_a)\, G - G\,(\nabla \cdot \nabla_a)\, \bar{F}, \end{aligned} \tag{12.116}$$

where $\nabla \cdot \nabla_a$ is analytically equal to the operator $\nabla_a \cdot \nabla$ defined by (12.115). By applying the divergence theorem to $\nabla \cdot \bar{f}$ we obtain a modified Green's theorem of the second kind in the form

$$\begin{aligned} &\iiint \left[F\,(\nabla_a \cdot \nabla)\, G - G\,(\nabla_a \cdot \nabla)\, F\right] dV \\ &= \oiint \hat{n} \cdot (F\nabla_a G - G\nabla_a F)\, dS. \end{aligned} \tag{12.117}$$

By considering three distinct scalar functions F_i with $i = (1, 2, 3)$ as the components of a vector function $\bar{F}$ in Cartesian coordinate system, a hybrid theorem involving a vector function $\bar{F}$ and a scalar function G can be derived; it has the form

$$\begin{aligned} &\iiint \left[\bar{F}\,(\nabla_a \cdot \nabla)\, G - G\,(\nabla_a \cdot \nabla)\, \bar{F}\right] dV \\ &= \oiint \hat{n} \cdot \left(\bar{F}\nabla_a G - G\nabla_a \bar{F}\right) dS. \end{aligned} \tag{12.118}$$

The function $\nabla_a \bar{F}$ is, of course, a dyadic function.

The technique of elevating (12.117) to (12.118) is discussed in Appendix D of the author's book on vector analysis [Tai, 1992] except that it is now extended to functions operated by ∇_a and $\nabla_a \cdot \nabla$. The theorem stated by (12.117) can be used to prove the symmetrical relation of the function $G_0^{(b)}$ that is needed later. We consider two functions with different values of $\bar{R}'$, denoted by $G_0^{(b)}(\bar{R}, \bar{R}_a)$ and $G_0^{(b)}(\bar{R}, \bar{R}_b)$; they satisfy the equations

$$\left(\nabla_a \cdot \nabla + k^2 a\right) G_0^{(b)}(\bar{R}, \bar{R}_A) = -\delta(\bar{R} - \bar{R}_A) \tag{12.119}$$

$$\left(\nabla_a \cdot \nabla + k^2 a\right) G_0^{(b)}(\bar{R}, \bar{R}_B) = -\delta(\bar{R} - \bar{R}_B). \tag{12.120}$$

In (12.117) we let $F = G_0^{(b)}(\bar{R}, \bar{R}_A)$ and $G = G_0^{(b)}(\bar{R}, \bar{R}_B)$, which yields, in view of the above two equations,

$$G_0^{(b)}(\bar{R}_B, \bar{R}_A) = G_0^{(b)}(\bar{R}_A, \bar{R}_B). \tag{12.121}$$

For an unbounded region the surface integral in (12.116) vanishes at infinity by assuming an appropriate radiation condition for the two functions. If we replace R_A and R_B by $\bar{R}$ and $\bar{R}'$ in (12.121), it becomes

$$G_0^{(b)}(\bar{R}', \bar{R}) = G_0^{(b)}(\bar{R}, \bar{R}'). \tag{12.122}$$

This is a symmetrical relation to be used later. It should be remarked that we are now treating $G_0^{(b)}(\bar{R}, \bar{R}')$ as the Fourier transform of an instantaneous function $G_0^{(b)}\left(\bar{R}, \bar{R}', \tau\right)$ defined in the τ-domain; that is,

$$G_0^{(b)}(\bar{R}, \bar{R}') = \int_{-\infty}^{\infty} G_0^{(b)}\left(\bar{R}, \bar{R}', \tau\right) e^{i\omega\tau} d\tau, \tag{12.123}$$

where $G_0^{(b)}\left(\bar{R}, \bar{R}', \tau\right)$ satisfies the differential equation

$$\begin{aligned} &\left[\nabla_a \cdot \nabla - a\left(\frac{n^2}{c^2}\right)\frac{\partial^2}{\partial \tau^2}\right] G_0^{(b)}\left(\bar{R}, \bar{R}', \tau\right) \\ &= -\delta(\bar{R} - \bar{R}')\delta\left(\tau - 0\right), \end{aligned} \tag{12.124}$$

where we purposely set the initial value of τ equal to zero.

A Fourier transform of (12.124) produces the equation for $G_0^{(b)}(\bar{R}, \bar{R}')$,

$$\left(\nabla_a \cdot \nabla + k^2 a\right) G_0^{(b)}(\bar{R}, \bar{R}') = -\delta(\bar{R} - \bar{R}'), \tag{12.125}$$

where $k = \omega n/c$. This equation is identical to (12.91) found in the formulation for monochromatically oscillating fields in moving medium. At present, the equation applies to the continuous spectrum in the ω-domain.

With this much explanation of the origin and the meaning of $G_0^{(b)}(\bar{R}, \bar{R}')$ in the present context we can find the integral solution for $\bar{A}(\bar{R}, \bar{R}')$ by identifying

$\bar{F} = \bar{A}(\bar{R})$ and $G = G_0^{(b)}(\bar{R}, \bar{R}')$ in (12.118) for a region extended to infinity or an open region. In view of the defining equations for these two functions, (12.114) and (12.124), we obtain

$$\bar{A}(\bar{R}') = \mu a \iiint \bar{J}(\bar{R}) G_0^{(b)}(\bar{R}, \bar{R}')\, dV. \tag{12.126}$$

The surface integral at infinity vanishes by assuming two proper radiation conditions for the two functions. The exact form of these conditions depend on the nature of the problem, particularly the value and the sign of the constant a. For the time being we merely accept the vanishing of the surface integral as a postulate. For the case of a stationary medium, $a = 1$, the radiation condition is certainly well known. By interchanging $\bar{R}$ and $\bar{R}'$ in (12.125), and making use of the symmetrical relation described by (12.122), we have

$$\bar{A}(\bar{R}) = \mu a \iiint \bar{J}(\bar{R}') G_0^{(b)}(\bar{R}, \bar{R}')\, dV'. \tag{12.127}$$

To find the instantaneous expression for $\bar{A}\left(\bar{R}, \tau\right)$ or eventually the same function expressed in terms of the original variables $(x, y, z; t)$ let us consider the case corresponding to $a > 0$ or $n\beta < 1$. The expression for $G_0^{(b)}(\bar{R}, \bar{R}')$ is then given by (12.92), namely,

$$G_0^{(b)}(\bar{R}, \bar{R}') = \frac{a^{\frac{1}{2}} e^{ika^{\frac{1}{2}} R_a}}{4\pi R_a},$$

where

$$\begin{aligned} R_a &= \left(r^2 + a\xi^2\right) \\ r^2 &= \left(x - x'\right)^2 + \left(y - y'\right)^2, \; \xi = z - z' \\ k &= \frac{\omega n}{c}. \end{aligned}$$

The reciprocal Fourier transform of (12.127) with respect to ω provides for the solution for $\bar{A}\left(\bar{R}, \tau\right)$; that is,

$$\begin{aligned} \bar{A}\left(\bar{R}, \tau\right) = & \frac{\mu a^{\frac{3}{2}}}{2\pi} \int_{-\infty}^{\infty} \Bigg\{ e^{-i\omega\tau} \\ & \cdot \iiint_V \left[\frac{\bar{J}(\bar{R}') e^{i\omega\tau_a}}{4\pi R_a} \right] dV' \Bigg\}\, d\omega \end{aligned} \tag{12.128}$$

where

$$\tau_a = nc^{-1} a^{\frac{1}{2}} R_a.$$

It is obvious from (12.124) that (12.128) has been derived under the condition that the intitial value of τ is equal to zero. If the initial value is chosen to be τ', the corresponding expression would be

$$\begin{aligned}\bar{A}(\bar{R},\tau) &= \frac{\mu a^{\frac{3}{2}}}{2\pi}\int_{-\infty}^{\infty}\Bigg\{e^{-i\omega(\tau-\tau')} \\ &\quad \cdot \iiint_V \left[\frac{\bar{J}(\bar{R}')e^{i\omega\tau_a}}{4\pi R_a}\right] dV'\Bigg\} d\omega \\ &= \frac{\mu a^{\frac{3}{2}}}{2\pi}\int_{-\infty}^{\infty}\Bigg\{e^{-i\omega\tau} \\ &\quad \cdot \iiint \left[\frac{\bar{J}(\bar{R}')e^{i\omega(\tau'+\tau_a)}}{4\pi R_a}\right] dV'\Bigg\} d\omega. \end{aligned} \tag{12.129}$$

The Fourier integral contained in (12.129) can now be evaluated by means of the convolution theorem in the theory of Fourier integral. If $\bar{g}_1(\omega)$ and $g_2(\omega)$ denote, respectively, the Fourier transform of two functions $\bar{f}_1(\tau)$ and $f_2(\tau)$ in the τ- domain, then the theorem states

$$\begin{aligned}&\frac{1}{2\pi}\int_{-\infty}^{\infty} \bar{g}_1(\omega)g_2(\omega)e^{-i\omega\tau}d\omega \\ &= \int_{-\infty}^{\infty} \bar{f}_1(\tau-\tau_0)f_2(\tau_0)d\tau_0. \end{aligned} \tag{12.130}$$

Because our simplified notation for the transformed function $\bar{J}(\bar{R}')$ does not reveal the explicit functional dependence on ω we should clearly identify $\bar{g}_1(\omega), g_2(\omega), \bar{f}_1(\tau)$, and $f_2(\tau)$ in our formulation. They are

$$\begin{aligned}\bar{g}_1(\omega) &= \bar{J}(\bar{R}') \\ \bar{f}_1(\tau) &= \bar{J}(\bar{R}',\tau) \\ g_2(\omega) &= e^{i\omega(\tau'+\tau_a)} \\ f_2(\tau) &= \frac{1}{2\pi}\int_{-\infty}^{\infty} g_2(\omega)e^{-i\omega\tau}d\omega \\ &= \delta(\tau-\tau'-\tau_a)\end{aligned}$$

By substituting these functions into (12.130), we find

$$\begin{aligned}&\frac{1}{2\pi}\int_{-\infty}^{\infty}\left[\bar{J}(\bar{R}')e^{i\omega(\tau'+\tau_a)}\right]e^{-i\omega\tau}d\omega \\ &= \bar{J}\left(\bar{R}',\tau-\tau'-\tau_a\right);\end{aligned}$$

hence

$$\bar{A}(\bar{R},\tau) = \mu a^{\frac{3}{2}}\iiint_V \left[\frac{\bar{J}(\bar{R}',\tau-\tau'-\tau_a)}{4\pi R_a}\right] dV', \tag{12.131}$$

which has the familiar characteristic of a retarded potential except it is expressed in the pseudo-time domain.

To convert $\bar{A}(\bar{R}, \tau)$ in the original variables $(\bar{R}, t)$ we merely put $\tau = t + \Omega z, \tau' = t' + \Omega z'$, then

$$\bar{A}(\bar{R}, t) = \mu a^{\frac{3}{2}} \iiint_V \left[\frac{\bar{J}(\bar{R}', t - t_d)}{4\pi R_a} \right] dV', \tag{12.132}$$

where

$$\begin{aligned} t - t_d &= t - t' + \Omega(z - z') - \tau_a \\ &= t - t' + \Omega(z - z') - nc^{-1} a^{\frac{1}{2}} R_a. \end{aligned} \tag{12.133}$$

By denoting $t - t' = \eta, z - z' = \xi$, then $R_a = \left(r^2 + a\xi^2\right)^{\frac{1}{2}}$; thus the locus of the wave front emitted from a point source located at $\bar{R}'$ corresponds to

$$\eta + \Omega\xi - nc^{-1} a^{\frac{1}{2}} \left(r^2 + a\xi^2\right)^{\frac{1}{2}} = 0, \tag{12.134}$$

where the constant a is defined by

$$a = \frac{(1 - \beta^2)}{(1 - n^2\beta^2)},$$

assuming to the positive $(n\beta < 1)$ in the present case. Equation (12.134) can be changed to

$$\frac{(\xi - \xi_c)^2}{A^2} + \frac{r^2}{B^2} = 1 \tag{12.135}$$

where

$$A = \frac{c\eta n \left(1 - \beta^2\right)}{(\eta^2 - \beta^2)}$$

$$B = c\eta \left[\frac{\left(1 - \beta^2\right)}{\left(n^2 - \beta^2\right)} \right]^{\frac{1}{2}}$$

$$\xi_c = c\eta \frac{\left(n^2 - 1\right)\beta}{\left(n^2 - \beta^2\right)}.$$

It represents an ellipsoid with center at $\xi = \xi_c, r = 0$. It can be proved that for the case considered here with $a > 0$, corresponding to $\eta\beta < 1$, the following inequalities hold true

$$\frac{A}{B} < 1, \frac{A}{\xi_c} > 1,$$

when $n = 1$ (air medium) or $\beta = 0$ (stationary medium), the ellipsoid degenerates to a spherical surface as it should be.

For the case with $a < 0$, corresponding to $n\beta > 1$, the Cerenkov condition prevails. The function $G_0^{(b)}$ to be used in (12.126) is given by (12.93). By following a similar analysis it can be shown that

$$\bar{A}\left(\bar{R}, t\right) = \mu \mid a \mid^{\frac{3}{2}} \iiint_V \left[\frac{\bar{J}\left(\bar{R}', t - t'_d\right)}{4\pi R'_a}\right] dV', \tag{12.136}$$

where

$$t - t'_d = \eta + \Omega\xi - \tau'_a,$$

with

$$\tau'_a = nc^{-1} \mid a \mid^{\frac{1}{2}} R'_a$$
$$R'_a = \left(\mid a \mid \xi^2 - r^2\right)^{\frac{1}{2}}.$$

The electromagnetic Mach cone is defined by $r = \mid a \mid^{\frac{1}{2}} \xi$. Equation (12.136), which is valid for a field point, lies inside the cone; outside of the cone the field vanishes. The situation is exactly like that of the monochromatically oscillating field. It turns out that the locus of the wave front is still given by (12.135) but with $A/\xi_c < 1$. The detached ellipsoidal wave front is inscribed inside the electromagnetic Mach cone. The physical interpretation of the wave fronts for both cases, $a \gtrless 0$, is found in Compton's work.

To find the expressions for $\bar{E}\left(\bar{R}, t\right)$ and $\bar{H}\left(\bar{R}, t\right)$ we have to go back to (12.109), (12.110), and (12.113). The expression for $\bar{H}\left(\bar{R}, t\right)$ can be found in a relatively simple manner. By taking a reciprocal Fourier transform of (12.109) with respect to ω, we have

$$\mu\bar{\bar{\alpha}} \cdot \bar{H}(\bar{R}, \tau) = \nabla \times \left[\bar{\bar{b}} \cdot \bar{A}(\bar{R}, \tau)\right]. \tag{12.137}$$

Now by treating these functions as implicit functions of $\bar{R}$ and t, one finds

$$\begin{aligned}&\nabla \times \left[\bar{\bar{b}} \cdot \bar{A}(\bar{R}, \tau)\right] \\ &= \nabla \times \left[\bar{\bar{b}} \cdot \bar{A}(\bar{R}, t)\right] - \bar{\Omega} \times \frac{\partial}{\partial t}\left[\bar{\bar{b}} \cdot \bar{A}(\bar{R}, t)\right].\end{aligned} \tag{12.138}$$

This transformation is similar to (12.101) but carried out in a reverse manner. Equation (12.137) is, therefore, equivalent to

$$\begin{aligned}&\mu\bar{\bar{\alpha}} \cdot \bar{H}(\bar{R}, t) \\ &= \nabla \times \left[\bar{\bar{b}} \cdot \bar{A}(\bar{R}, t)\right] - \bar{\Omega} \times \frac{\partial}{\partial t}\left[\bar{\bar{b}} \cdot \bar{A}(\bar{R}, t)\right],\end{aligned} \tag{12.139}$$

which is the desired formula to calculate $\bar{H}(\bar{R}, t)$ with $\bar{A}(\bar{R}, t)$ given by (12.132) or (12.136). The calculation of $\bar{E}(\bar{R}, t)$ is more involved. We leave it as an exercise in Appendix D with sufficient hint to execute it. It should be mentioned that the equations for $\overline{E}(\overline{R},\tau)$ and $\overline{H}(\overline{R},\tau)$ described by (12.103) and (12.104) can also be solved by introducing two dyadic Green functions $\overline{\overline{G}}_{eo}(\overline{R},\tau)$ and $\overline{\overline{G}}_{mo}(\overline{R},\tau)$ defined in the $(\overline{R},\tau)$ domain. An exercise is also assigned for the readers to practice this approach.

In conclusion, we have presented here a new method of formulating and solving the field equations in a moving medium by introducing a pseudo-time variable that facilitates considerably the handling of the original differential equations. The theory is based on the technique of functional mapping in mathematic analysis. The method appears to be simpler than the direct method used by Compton with the introduction of a four-dimensional space and time operator because only existing formulas, already available in the theory of monochromatic field, are needed. As far as the result is concerned, the solution can also be obtained by considering the monochromatic solution as the Fourier transform of a time-domain field. The inverse Fourier transform would yield directly the time-domain dyadic Green function.

12-6 RECTANGULAR WAVEGUIDE WITH A MOVING MEDIUM

The method of eigenfunction expansion can also be applied to find the dyadic Green functions for waveguides filled with a moving isotropic medium. In view of (12.66) and (12.67) we define a pair of electric and magnetic dyadic Green functions satisfying the following two equations for monochromatic oscillating fields:

$$\nabla\times\left[\bar{\bar{b}}\cdot\overline{\overline{G}}_e^{(b)}\right]=\overline{\overline{G}}_m^{(b)} \tag{12.140}$$

$$\nabla\times\left[\bar{\bar{b}}\cdot\overline{\overline{G}}_m^{(b)}\right]=\bar{\bar{I}}\delta(\overline{R}-\overline{R}')+k^2\overline{\overline{G}}_e^{(b)}. \tag{12.141}$$

The wave equations for these two functions are

$$\nabla\times\left[\bar{\bar{b}}\cdot\nabla\times\left(\bar{\bar{b}}\cdot\overline{\overline{G}}_e^{(b)}\right)\right]-k^2\overline{\overline{G}}_e^{(b)}=\bar{\bar{I}}\delta(\overline{R}-\overline{R}') \tag{12.142}$$

$$\nabla\times\left[\bar{\bar{b}}\cdot\nabla\times\left(\bar{\bar{b}}\cdot\overline{\overline{G}}_m^{(b)}\right)\right]-k^2\overline{\overline{G}}_m^{(b)}=\nabla\times\left[\bar{\bar{b}}\delta(\overline{R}-\overline{R}')\right]. \tag{12.143}$$

To find the eigenfunction expansions for the two functions for a rectangular waveguide, we need $\overline{\overline{G}}_{e1}^{(b)}$ and $\overline{\overline{G}}_{m2}^{(b)}$. They satisfy the boundary conditions at the inner surface of the waveguide, corresponding to $x=0$ and x_0, $y=0$ and y_0,

$$\hat{n}\times\left[\bar{\bar{b}}\cdot\overline{\overline{G}}_{e1}^{(b)}\right]=0 \tag{12.144}$$

$$\hat{n}\times\nabla\times\left[\bar{\bar{b}}\cdot\overline{\overline{G}}_{m2}^{(b)}\right]=0. \tag{12.145}$$

In Chapter 5, we used "a" and "b" for x_0 and y_0. In this chapter the letters "a" and "b" have been used to denote two parameters in the constitutive relations for moving medium; hence a change of notations for the widths of the waveguide is necessary.

The solenoidal vector wave functions needed to construct $\bar{\bar{b}}\cdot\overline{\overline{G}}_{e1}^{(b)}$ and $\bar{\bar{b}}\cdot\overline{\overline{G}}_{m2}^{(b)}$ are solutions of the following homogeneous differential equation

$$\nabla \times \left[\bar{\bar{b}} \cdot \nabla \times \left(\bar{\bar{b}} \cdot \overline{F}\right)\right] - \kappa^2 \overline{F} = 0; \tag{12.146}$$

they are

$$\overline{M}_{\substack{e\\o}mn}(h) = \nabla \times \left[\Phi_{\substack{e\\o}mn}(h)\hat{z}\right] \tag{12.147}$$

and

$$\begin{aligned}\overline{N}_{\substack{e\\o}mn}(h) &= \frac{1}{\kappa}\nabla \times \left[\bar{\bar{b}} \cdot \overline{M}_{\substack{e\\o}mn}(h)\right]\\ &= \frac{1}{\kappa a}\nabla \times \nabla \times \left[\Phi_{\substack{e\\o}mn}(h)\hat{z}\right],\end{aligned} \tag{12.148}$$

where the scalar functions $\Phi_{\substack{e\\o}mn}(h)$ satisfy the following equation:

$$\left(\frac{\partial^2}{\partial x^2} + \frac{\partial^2}{\partial y^2} + \frac{1}{a}\frac{\partial^2}{\partial z^2} + \kappa^2 a\right)\Phi_{\substack{e\\o}mn}(h) = 0. \tag{12.149}$$

By means of (12.146)–(12.148), it can readily be shown that

$$\overline{M}_{\substack{e\\o}mn}(h) = \frac{1}{\kappa}\nabla \times \left[\bar{\bar{b}} \cdot \overline{N}_{\substack{e\\o}mn}(h)\right]. \tag{12.150}$$

By applying method of separation of variables to (12.149), one finds that the resulting eigenfunctions are

$$\Phi_{\substack{e\\o}mn}(h) = \begin{pmatrix} C_x C_y \\ S_x S_y \end{pmatrix} e^{ihz}, \tag{12.151}$$

where

$$\begin{aligned}C_x &= \cos\left(\frac{m\pi x}{x_0}\right)\\ S_x &= \sin\left(\frac{m\pi x}{x_0}\right)\\ C_y &= \cos\left(\frac{n\pi y}{y_0}\right)\\ S_y &= \sin\left(\frac{n\pi y}{y_0}\right),\end{aligned}$$

with

$$(m, n) = 0, 1, 2, \ldots$$

and

$$\begin{aligned}\kappa^2 a^2 &= h^2 + a k_c^2 \\ k_c^2 &= \left(\frac{m\pi}{x_0}\right)^2 + \left(\frac{n\pi}{y_0}\right)^2.\end{aligned} \tag{12.152}$$

The trigonometrical functions and the cut-off wave number k_c occurred previously in the theory of rectangular waveguide with stationary medium. The relation between the eigenvalues stated by (12.152), however, is quite different. The constant "a" in that equation, defined in Sec. 12-4, is given by

$$a = \frac{(1-\beta^2)}{(1-n^2\beta^2)},$$

with $\beta = v/c, n = (\mu\epsilon/\mu_0\epsilon_0)$. It is recalled that "$a$" could be either positive or negative, depending upon the value of $n\beta$.

It can be verified that the normalization relations of the vector wave functions thus introduced are

$$\iiint \overline{M}_{\substack{e\\o}mn}(h) \cdot \overline{N}_{\substack{e\\o}m'n'}(-h')\, dV = 0$$

$$\iiint \overline{M}_{\substack{e\\o}mn}(h) \cdot \overline{M}_{\substack{e\\o}m'n'}(-h')\, dV = \begin{cases} 0, m \neq m', n \neq n' \\ \frac{1}{2}(1+\delta_0)\pi k_c^2 x_0 y_0 \delta(h-h') \end{cases} \tag{12.153}$$

$$\iiint \overline{N}_{\substack{e\\o}mn}(h) \cdot \overline{N}_{\substack{e\\o}m'n'}(-h')\, dV = \begin{cases} 0, m \neq m', n \neq n' \\ \frac{1}{2\kappa^2 a^2}(1+\delta_0)\pi k_c^2 x_0 y_0 \delta(h-h')(h^2+k_c^2), \end{cases} \tag{12.154}$$

where

$$\delta_0 = \begin{cases} 0, m \text{ or } n = 0 \\ 1, m \neq 0, n \neq 0 \end{cases}$$

According to the method of magnetic dyadic Green function, we let

$$\nabla \times \left[\bar{\bar{I}}\delta(\overline{R}-\overline{R}')\right] = \int_{-\infty}^{\infty} dh \sum_{m,n} \left[\overline{M}_{omn}(h)\overline{A}_{omn}(h) + \overline{N}_{emn}(h)\overline{B}_{emn}(h)\right]. \tag{12.155}$$

As a result of the normalization relations of the vector wave functions we find

$$\overline{A}_{omn}(h) = \frac{(2-\delta_0)\,\kappa a \bar{\bar{I}} \cdot \overline{N}'_{omn}(-h)}{\pi k_c^2 x_0 y_0}$$

$$\overline{B}_{emn}(h) = \frac{(2-\delta_0)\,\kappa a \bar{\bar{I}} \cdot \overline{M}'_{emn}(-h)}{\pi k_c^2 x_0 y_0}.$$

The primed functions are defined with respect to (x', y', z'), the coordinate variables of $\overline{R}'$. As a result of (12.142), (12.148), and (12.149), the function $\overline{\overline{G}}_{m2}^{(b)}$

must have the following form:

$$\overline{\overline{G}}_{m2}^{(b)}(\overline{R},\overline{R}') = \int_{-\infty}^{\infty} dh \sum_{m,n} \kappa D \frac{\left[\overline{M}_0(h)\bar{\bar{b}}\cdot\overline{N}_0'(-h) + \overline{N}_e(h)\bar{\bar{b}}\cdot\overline{M}_e(-h)\right]}{(\kappa^2-k^2)}, \tag{12.156}$$

where

$$D = \frac{a\,(2-\delta_0)}{\pi k_c^2 x_0 y_0}$$

$$\kappa^2 - k^2 = \frac{\left(h^2 - a^2k^2 + ak_c^2\right)}{a^2}. \tag{12.157}$$

For simplicity in writing, we have omitted the subscript 'mn' for the wave functions. Unlike in the theory of ordinary rectangular waveguide, there are two cases to be considered.

Case 1. $a = \left(1-\beta^2\right)\left(1-n^2\beta^2\right) > 0 \quad \text{or} \quad n\beta < 1.$

In this case, there are two poles of the integrand in (12.154), resulting from $\kappa^2 - k^2 = 0$; they are

$$h = \pm k_g = \pm\left(a^2k^2 - ak_c^2\right)^{\frac{1}{2}}. \tag{12.158}$$

By applying the residue theorem to (12.156) in the h-plane, we obtain

$$\overline{\overline{G}}_{m2}^{(b)\pm}(\overline{R},\overline{R}') = \sum_{m,n}\frac{i\pi k a^2}{k_g} D\left[\overline{M}_0\left(\pm k_g\right)\bar{\bar{b}}\cdot\overline{N}_0'\left(\mp k_g\right)\right.$$
$$\left. + \overline{N}_e\left(\pm k_g\right)\bar{\bar{b}}\cdot\overline{M}_e\left(\mp k_g\right)\right],\ z \gtrless z', \tag{12.159}$$

where k_g denotes the guided wave number.

The discontinuity condition for $\overline{\overline{G}}_{m2}^{(b)}$ in the plane $z = z'$, derivable from (12.141), is given by

$$\hat{z}\times\left[\bar{\bar{b}}\cdot\left(\overline{\overline{G}}_{m2}^{(b)+} - \overline{\overline{G}}_{m2}^{(b)-}\right)\right] = \bar{\bar{I}}_t\delta(\overline{R}-\overline{R}'), \tag{12.160}$$

where

$$\bar{\bar{I}}_t = \hat{x}\hat{x} + \hat{y}\hat{y},$$
$$\delta(\overline{R}-\overline{R}') = \delta(x-x')\delta(y-y').$$

Subsequently, one finds

$$\overline{\overline{G}}_{e1}^{(b)} = \frac{1}{k^2}\left[\nabla\times\left(\bar{\bar{b}}\cdot\overline{\overline{G}}_{m2}^{(b)+}\right)U\left(z-z'\right)\right.$$
$$\left. + \nabla\times\left(\bar{\bar{b}}\cdot\overline{\overline{G}}_{m2}^{(b)-}\right)U(z'-z) - \hat{z}\hat{z}\delta(\overline{R}-\overline{R}')\right]. \tag{12.161}$$

By substituting the expressions for $\overline{\overline{G}}_{m2}^{(b)\pm}$ from (12.159) to (12.161), we obtain the desired expression for $\overline{\overline{G}}_{e1}^{(b)}$:

$$\overline{\overline{G}}_{e1}^{(b)}(\overline{R},\overline{R}') = -\frac{1}{k^2}\hat{z}\hat{z}\delta(\overline{R}-\overline{R}') + \sum_{m,n} C\left[\overline{M}_e(\pm k_g)\bar{\bar{b}}\cdot\overline{M}'_e(\mp k_g) + \overline{N}_0(\pm k_g)\bar{\bar{b}}\cdot\overline{N}'_0(\mp k_g)\right],\quad z \gtrless z', \tag{12.162}$$

where

$$C = \frac{ia^3(2-\delta_0)}{k_c^2 k_g x_0 y_0}.$$

Equations (12.159) and (12.162) have been derived under the condition $a > 0$ or $n\beta < 1$. When $a = 1$, they reduce to the expressions for a waveguide with a stationary medium.

Case 2. $$a = \frac{(1-\beta^2)}{(1-n^2\beta^2)} < 0 \quad \text{or} \quad n\beta > 1.$$

In this case, the poles of the integrand of (12.154) become real. They are given by

$$h = \pm k'_g = \pm\left(a^2k^2 + \mid a \mid k_c^2\right)^{\frac{1}{2}}. \tag{12.163}$$

Since the primary field, in the absence of the waveguide, is confined within an electromagnetic Mach cone or the Cerenkov cone, the contour of integration must exclude these poles for $z > z'$ and enclose both poles for $z > z'$. The expression for $\overline{\overline{G}}_{m2}^{(b)+}$ is then given by

$$\overline{\overline{G}}_{m2}^{(b)}(\overline{R},\overline{R}') = \sum_{mn}\frac{i\pi k a^2}{k'_g}D \cdot\left[\overline{M}_0(k_g)\bar{\bar{b}}\cdot\overline{N}'_0(-k_g) + \overline{N}_e(k_g)\bar{\bar{b}}\cdot\overline{M}'_e(-k_g) - \overline{M}_0(-k_g)\bar{\bar{b}}\cdot\overline{N}'_0(k_g) - \overline{N}_e(-k_g)\bar{\bar{b}}\cdot\overline{M}'_e(k_g)\right],\quad z > z'. \tag{12.164}$$

The function $\overline{\overline{G}}_{m2}^{(b)-}$ vanishes for $z < z'$. The corresponding expression for $\overline{\overline{G}}_{e1}^{(b)}(\overline{R},\overline{R}')$ becomes

$$\begin{aligned}\overline{\overline{G}}_{e1}^{(b)}(\overline{R},\overline{R}') = &-\frac{1}{k^2}\hat{z}\hat{z}\delta(\overline{R}-\overline{R}') \\ &+\sum_{m,n} C'\Big[\overline{M}_e(k_g)\bar{\bar{b}}\cdot\overline{M}'_e(-k_g)+\overline{N}_0(k_g)\bar{\bar{b}}\cdot\overline{N}'_0(-k_g) \\ &-\overline{M}_e(-k_g)\bar{\bar{b}}\cdot\overline{M}'_e(k_g)-\overline{N}_0(-k_g)\bar{\bar{b}}\cdot\overline{N}'_0(k_g)\Big], \; z>z',\end{aligned} \tag{12.165}$$

where

$$C' = ia^3(2-\delta_0)'k_c^2k_g'x_0y_0.$$

For $z < z'$, the function $\overline{\overline{G}}_{e1}^{-}$ vanishes. The structure of $\overline{\overline{G}}_{e1}^{(b)+}$ and $\overline{\overline{G}}_{m2}^{(b)-}$ indicates that standing waves are existing within the original Cerenkov cone. This phenomenon resembles the solution for $\overline{\overline{G}}_{eo}^{(b)}$ in an infinite region given by (12.93).

12-7 CYLINDRICAL WAVEGUIDE WITH A MOVING MEDIUM

Without going through the detailed derivation that is very similar to the previous treatment, we merely give the answers for $\overline{\overline{G}}_{m2}^{(b)}$ and $\overline{\overline{G}}_{e1}^{(b)}$ as follows:

Case 1. $$a = \frac{(1-\beta^2)}{(1-n^2\beta^2)} > 0$$

$$\begin{aligned}\overline{\overline{G}}_{m2}^{(b)}(\overline{R},\overline{R}') = &\frac{ika^3}{4\pi}\sum_n(2-\delta_0) \\ &\cdot\left\{\sum_\lambda\frac{1}{\lambda^2I_\lambda k_\lambda}\left[\overline{M}_\lambda(\pm k_\lambda)\bar{\bar{b}}\cdot\overline{N}'_\lambda(\mp k_\lambda)\right]\right. \\ &\left.+\sum_\mu\frac{1}{\mu^2I_\mu k_\mu}\left[\overline{M}_\mu(\pm k_\mu)\bar{\bar{b}}\cdot\overline{M}'_\mu(\mp k_\mu)\right]\right\}, \; z\gtrless z'\end{aligned} \tag{12.166}$$

$$\begin{aligned}\overline{\overline{G}}_{e1}^{(b)}(\overline{R},\overline{R}') = &-\frac{1}{k^2}\hat{z}\hat{z}\delta(\overline{R}-\overline{R}')+\frac{ia^3}{4\pi}\sum_n(2-\delta_0) \\ &\cdot\left\{\sum_\lambda\frac{1}{\lambda^2I_\mu k_\lambda}\left[\overline{N}_\lambda(\pm k_\lambda)\bar{\bar{b}}\cdot\overline{N}'_\lambda(\mp k_\lambda)\right]\right. \\ &\left.+\sum_\mu\frac{1}{\mu^2I_\mu k_\mu}\left[\overline{M}_\mu(\pm k_\mu)\bar{\bar{b}}\cdot\overline{M}'_\mu(\mp k_\mu)\right]\right\}, \; z\gtrless z',\end{aligned} \tag{12.167}$$

where

$$\overline{M}_\lambda(k_\lambda) = \overline{M}_{\substack{e\\o}n\lambda}(k_\lambda) = \nabla \times \left[\psi_{\substack{e\\o}n\lambda}(k_\lambda)\hat{z}\right]$$
$$\overline{N}_\lambda(k_\lambda) = \overline{N}_{\substack{e\\o}n\lambda}(k_\lambda) = \frac{1}{k}\nabla \times \left[\bar{\bar{b}}\cdot \overline{M}_\lambda(k_\lambda)\right]$$
$$\overline{M}_\mu(k_\mu) = \overline{M}_{\substack{e\\o}n\mu}(k_\mu) = \nabla \times \left[\psi_{\substack{e\\o}n\mu}(k_\mu)\hat{z}\right]$$
$$\overline{N}_\lambda(k_\mu) = \overline{N}_{\substack{e\\o}n\mu}(k_\mu) = \frac{1}{k}\nabla \times \left[\bar{\bar{b}}\cdot \overline{M}_\mu(k_\mu)\right]$$
$$J_n(\lambda r_0) = 0,$$
$$J_n'(\mu r_0) = 0, J_n'(\mu r_0) = dJ_n(\mu r_0)/d(\mu r_0)$$
$$r_0 = \text{radius of the cylindrical waveguide}$$
$$k_\lambda = (a^2k^2 - a\lambda^2)^{\frac{1}{2}}$$
$$k_\mu = (a^2k^2 - a\mu^2)^{\frac{1}{2}}$$
$$I_\lambda = \frac{r_0^2}{2}\left[J_n'(\lambda r_0)\right]^2, J_n'(\lambda r_0) = \frac{dJ_n(\lambda r_0)}{d(\lambda r_0)}$$
$$I_\mu = \frac{1}{2\mu^2}(\mu^2 r_0^2 - n^2)J_n^2(\mu r_0)$$
$$\psi_{\substack{e\\o}n\lambda}(k_\lambda) = J_n(\lambda r)\,{}^{\cos}_{\sin}\, n\phi e^{ik_\lambda z}$$
$$\psi_{\substack{e\\o}n\mu}(k_\mu) = J_n(\mu r)\,{}^{\cos}_{\sin}\, n\phi e^{ik_\mu z}.$$

Case 2. $a = \dfrac{(1-\beta^2)}{(1-n^2\beta^2)} < 0$

$$\begin{aligned}\bar{\bar{G}}_{m2}^{(b)}(\overline{R}, \overline{R}') =& \frac{ika^3}{2\pi}\sum_n (2-\delta_0)\\ &\cdot\left\{\sum_\lambda \frac{1}{\lambda^2 I_\lambda k_\lambda'}\left[\overline{M}_\lambda(k_\lambda')\bar{\bar{b}}\cdot\overline{M}_\lambda'(-k_\lambda')\right.\right.\\ &\left. -\overline{M}_\lambda(-k_\lambda')\bar{\bar{b}}\cdot\overline{M}_\lambda(k_\lambda')\right] + \sum_\mu \frac{1}{\mu^2 I_\mu k_\mu'}\\ &\left.\cdot\left[\overline{N}_\mu(k_\mu'\bar{\bar{b}}\cdot\overline{N}_\mu'(-k_\mu') - \overline{M}_\mu(-k_\mu')\bar{\bar{b}}\cdot\overline{N}_\mu'(k_\mu')\right]\right\}, \ z > z'\end{aligned} \tag{12.168}$$

$$\bar{\bar{G}}_{m2}^{(b)}(\overline{R}, \overline{R}') = 0, z < z'$$
$$\begin{aligned}\bar{\bar{G}}_{e1}^{(b)}(\overline{R}, \overline{R}') =& -\frac{1}{k^2}\hat{z}\hat{z}\delta(\overline{R}-\overline{R}')\\ &+\frac{ia^3}{2\pi}\sum(2-\delta_0)\left\{\sum_\lambda \frac{1}{\lambda^2 I_\lambda k_\lambda'}\left[\overline{N}_\lambda(h_\lambda')\bar{\bar{b}}\cdot\overline{N}_\lambda'(-h_\lambda')\right.\right.\\ &\left.-\overline{N}_\lambda(-h_\lambda')\bar{\bar{b}}\cdot\overline{N}_\lambda'(h_\lambda')\right] + \sum_\mu \frac{1}{\mu^2 I_\mu k_\mu'}\end{aligned}$$

$$\cdot \left[\overline{M}_{\mu}(h'_{\mu})\bar{\bar{b}} \cdot \overline{M}'_{\mu}(-h'_{\mu}) \right.$$
$$\left. - \overline{M}_{\mu}(-h'_{\mu})\bar{\bar{b}} \cdot \overline{M}_{\mu}(h'_{\mu})\right]\Big\}, \; z > z' \tag{12.169}$$

$$\overline{\overline{G}}_{e1}^{(b)-}(\overline{R}, \overline{R}') = 0, \; z < z',$$

where

$$k'_{\lambda} = \left(a^2k^2 + \mid a \mid \lambda^2\right)^{\frac{1}{2}}$$
$$k'_{\mu} = \left(a^2k^2 + \mid a \mid \mu^2\right)^{\frac{1}{2}}.$$

The functions and other parameters are identical to the ones defined in Case 1.

It should be mentioned that an early work on this subject [Stubenrauch and Tai, 1971] contained a mistake resulting from the direct synthesis of $\overline{\overline{G}}_{e1}^{(b)}$ without the longitudinal functions. This mistake is now amended by way of the method of $\overline{\overline{G}}_{m2}^{(b)}$.

12-8 INFINITE CONDUCTING CYLINDER IN A MOVING MEDIUM

The problem under consideration has the same geometry as the one discussed in Chapter 7 except that the medium surrounding the conducting cylinder is an isotropic medium moving in the z-direction with velocity $\bar{v} = v\hat{z}$. The functions which we are interested in are again $\overline{\overline{G}}_{m2}$ and $\overline{\overline{G}}_{e1}$. Only the case for $a > 0$ will be treated. The other case corresponding to $a < 0$ has no practical significance in terrestrial problems. From the point of view of the theory of relativity, this problem is equivalent to that of a conducting cylinder moving with a velocity $-v\hat{z}$ in a stationary medium, and the field under investigation is expressed in a frame stationary with respect to the moving cylinder.

To resolve this problem we need a Fourier integral representation of the functions $\overline{\overline{G}}_{m0}^{(b)}$ and $\overline{\overline{G}}_{e0}^{(b)}$. The vector wave functions to be used in the eigenfunction expansion are functions defined in a continuous spectrum both in the λ and h domain; they are

$$\overline{M}_{{e \atop o}mn}(h) = \nabla \times \left[J_n(\lambda r) {\cos \atop \sin} n\phi e^{ihz}\hat{z}\right]$$
$$\overline{N}_{{e \atop o}mn}(h) = \frac{1}{\kappa}\nabla \times \left[\bar{\bar{b}} \cdot \overline{M}_{{e \atop o}mn}(h)\right],$$

with

$$\kappa^2 a^2 = \lambda^2 a^2 + h^2.$$

Furthermore,

$$\overline{M}_{{e \atop o}mn}(h) = \frac{1}{\kappa}\nabla \times \left[\bar{\bar{b}} \cdot \overline{N}_{{e \atop o}mn}(h)\right].$$

The orthogonal relations of these functions are

$$\iiint \overline{M}_{\substack{e\\o}mn}(h) \cdot \overline{M}_{\substack{e\\o}m'n'}(-h')\, dV$$
$$= \begin{cases} 0, m \neq m', n \neq n' \\ 2\pi^2(1+\delta_0)\lambda\delta(h-h')\delta(\lambda-\lambda'), m = m', n = n' \end{cases}$$
$$\iiint \overline{N}_{\substack{e\\o}mn}(h) \cdot \overline{N}_{\substack{e\\o}m'n'}(-h')\, dV$$
$$= \begin{cases} 0, m \neq m', n \neq n' \\ \frac{2\pi^2(1+\delta_0)\lambda\delta(h-h')\delta(\lambda-\lambda')(h^2+\lambda^2)}{\kappa^2 a^2} \end{cases}$$
$$\iiint \overline{M}_{\substack{e\\o}mn}(h) \cdot \overline{N}_{\substack{e\\o}m'n'}(-h')\, dV = 0.$$

With the aid of these relations we find

$$\nabla \times \left[\bar{\bar{b}}\delta(\overline{R} - \overline{R}')\right] = \int_{-\infty}^{\infty} dh \int_0^{\infty} d\lambda$$
$$\cdot \sum_n \frac{(2-\delta_0)\kappa a}{4\pi^2\lambda} \left[\overline{M}\bar{\bar{b}} \cdot \overline{N}' + \overline{N}\bar{\bar{b}} \cdot \overline{M}'\right] \tag{12.170}$$

For simplicity, the subscript $\substack{e\\o}mn$ attached to the vector wave functions has been omitted; that is,

$$\overline{N} = \overline{N}_{\substack{e\\o}mn}(h)$$
$$\overline{M} = \overline{M}_{\substack{e\\o}mn}(h).$$

The primed functions in (12.170) are defined with respect to $\overline{R}'$. The magnetic dyadic Green function $\overline{\overline{G}}{}_{mo}^{(b)}$ in an unbounded region in view of (12.143) must have the form

$$\overline{\overline{G}}{}_{mo}^{(b)}(\overline{R}, \overline{R}') = \int_{-\infty}^{\infty} dh \int_0^{\infty} d\lambda$$
$$\cdot \sum_n \frac{(2-\delta_0)\kappa a}{4\pi^2\lambda(\kappa^2 - k^2)} \left[\overline{M}\bar{\bar{b}} \cdot \overline{N}' + \overline{N}\bar{\bar{b}} \cdot \overline{M}'\right]. \tag{12.171}$$

By eliminating the λ-integration with the aid of the dyadic operational method described in Sec. 6-2, we obtain

$$\overline{\overline{G}}{}_{mo}^{(b)}(\overline{R}, \overline{R}') = \int_{-\infty}^{\infty} dh \sum_n \frac{i(2-\delta_0)ka^2}{8\pi\eta^2}$$
$$\cdot \begin{bmatrix} \overline{M}^{(1)}\bar{\bar{b}} \cdot \overline{N}' + \overline{N}^{(1)}\bar{\bar{b}} \cdot \overline{M}' \\ \overline{M}\bar{\bar{b}} \cdot \overline{N}'^{(1)} + \overline{N}\bar{\bar{b}} \cdot \overline{M}'^{(1)} \end{bmatrix}, \quad z \gtrless z', \tag{12.172}$$

where

$$\eta^2 = \frac{1}{a}(k^2a^2 - h^2).$$

Subsequently, by the method of $\overline{\overline{G}}_m$ one finds

$$\overline{\overline{G}}_{eo}^{(b)}(\overline{R}, \overline{R}') = -\frac{1}{k^2}\hat{z}\hat{z}\delta(\overline{R} - \overline{R}')$$

$$+ \int_{-\infty}^{\infty} dh \sum_n C_\eta \begin{bmatrix} \overline{M}^{(1)}\bar{\bar{b}} \cdot \overline{M}' + \overline{N}^{(1)}\bar{\bar{b}} \cdot \overline{N}' \\ \overline{M}\bar{\bar{b}} \cdot \overline{M}'^{(1)} + \overline{N}\bar{\bar{b}} \cdot \overline{N}'^{(1)} \end{bmatrix}, \; z \gtrless z', \tag{12.173}$$

where

$$C_\eta = \frac{ia^2(2 - \delta_0)}{8\pi\eta^2}.$$

Vector wave functions with superscript "(1)" are defined with respect to Hankel functions of the first kind in the generating function. In the presence of a conducting cylinder with radius r_0, the method of scattering superposition yields

$$\overline{\overline{G}}_{e1}^{(b)}(\overline{R}, \overline{R}') = \overline{\overline{G}}_{eo}^{(b)}(\overline{R}, \overline{R}') + \bar{\bar{b}}_{es}^{(b)}(\overline{R}, \overline{R}'), \tag{12.174}$$

where the scattered term is given by

$$\overline{\overline{G}}_{es}^{(b)}(\overline{R}, \overline{R}') = \int_{-\infty}^{\infty} dh \sum_n C_\eta$$

$$\cdot \left[\alpha \overline{M}^{(1)}\bar{\bar{b}}\overline{M}'^{(1)} + \beta \overline{N}^{(1)}\bar{\bar{b}} \cdot \overline{N}'^{(1)}\right], \tag{12.175}$$

with

$$\alpha = -\frac{\frac{\partial J_n(x)}{\partial x}}{\frac{\partial H_n^{(1)}(x)}{\partial x}}, x = \eta r_0$$

$$\beta = \frac{-J_n(\eta r_0)}{H_n^{(1)}(\eta r_0)}.$$

Once $\overline{\overline{G}}_{e1}^{(b)}$ is known, the electric field due to a current distribution placed outside of the conducting cylinder can be calculated using the formula

$$\overline{E}(\overline{R}') = i\omega\mu \iiint \bar{\bar{b}} \cdot \overline{\overline{G}}_{e1}(\overline{R}, \overline{R}') \cdot \overline{J}(\overline{R}')e^{-i\omega\Omega(z-z')}dV'. \tag{12.176}$$

Problems of similar nature, like a moving dielectric cylinder in air with a current source placed either inside or outside of the cylinder, can be formulated accordingly [Stubenrauch, 1972].

In conclusion, we have compiled in this book many basic formulas for the dyadic Green functions of various canonical problems which could be used to formulate boundary-value problems in electromagnetic theory. Some topics which are not covered in the text are absorbed in the exercises with enough hints so that the readers may wish to broaden the scope of study and to gain confidence in digesting and applying the method so introduced.

Appendix A

Mathematical Formulas

A-1 GRADIENT, DIVERGENCE, AND CURL IN ORTHOGONAL SYSTEMS

$$\text{coordinate variables: } v_1, v_2, v_3$$
$$\text{unit vectors: } \hat{u}_1, \hat{u}_2, \hat{u}_3$$
$$\text{metric coefficients: } h_1, h_2, h_3, \ \ \Omega = h_1 h_2 h_3$$

Gradient of f

$$\nabla f = \sum_i \frac{\hat{u}_i \partial f}{h_i \partial v_i}. \tag{A.1}$$

Divergence of $\overline{F}$

$$\begin{aligned} \nabla \cdot \overline{F} &= \sum_i \frac{\hat{u}_i}{h_i} \cdot \frac{\partial \overline{F}}{\partial v_i} \\ &= \sum_i \frac{1}{\Omega} \frac{\partial}{\partial v_i} \left(\frac{\Omega}{h_i} F_i \right). \end{aligned} \tag{A.2}$$

Curl of $\overline{F}$

$$\begin{aligned} \nabla \times \overline{F} &= \sum_i \frac{\hat{u}_i}{h_i} \times \frac{\partial \overline{F}}{\partial v_i} \\ &= \sum_i \frac{h_i \hat{u}_i}{\Omega} \left[\frac{\partial (h_k F_k)}{\partial v_j} - \frac{\partial (h_j F_j)}{\partial v_k} \right] \\ & (i, j, k) = (1, 2, 3) \text{ in cyclic order.} \end{aligned} \tag{A.3}$$

Laplacian of f

$$\nabla^2 f = \sum_i \frac{1}{\Omega}\frac{\partial}{\partial v_i}\left(\frac{\Omega}{h_i^2}\frac{\partial f}{\partial v_i}\right) = \nabla \cdot \nabla f. \tag{A.4}$$

Laplacian of $\overline{F}$

$$\begin{aligned}\nabla^2 \overline{F} &= \sum_i \frac{1}{\Omega}\frac{\partial}{\partial v_i}\left(\frac{\Omega}{h_i^2}\frac{\partial \overline{F}}{\partial v_i}\right)\\ &= \nabla\nabla \cdot \overline{F} - \nabla \times \nabla \times \overline{F}.\end{aligned} \tag{A.5}$$

Derivatives of Unit Vectors

$$\frac{\partial \hat{u}_j}{\partial v_k} = \frac{1}{h_j}\frac{\partial h_k}{\partial v_j}\hat{u}_k, \quad j \neq k \tag{A.6}$$

$$\frac{\partial \hat{u}_i}{\partial v_i} = -\left(\frac{1}{h_j}\frac{\partial h_i}{\partial v_j}\hat{u}_j + \frac{1}{h_k}\frac{\partial h_i}{\partial v_k}\hat{u}_k\right), \tag{A.7}$$

$i, j, k = 1, 2, 3,$ in cyclic order.

Cartesian System

$$\begin{aligned} v_1, v_2, v_3 &= x, y, z\\ h_1, h_2, h_3 &= 1, 1, 1\\ \hat{u}_1, \hat{u}_2, \hat{u}_3 &= \hat{x}, \hat{y}, \hat{z}.\end{aligned}$$

Cylindrical System

$$\begin{aligned} v_1, v_2, v_3 &= r, \phi, z\\ h_1, h_2, h_3 &= 1, r, 1\\ \hat{u}_1, \hat{u}_2, \hat{u}_3 &= \hat{r}, \hat{\phi}, \hat{z}.\end{aligned}$$

Spherical System

$$\begin{aligned} v_1, v_2, v_3 &= R, \theta, \phi\\ h_1, h_2, h_3 &= 1, R, R\sin\theta\\ \hat{u}_1, \hat{u}_2, \hat{u}_3 &= \hat{R}, \hat{\theta}, \hat{\phi}.\end{aligned}$$

Elliptical Cylinder

$$\begin{aligned} v_1, v_2, v_3 &= u, v, z\\ h_1, h_2, h_3 &= c(ch^2 u - \cos^2 v), c(ch^2 u - \cos^2 v), 1\\ \hat{u}_1, \hat{u}_2, \hat{u}_3 &= \hat{u}, \hat{v}, \hat{z}.\end{aligned}$$

Invariance of Differential Operators

$$\sum_i \frac{\hat{u}_i}{h_i}\frac{\partial}{\partial v_i} = \sum_i \frac{\hat{u}'_i}{h'_i}\frac{\partial}{\partial v'_i} \quad \text{(gradient operator)}$$

$$\sum_i \frac{\hat{u}_i}{h_i}\cdot\frac{\partial}{\partial v_i} = \sum_i \frac{\hat{u}'_i}{h'_i}\cdot\frac{\partial}{\partial v'_i} \quad \text{(divergence operator)}$$

$$\sum_i \frac{\hat{u}_i}{h_i}\times\frac{\partial}{\partial v_i} = \sum_i \frac{\hat{u}'_i}{h'_i}\times\frac{\partial}{\partial v'_i} \quad \text{(curl operator)}$$

applicable to any two orthogonal sets.

A-2 VECTOR IDENTITIES

$$\bar{a}\cdot(\bar{b}\times\bar{c}) = \bar{b}\cdot(\bar{c}\times\bar{a}) = \bar{c}\cdot(\bar{a}\times\bar{b}) \quad (A.8)$$

$$\bar{a}\times(\bar{b}\times\bar{c}) = (\bar{a}\cdot\bar{c})\bar{b} - (\bar{a}\cdot\bar{b})\bar{c} \quad (A.9)$$

$$\nabla(ab) = a\nabla b + b\nabla a \quad (A.10)$$

$$\nabla\cdot(a\bar{b}) = a\nabla\cdot\bar{b} + \bar{b}\cdot\nabla a \quad (A.11)$$

$$\nabla\times(a\bar{b}) = a\nabla\times\bar{b} - \bar{b}\times\nabla a \quad (A.12)$$

$$\nabla\cdot(\bar{a}\times\bar{b}) = \bar{b}\cdot\nabla\times\bar{a}\cdot\nabla\times\bar{b} \quad (A.13)$$

$$\begin{aligned}\nabla(\bar{a}\cdot\bar{b}) &= \bar{a}\times\nabla\times\bar{b}\times\nabla\times\bar{a} \\ &\quad + (\bar{a}\cdot\nabla)\bar{b} + (\bar{b}\cdot\nabla)\bar{a}\end{aligned} \quad (A.14)$$

$$\begin{aligned}\nabla\times(\bar{a}\times\bar{b}) &= \bar{a}\nabla\cdot\bar{b} - \bar{b}\nabla\cdot\bar{a} \\ &\quad - (\bar{a}\cdot\nabla)\bar{b} + (\bar{b}\cdot\nabla)\bar{a}\end{aligned} \quad (A.15)$$

$$\nabla\cdot(\nabla a) = \nabla^2 a \quad (A.16)$$

$$\nabla\cdot(\nabla\bar{a}) = \nabla^2\bar{a} \quad (A.17)$$

$$\nabla\times(\nabla\times\bar{a}) = \nabla(\nabla\cdot\bar{a}) - \nabla^2\bar{a} \quad (A.18)$$

$$\nabla\times(\nabla a) = 0 \quad (A.19)$$

$$\nabla\cdot(\nabla\times\bar{a}) = 0. \quad (A.20)$$

A-3 DYADIC IDENTITIES

$$\bar{a}\cdot(\bar{b}\times\bar{\bar{c}}) = -\bar{b}\cdot(\bar{a}\times\bar{\bar{c}}) = (\bar{a}\times\bar{b})\cdot\bar{\bar{c}} \quad (A.21)$$

$$\bar{a}\times(\bar{b}\times\bar{\bar{c}}) = \bar{b}\cdot(\bar{a}\times\bar{\bar{c}}) - (\bar{a}\cdot\bar{b})\bar{\bar{c}} \quad (A.22)$$

$$\nabla(a\bar{b}) = a\nabla\bar{b} + (\nabla a)\bar{b} \quad (A.23)$$

$$\nabla\cdot(a\bar{\bar{b}}) = a\nabla\cdot\bar{\bar{b}} + (\nabla a)\cdot\bar{\bar{b}} \quad (A.24)$$

$$\nabla\times(a\bar{\bar{b}}) = a\nabla\times\bar{\bar{b}} + (\nabla a)\times\bar{\bar{b}} \quad (A.25)$$

$$\nabla\times(\nabla\times\bar{\bar{a}}) = \nabla(\nabla\cdot\bar{\bar{a}}) - \nabla^2\bar{\bar{a}} \quad (A.26)$$

$$\nabla \cdot (\nabla \times \overline{\overline{a}}) = 0 \tag{A.27}$$

$$\overline{a} \cdot \overline{\overline{b}} = (\overline{\overline{b}})^T \cdot \overline{a} \tag{A.28}$$

$$\overline{a} \times \overline{\overline{b}} = -\left[(\overline{\overline{b}})^T \times \overline{a}\right]^T \tag{A.29}$$

$$(\overline{c})^T \cdot (\overline{a} \times \overline{\overline{b}}) = -(\overline{a} \times \overline{c})^T \cdot \overline{\overline{b}}. \tag{A.30}$$

A-4 INTEGRAL THEOREMS

Gauss Theorem or Divergence Theorem

$$\iiint \nabla \cdot \overline{F} dV = \oiint \left(\hat{n} \cdot \overline{F}\right) dS. \tag{A.31}$$

Curl Theorem

$$\iiint \nabla \times \overline{F} dV = \oiint \left(\hat{n} \times \overline{F}\right) dS. \tag{A.32}$$

Gradient Theorem

$$\iiint \nabla f dV = \oiint \hat{n} f dS. \tag{A.33}$$

Surface Divergence Theorem

$$\iint \nabla_s \cdot \overline{F} dS = \oint \hat{m} \cdot \overline{F} d\ell$$

$$\left(\hat{m} = \hat{\ell} \times \hat{n}\right). \tag{A.34}$$

Surface Curl Theorem

$$\iint \nabla_s \times \overline{F} dS = \oint \left(\hat{m} \times \overline{F}\right) d\ell. \tag{A.35}$$

Surface Gradient Theorem

$$\iint \nabla_s f dS = \oint \hat{m} f d\ell. \tag{A.36}$$

Stokes Theorem

$$\iint \hat{n} \cdot \nabla \times \overline{F} dS = \oint \overline{F} \cdot d\overline{\ell}. \tag{A.37}$$

Cross-Gradient Theorem

$$\iint \hat{n} \times \nabla f dS = \oint f d\overline{\ell}. \tag{A.38}$$

Cross-Del-Cross Theorem

$$\iint (\hat{n} \times \nabla) \times \overline{F} dS = -\oint \overline{F} \times d\bar{\ell}. \tag{A.39}$$

First Scalar Green's Theorem

$$\iiint \left(f_1 \nabla^2 f_2 + \nabla f_1 \cdot \nabla f_2\right) dV = \oint\!\!\!\oint f_1 \nabla f_2 \cdot d\overline{S}. \tag{A.40}$$

Second Scalar Green's Theorem

$$\iiint \left(f_1 \nabla^2 f_2 - f_2 \nabla^2 f_1\right) dV = \iint \left(f_1 \nabla f_2 - f_2 \nabla f_1\right) \cdot d\overline{S}. \tag{A.41}$$

First Vector Green's Theorem

$$\iiint \left[(\nabla \times \overline{F}_1) \cdot (\nabla \times \overline{F}_2) - \overline{F}_1 \cdot \nabla \times \nabla \times \overline{F}_2\right] dV$$

$$= \oint\!\!\!\oint \hat{n} \cdot \left(\overline{F}_1 \times \nabla \times \overline{F}_2\right) dS. \tag{A.42}$$

Second Vector Green's Theorem

$$\iiint \left(\overline{F}_1 \cdot \nabla \times \nabla \times \overline{F}_2 - \overline{F}_2 \cdot \nabla \times \nabla \times \overline{F}_1\right) dV$$

$$= \oint\!\!\!\oint \hat{n} \cdot \left(\overline{F}_2 \times \nabla \times \overline{F}_1 - \overline{F}_1 \times \nabla \times \overline{F}_2\right) dS. \tag{A.43}$$

First Vector-Dyadic Green's Theorem

$$\iiint \left[(\nabla \times \overline{P}) \cdot \nabla \times \overline{\overline{Q}} - \overline{P} \cdot \nabla \times \nabla \times \overline{\overline{Q}}\right] dV$$

$$= \oint\!\!\!\oint \hat{n} \cdot \left(\overline{P} \times \nabla \times \overline{\overline{Q}}\right) dS. \tag{A.44}$$

Second Vector-Dyadic Green's Theorem

$$\iiint \left[(\nabla \times \nabla \times \overline{P}) \cdot \overline{\overline{Q}} - \overline{P} \cdot \nabla \times \nabla \times \overline{\overline{Q}}\right] dV$$

$$= \oint\!\!\!\oint \hat{n} \cdot \left[\overline{P} \times \nabla \times \overline{\overline{Q}} + (\nabla \times \overline{P}) \times \overline{\overline{Q}}\right] dS. \tag{A.45}$$

First Dyadic-Dyadic Green's Theorem

$$\iiint \left[\left(\nabla \times \overline{\overline{Q}} \right)^T \cdot \nabla \times \overline{\overline{P}} - \left(\nabla \times \nabla \times \overline{\overline{Q}} \right)^T \cdot \overline{\overline{P}} \right] dV$$
$$= \oiint \left(\nabla \times \overline{\overline{Q}} \right)^T \cdot \left(\hat{n} \times \overline{\overline{P}} \right) dS. \tag{A.46}$$

Second Dyadic-Dyadic Green's Theorem

$$\iiint \left[(\overline{\overline{Q}})^T \cdot \nabla \times \nabla \times \overline{\overline{P}} - \left(\nabla \times \nabla \times \overline{\overline{Q}} \right)^T \cdot \overline{\overline{P}} \right] dV$$
$$= \oiint \left[\left(\nabla \times \overline{\overline{Q}} \right)^T \cdot \left(\hat{n} \times \overline{\overline{P}} \right) + (Q)^T \cdot \left(\hat{n} \times \nabla \times \overline{\overline{P}} \right) \right] dS. \tag{A.47}$$

In the above formulas $\hat{n}$ denotes the outward unit normal vector for a closed surface. For an open surface, $\hat{\ell}$ denotes the tangential unit vector to the edge of the surface, and $\hat{n}$ follows the right-hand screw rule by turning $\hat{\ell}$. The unit vector $\hat{m}$ is perpendicular to the edge but tangential to the open surface; it is defined by $\hat{m} = \hat{\ell} \times \hat{n}$.

Appendix B

Vector Wave Functions and Their Mutual Relationships

B-1 RECTANGULAR VECTOR WAVE FUNCTIONS

$$
\begin{aligned}
&\psi_{\substack{e\\o}mn}(h) = \begin{Bmatrix} C_x & C_y \\ S_x & S_y \end{Bmatrix} e^{ihx}, \quad m,n = 0,1,2,\ldots \\
&C_x = \cos\frac{m\pi x}{a} = \cos k_x x \\
&S_x = \sin\frac{m\pi x}{a} = \sin k_x x \\
&C_y = \cos\frac{n\pi y}{b} = \cos k_y y \\
&S_y = \sin\frac{n\pi y}{b} = \sin k_y y \\
&k_x = \frac{m\pi}{a}, k_y = \frac{n\pi}{b} \\
&\nabla^2\psi_{\substack{e\\o}mn}(h) + \kappa^2\psi_{\substack{e\\o}mn}(h) = 0 \\
&\kappa^2 = k_x^2 + k_y^2 + h^2 = k_c^2 + h^2 \\
&k_c^2 = k_x^2 + k_y^2 \\
&\overline{M}_{\substack{e\\o}mn}(h) = \nabla \times \left[\psi_{\substack{e\\o}mn}(h)\hat{z}\right] \\
&\qquad = \left[\mp k_y \begin{pmatrix} C_x & S_y \\ S_x & C_y \end{pmatrix} \hat{x} \pm k_y \begin{pmatrix} S_x & C_y \\ C_x & S_y \end{pmatrix} \hat{y}\right] e^{ihz}
\end{aligned} \tag{B.1}
$$

$$
\begin{aligned}
\overline{N}_{\substack{e\\o}mn}(h) &= \frac{1}{\kappa}\nabla\times\nabla\times\left[\psi_{\substack{e\\o}mn}(h)\hat{z}\right] \\
&= \frac{1}{\kappa}\left[\mp ihk_x\begin{pmatrix} S_x & C_y \\ C_x & S_y \end{pmatrix}\hat{x} \mp ihk_y\begin{pmatrix} C_x & S_y \\ S_x & C_y \end{pmatrix}\hat{y}\right. \\
&\quad \left. + k_c^2\begin{pmatrix} C_x & C_y \\ S_x & S_y \end{pmatrix}\hat{z}\right]e^{ihz} \qquad (B.2)
\end{aligned}
$$

$$
\nabla\times\overline{M}_{\substack{e\\o}mn}(h) = \kappa\overline{N}_{\substack{e\\o}mn}(h) \qquad (B.3)
$$

$$
\nabla\times\overline{N}_{\substack{e\\o}mn}(h) = \kappa\overline{M}_{\substack{e\\o}mn}(h) \qquad (B.4)
$$

$$
\begin{aligned}
\overline{M}_{mn}^{(x)}(h) &= \nabla\times\left(S_xC_ye^{ihz}\hat{x}\right) \\
&= \left(ihS_xC_y\hat{y} + k_yS_xS_y\hat{z}\right)e^{ihz} \qquad (B.5)
\end{aligned}
$$

$$
\begin{aligned}
\overline{M}_{mn}^{(y)}(h) &= \nabla\times\left(C_xS_ye^{ihz}\hat{y}\right) \\
&= \left(-ihC_xS_y\hat{x} - k_xS_xS_y\hat{z}\right)e^{ihz} \qquad (B.6)
\end{aligned}
$$

$$
\begin{aligned}
\overline{N}_{mn}^{(x)}(h) &= \frac{1}{\kappa}\nabla\times\nabla\times\left(C_xS_ye^{ihz}\hat{x}\right) \\
&= \frac{1}{\kappa}\left[\left(\kappa^2 - k_x^2\right)C_xS_y\hat{x} - k_xk_yS_xC_y\hat{y}\right. \\
&\quad \left. -ihk_xS_xS_y\hat{z}\right]e^{ihz} \qquad (B.7)
\end{aligned}
$$

$$
\begin{aligned}
\overline{N}_{mn}^{(y)}(h) &= \frac{1}{\kappa}\nabla\times\nabla\times\left(S_xC_ye^{ihz}\hat{y}\right) \\
&= \frac{1}{\kappa}\left[-k_xk_yC_xS_y\hat{x} + \left(\kappa^2 - k_y^2\right)S_xC_y\hat{y}\right. \\
&\quad \left. -ihk_yS_xS_y\hat{z}\right]e^{ihz} \qquad (B.8)
\end{aligned}
$$

$$
\overline{M}_{mn}^{(x)}(h) = \frac{1}{k_c^2}\left[\kappa k_y\overline{N}_{omn}(h) + ihk_x\overline{M}_{emn}(h)\right] \qquad (B.9)
$$

$$
\overline{M}_{mn}^{(y)}(h) = \frac{1}{k_c^2}\left[-\kappa k_x\overline{N}_{omn}(h) + ihk_y\overline{M}_{emn}(h)\right] \qquad (B.10)
$$

$$
\overline{N}_{mn}^{(x)}(h) = \frac{1}{k_c^2}\left[-ihk_x\overline{N}_{omn}(h) - \kappa k_y\overline{M}_{emn}(h)\right] \qquad (B.11)
$$

$$
\overline{N}_{mn}^{(y)}(h) = \frac{1}{k_c^2}\left[-ihk_y\overline{N}_{omn}(h) + \kappa k_x\overline{M}_{emh}(h)\right]. \qquad (B.12)
$$

It is to be noticed that $\overline{N}_{mn}^{(x)}$ is not equal to $\frac{1}{\kappa}\nabla\times\overline{M}_{mn}^{(x)}$ because the generating functions used to define them are different. The functions defined by (*B*.5) to (*B*.8) all satisfy the vector Dirichlet boundary condition at the side walls of the rectangular guide, corresponding to $x = 0$ and $a, y = 0$ and b. Another four functions that satisfy the vector Neumann boundary condition can be found in a similar manner.

B-2 CYLINDRICAL VECTOR WAVE FUNCTIONS WITH DISCRETE EIGENVALUES

$$\psi_{\substack{e\\o}n\lambda}(h) = J_n(\lambda r) \frac{\cos}{\sin} n\phi e^{ihz}, \quad n = 0, 1, 2, \ldots$$

$$J_n(\lambda_{nm} a) = 0 \text{ or } \lambda_{nm} a = p_{nm}, \quad m = 1, 2, 3, \ldots$$

$$\psi_{\substack{e\\o}n\mu}(h) = J_n(\mu r) \frac{\cos}{\sin} n\phi e^{ihz}$$

$$\frac{dJ_n(\mu_{mn} a)}{d(\mu_{nm} a)} = 0 \text{ or } \mu_{nm} a = q_{nm}$$

$$\nabla^2 \psi_{\substack{e\\o}n\mu} + \kappa_\mu^2 \psi_{\substack{e\\o}n\mu}(h) = 0, \kappa_\mu^2 = \mu^2 + h^2$$

$$\nabla^2 \psi_{\substack{e\\o}n\lambda} + \kappa_\lambda^2 \psi_{\substack{e\\o}n\lambda}(h) = 0, \kappa_\lambda^2 = \lambda^2 + h^2$$

$$\overline{M}_{\substack{e\\o}n\mu}(h) = \nabla \times \left[\psi_{\substack{e\\o}n\mu}(h)\hat{z}\right]$$

$$= \left[\mp \frac{nJ_n(\mu r)}{r} \frac{\sin}{\cos} n\phi \hat{r} - \frac{\partial J_n(\mu r)}{\partial r} \frac{\cos}{\sin} n\phi \hat{\phi}\right] e^{ihz} \tag{B.13}$$

$$\overline{N}_{\substack{e\\o}n\lambda}(h) = \frac{1}{\kappa_\lambda} \nabla \times \nabla \times \left[\psi_{\substack{e\\o}n\lambda}(h)\hat{z}\right] = \frac{1}{\kappa_\lambda}\left[ih \frac{\partial J_n(\lambda r)}{\partial r} \frac{\cos}{\sin} n\phi \hat{r} \right.$$

$$\left. \mp \frac{ihn}{r} J_n(\lambda r) \frac{\sin}{\cos} n\phi \hat{\phi} + \lambda^2 J_n(\lambda r) \frac{\cos}{\sin} n\phi \hat{z}\right] e^{ihz}. \tag{B.14}$$

The functions $\overline{M}_{\substack{e\\o}n\lambda}$ and $\overline{N}_{\substack{e\\o}n\mu}$ are obtained, respectively, by replacing μ by λ in (B.13) and λ by μ in (B.14).

When the eigenfunctions are continuous, we discard the constraints on λ and μ so that they may assume any value. In that case, we have only two sets of functions, namely, $\overline{M}_{\substack{e\\o}n\lambda}$ and $\overline{N}_{\substack{e\\o}n\lambda}$, instead of four sets. Other types of cylindrical vector wave functions can also be used to represent the electromagnetic fields. Thus we have

$$\overline{M}^{(x)}_{\substack{e\\o}n\lambda}(h) = \nabla \times \left[J_n(\lambda r) \frac{\cos}{\sin} n\phi e^{ihz} \hat{x}\right] \tag{B.15}$$

$$\overline{N}^{(x)}_{\substack{e\\o}n\lambda}(h) = \frac{1}{\kappa} \nabla \times \nabla \times \left[J_n(\lambda r) \frac{\cos}{\sin} n\phi e^{ihz} \hat{x}\right] \tag{B.16}$$

$$\overline{M}^{(x)}_{\substack{e\\o}n\lambda}(h) = \frac{1}{2}\left\{ \frac{-ih}{\lambda} \left[\overline{M}_{\substack{e\\o}(n+1)\lambda}(h) - \overline{M}_{\substack{e\\o}(n-1)\lambda}(h)\right] \right.$$

$$\left. \pm \frac{\kappa}{\lambda} \left[\overline{N}_{\substack{o\\e}(n+1)\lambda}(h) + \overline{N}_{\substack{o\\e}(n-1)\lambda}(h)\right] \right\} \tag{B.17}$$

$$\overline{N}^{(x)}_{\substack{e\\o}n\lambda}(h) = \frac{1}{2}\left\{ \frac{-ih}{\lambda}\left[\overline{N}_{\substack{e\\o}(n+1)\lambda}(h) - \overline{N}_{\substack{e\\o}(n-1)\lambda}(h)\right] \right.$$
$$\left. \pm\frac{\kappa}{\lambda}\left[\overline{M}_{\substack{o\\e}(n+1)\lambda}(h) + \overline{M}_{\substack{o\\e}(n-1)\lambda}(h)\right]\right\}. \tag{B.18}$$

B-3 SPHERICAL VECTOR WAVE FUNCTIONS

$$\psi_{\substack{e\\o}mn}(\kappa) = j_n(\kappa R)P_n^m(\cos\theta)\,{}^{\cos}_{\sin}\, m\phi$$
$$\nabla^2\psi_{\substack{e\\o}mn}(\kappa) + \kappa^2\psi_{\substack{e\\o}mn}(\kappa) = 0$$
$$\overline{M}_{\substack{e\\o}mn}(\kappa) = \nabla\times\left[\psi_{\substack{e\\o}mn}(\kappa)\overline{R}\right]$$
$$= j_n(\kappa R)\left[\mp\frac{m}{\sin\theta}P_n^m(\cos\theta)\,{}^{\sin}_{\cos}\, m\phi\hat{\theta} - \frac{\partial P_n^m(\cos\theta)}{\partial\theta}\,{}^{\cos}_{\sin}\, m\phi\hat{\phi}\right] \tag{B.19}$$
$$\overline{N}_{\substack{e\\o}mn}(\kappa) = \frac{1}{\kappa}\nabla\times\nabla\times\left[\psi_{\substack{e\\o}mn}(\kappa)\overline{R}\right]$$
$$= \frac{n(n+1)}{\kappa R}j_n(\kappa R)P_n^m(\cos\theta)m\phi\hat{R} + \frac{1}{\kappa R}\frac{\partial}{\partial R}[Rj_n(\kappa R)]\left[\frac{\partial P_n^m}{\partial\theta}\,{}^{\cos}_{\sin}\, m\phi\hat{\theta} \mp\frac{m}{\sin\theta}P_n^m(\cos\theta)\,{}^{\sin}_{\cos}\, m\phi\hat{\phi}\right] \tag{B.20}$$
$$\overline{M}^{(x)}_{\substack{e\\o}mn}(\kappa) = \frac{1}{\kappa}\nabla\times\left[\psi_{\substack{e\\o}mn}(\kappa)\hat{x}\right]$$
$$= \frac{1}{\kappa}\nabla\times\left[\psi_{\substack{e\\o}mn}(\kappa)\left(\sin\theta\cos\phi\hat{R} + \cos\theta\cos\phi\hat{\theta} - \sin\phi\hat{\phi}\right)\right]$$
$$= \pm\frac{1}{2n(n+1)}\left[\overline{N}_{\substack{o\\e}(m+1)n} + (n+m)(n-m+1)\overline{N}_{\substack{o\\e}(m-1)n}\right]$$
$$+\frac{1}{2(n+1)(2n+1)}\left[\overline{M}_{\substack{e\\o}(m+1)(n+1)} - (n-m+1)(n-m+2)\overline{M}_{\substack{e\\o}(m-1)(n+1)}\right]$$
$$-\frac{1}{2n(2n+1)}\left[\overline{M}_{\substack{e\\o}(m+1)(n-1)} - (n+m-1)(n+m)\overline{M}_{\substack{e\\o}(m-1)(n-1)}\right], \tag{B.21}$$

where the functional dependence (κ) for $\overline{N}$ and $\overline{M}$ is implied.

$$
\begin{aligned}
\overline{M}^{(y)}_{\substack{e\\o}mn}(\kappa) &= \frac{1}{\kappa}\nabla\times\left[\psi_{\substack{e\\o}mn}(\kappa)\hat{y}\right] \\
&= \frac{1}{\kappa}\nabla\times\left[\psi_{\substack{e\\o}mn}(\kappa)\left(\sin\theta\sin\phi\hat{R}\right.\right. \\
&\quad \left.\left.+\cos\theta\sin\phi\hat{\theta}+\cos\phi\hat{\phi}\right)\right] \\
&= -\frac{1}{2n(n+1)}\left[\overline{N}_{\substack{e\\o}(m+1)n}\right. \\
&\quad \left.-(n+m)(n-m+1)\overline{N}_{\substack{e\\o}(m-1)n}\right] \\
&\quad \pm\frac{1}{2(n+1)(2n+1)}\left[\overline{M}_{\substack{o\\e}(m+1)(n+1)}\right. \\
&\quad \left.+(n-m+1)(n-m+2)\overline{M}_{\substack{o\\e}(m-1)(n+1)}\right] \\
&\quad \mp\frac{1}{2n(2n+1)}\left[\overline{M}_{\substack{o\\e}(m+1)(n-1)}\right. \\
&\quad \left.+(n+m-1)(n+m)\overline{M}_{\substack{o\\e}(m-1)(n-1)}\right]
\end{aligned}
\tag{B.22}
$$

$$
\begin{aligned}
\overline{M}^{(z)}_{\substack{e\\o}mn}(\kappa) &= \frac{1}{\kappa}\nabla\times\left[\psi_{\substack{e\\o}mn}(\kappa)\hat{z}\right] \\
&= \frac{1}{\kappa}\nabla\times\left[\psi_{\substack{e\\o}mn}(\kappa)\left(\cos\theta\hat{R}-\sin\theta\hat{\theta}\right)\right] \\
&= \mp\frac{m}{n(n+1)}\overline{N}_{\substack{o\\e}mn}+\frac{1}{2n+1} \\
&\qquad \left[\frac{n-m+1}{n+1}\overline{M}_{\substack{e\\o}m(n+1)}+\frac{n+m}{n}\overline{M}_{\substack{e\\o}m(n-1)}\right]\cdot
\end{aligned}
\tag{B.23}
$$

The relations between $\overline{N}^{(x)}_{\substack{e\\o}mn}, \overline{N}^{(y)}_{\substack{e\\o}mn}, \overline{N}^{(z)}_{\substack{e\\o}mn}$ and $\overline{M}_{\substack{e\\o}mn}, \overline{N}_{\substack{e\\o}mn}$ can be obtained by taking the curl of $(B.21)$, $(B.22)$, and $(B.23)$. The result is the same by interchanging the roles of $\overline{M}$ and $\overline{N}$ in $(B.21)$–$(B.23)$. For example, we have

$$
\begin{aligned}
\overline{N}^{(z)}_{\substack{e\\o}mn}(\kappa) &= \mp\frac{m}{n(n+1)}\overline{M}_{\substack{o\\e}mn}+\frac{1}{2n+1} \\
&\qquad \left[\frac{n-m+1}{n+1}\overline{N}_{\substack{e\\o}m(n+1)}+\frac{n+m}{n}\overline{N}_{\substack{e\\o}m(n-1)}\right]\cdot
\end{aligned}
\tag{B.24}
$$

B-4 CONICAL VECTOR WAVE FUNCTIONS

$$
\psi_{\substack{e\\o}m\mu}(\kappa) = j_\mu(\kappa R)P_\mu^m(\cos\theta)\,{\cos \atop \sin}\, m\phi.
$$

Characteristic equation for μ:

$$\left.\frac{\partial P_\mu^m(\cos\theta)}{\partial\theta}\right|_{\theta=\theta_0} = 0$$

$$\Psi_{\substack{e\\o}m\mu}(\kappa) = j_\nu(\kappa R)P_\nu^m(\cos\theta)\begin{matrix}\cos\\ \sin\end{matrix} m\phi.$$

Characteristic equation for ν:

$$P_\nu^m(\cos\theta_0) = 0$$

$$(\nabla^2 + \kappa^2)\begin{pmatrix}\psi_{\substack{e\\o}m\mu}\\ \Psi_{\substack{e\\o}m\mu}\end{pmatrix} = 0\cdot$$

The expressions for $\overline{M}_{\substack{e\\o}m\mu}$, $\overline{N}_{\substack{e\\o}m\mu}$, $\overline{M}_{\substack{e\\o}n\nu}$, and $\overline{M}_{\substack{e\\o}n\nu}$ have the same form as the spherical vector wave functions defined by (B.19) and (B.20) with n replaced by μ or ν.

Appendix C

Exercises

Chapter 1

C-1.1

The method of gradient [Tai, 1992, p. 60] states that the gradient operator is invariant to the coordinate system; that is,

$$\sum_{i=1}^{3} \frac{\hat{u}_i}{h_i}\frac{\partial f}{\partial v_i} = \sum_{i=1}^{3} \frac{\hat{u}_i'}{h_i'}\frac{\partial f}{\partial v_i'},$$

where the unprimed and the primed quantities are defined in two arbitrary curvilinear orthogonal systems.

Find the relations between the unit vectors $\hat{x}, \hat{y}, \hat{z}$ and $\hat{R}, \hat{\theta}, \hat{\phi}$ as tabulated in Table 1-2 by taking the gradient of an appropriate scalar quantity expressed in the two coordinate systems. For example, let

$$f = x = R\sin\theta\cos\phi;$$

then

$$\nabla x = \hat{x}$$
$$\nabla\left(R\sin\theta\cos\phi\right) = \sin\theta\cos\phi\hat{R} + \cos\theta\cos\phi\hat{\theta} - \sin\phi\hat{\phi}.$$

Hence

$$\hat{x} = \sin\theta\cos\phi\hat{R} + \cos\theta\cos\phi\hat{\theta} - \sin\phi\hat{\phi}.$$

Do the rest to verify the results given in Table 1-2.

C-1.2

Find the Fourier series representation of a one-dimensional delta function $\delta(x-x')$ in terms of the sine and cosine functions with periodicity 2a, that is, let

$$\delta(x-x') = \sum_n A_n \cos\left(\frac{n\pi x}{a}\right) + B_n \sin\left(\frac{n\pi x}{a}\right);$$

then determine A_n and B_n.

C-1.3

A two-dimensional dyadic delta function in the form of

$$(\hat{x}\hat{x} + \hat{y}\hat{y})\,\delta(x-x')\,\delta(y-y')$$

can be represented by a two-dimensional Fourier series using two orthogonal sets of functions defined by

$$\begin{aligned} \overline{M}_e &= \nabla \times (C_x C_y \hat{z}) \\ &= -k_y C_x S_y \hat{x} + k_x S_x C_y \hat{y} \\ \overline{N}_0 &= \hat{z} \times \overline{M}_0 \\ &= \hat{z} \times \nabla \times (S_x S_y \hat{z}) \\ &= k_x C_x S_y \hat{x} + k_y S_x C_y \hat{y} \end{aligned}$$

where

$$\begin{aligned} C_x &= \cos k_x x, & S_x &= \sin k_x x \\ C_y &= \cos k_y y, & S_y &= \sin k_y y \\ k_x &= \frac{m\pi}{a}, & k_y &= \frac{n\pi}{b} \end{aligned}$$

The series can be written in the form

$$\begin{aligned} &(\hat{x}\hat{x} + \hat{y}\hat{y})\,\delta(x-x')\,\delta(y-y') \\ &= \sum_{m,n} \left(\overline{M}_e \overline{A}_e + \overline{N}_0 \overline{B}_0\right). \end{aligned}$$

Determine the unknown coefficients $\overline{A}_e$ and $\overline{B}_0$. It is observed that the unknown coefficients must be placed at the posterior position.

C-1.4

According to the theory of generalized functions [Gelfand and Shilov, 1964, pp. 4, 18], the one-dimensional delta function $\delta(x-x')$ and its derivative are defined by

$$\int_{x\epsilon x'} f(x)\delta(x-x')dx = f(x')$$

$$\int_{x\epsilon x'} f(x)\frac{d}{dx}\left[\delta(x-x')\right]dx$$

$$= -\frac{df(x)}{dx}\bigg]_{x=x'} = \frac{df(x')}{dx'}.$$

Based on these relations, prove that

$$\iiint_{\overline{R}\epsilon\overline{R}'} \overline{F}(\overline{R})\cdot\overline{\overline{I}}\delta(\overline{R}-\overline{R}')\,dV = \overline{F}(\overline{R}')$$

and

$$\iiint_{\overline{R}\epsilon\overline{R}'} \overline{F}(\overline{R})\cdot\nabla\times\left[\overline{\overline{I}}\delta(\overline{R}-\overline{R}')\right]dV$$

$$= \nabla\times\overline{F}(\overline{R})_{\overline{R}=\overline{R}'} = \nabla'\times\overline{F}(\overline{R}'),$$

where $\overline{\overline{I}}$ denotes the idem factor and $\delta(\overline{R}-\overline{R}')$, the three-dimensional delta function, is defined by

$$\iiint_{\overline{R}\epsilon\overline{R}} f(\overline{R})\delta(\overline{R}-\overline{R}')\,dV = f(\overline{R}').$$

Chapter 2

C-2.1

The differential equations governing the current and the voltage on a transmission line excited by a shunt current are given by (2.1) and (2.2); that is,

$$\frac{dV}{dx} = i\omega LI(x)$$

$$\frac{dI(x)}{dx} = i\omega CV(x) + K(x).$$

We can define two scalar Green functions $g_v(x,x')$ and $g_i(x,x')$ such that

$$\frac{dg_v(x,x')}{dx} = g_i(x,x') \tag{1}$$

$$\frac{dg_i(x,x')}{dx} = -k^2 g_v(x,x') - \delta(x-x'), \tag{2}$$

where g_v and g_i satisfy, respectively, the equations

$$\frac{d^2 g_v(x,x')}{d^2 x} + k^2 g_v(x,x') = -\delta(x-x') \tag{3}$$

$$\frac{d^2 g_i(x,x')}{dx^2} + k^2 g_i(x,x') = -\frac{d}{dx}\delta(x-x'). \tag{4}$$

In Chapter 3, we have applied the Ohm-Rayleigh method to solve (3) for an infinite line where the result is given by (2.17) with $g_v(x, x')$ denoted by $g_0(x, x')$ therein. An alternative approach is to solve (4) by applying the same method, and then determine g_v with the aid of (2).

Hint: By definition, the generalized function $\frac{d}{dx}\delta(x - x')$ is defined by

$$\int_{x'-\Delta}^{x'+\Delta} f(x)\frac{d}{dx}\delta(x - x')dx = -\frac{df(x')}{dx'} .$$

For a discontinuous function described by

$$f(x) = f^+(x)U(x - x') + f^-(x)U(x' - x),$$

where

$$U(x - x') = \begin{cases} 1, & x > x' \\ 0, & x < x' \end{cases}$$

$$U(x' - x) = \begin{cases} 1, & x < x' \\ 0, & x > x', \end{cases}$$

$$\frac{d}{dv}\left[f^+(x)U(x - x') + f^-(x)U(x' - x)\right]$$

$$= \frac{df^+(x)}{dx}U(x - x') + \frac{df^-(x)}{dx}U(x' - x)$$

$$+ \left[f^+(x') - f^-(x')\right]\delta(x - x').$$

This exercise is a scalar version of the method of $\overline{\overline{G}}_m$ to be used to find the eigenfunction expansion of the electric dyadic Green function in subsequent chapters. A clear understanding of the present exercise would help to understand the foundation of the method of $\overline{\overline{G}}_m$.

C-2.2

For a three-dimensional scalar wave equation with spherical symmetry, the corresponding free-space Green function $G_0(\overline{R})$ satisfies the equation

$$\frac{1}{R^2}\frac{d}{dR}\left[R^2\frac{dG_0(R)}{dR}\right] + \kappa^2 G_0(R) = -\frac{\delta(R - 0)}{4\pi R^2}.$$

Show that the solution for $G_0(R)$ can be obtained by the method of spherical Hankel transform.

Hint: The spherical Hankel transform of the singular function $\delta(R - 0)/4\pi R^2$ according to (1.84) with $n = 0$ and $R' = 0$ is given by

$$\frac{\delta(R - 0)}{4\pi R^2} = \frac{1}{2\pi^2}\int_0^\infty j_0(\lambda R)\lambda^2\, d\lambda$$

because $j_0(0) = 1$ [Stratton, 1941, p. 405]. Let

$$G_0(R) = \frac{1}{2\pi^2} \int_0^\infty g(\lambda) j_0(\lambda R) \lambda^2 \, d\lambda.$$

Determine $g(\lambda)$ by substituting the above two expressions into the differential equation. With the aid of (1.113), the circulation relation of the spherical Bessel functions with $\nu = 0, R' = 0$, the function $G_0(R)$ can readily be determined. The result should be the same as (2.87) with R_1 therein replaced by R.

C-2.3

An infinite line is excited by a shunt current source described by

$$K(x) = \begin{cases} K_0, & |x| \leq x_0 \\ 0, & |x| > x_0, \end{cases}$$

where K_0 is a constant. Apply the Green function technique to determine $V(x)$ along the line and verify your result by the classical method based on the theory of differential equations. This exercise will show that (2.10) is valid for the point of observation, x, located either outside or inside the source region.

C-2.4

Derive the expression for the two-dimensional free-space Green function given by (2.89) by the method of Fourier transform.

Hint: The differential equation for $G_0(\overline{r}, r')$ is

$$\frac{\partial^2 G_0}{\partial x^2} + \frac{\partial^2 G_0}{\partial y^2} + k^2 G_0 = -\delta(x - x')\delta(y - y').$$

Apply a two-dimensional Fourier integral transform to this equation; then evaluate the inverse Fourier transform by the method of contour integration.

Chapter 3

C-3.1

Show that for two lossy dielectrics in contact with complex dielectric constants

$$\epsilon_{1c} = \epsilon_1 \left(\frac{1 + i\sigma_1}{\omega\epsilon_1} \right)$$

$$\epsilon_{2c} = \epsilon_2 \left(\frac{1 + i\sigma_2}{\omega\epsilon_2} \right).$$

The boundary condition for the normal components of the electric fields is

$$\hat{n} \cdot \left(\epsilon_{1c}\overline{E}_1 - \epsilon_{2c}\overline{E}_2 \right) = 0.$$

Find the surface charge density at the interface in terms of the normal component of $\overline{E}_1$ or $\overline{E}_2$ [Stratton, 1941, p. 483].

C-3.2

The Hertzian potential function $\overline{A}$ obeys the gauge condition $\nabla \cdot \overline{A} = i\omega\mu_0\epsilon_0\psi$ as described by (3.71). The Debye potential function was introduced under the assumption that $\overline{A} = A_R\hat{R}$ and the gauge condition

$$\frac{\partial A_R}{\partial R} = i\omega\mu_0\epsilon_0\Phi \cdot$$

Find the differential equations for A_R and Φ for the case that $\overline{J} = J_R\hat{R}$. Show that the electromagnetic field of a Hertzian dipole can be found by using A_R instead of A_z. This exercise demonstrates very clearly the nonuniqueness of the potential functions as far as the solutions for $\overline{E}$ and $\overline{H}$ are concerned.

C-3.3

Show that the integral solutions for $\overline{A}$ and ψ given by (3.74) and (3.75) indeed satisfy the gauge condition postulated by (3.71).

Hint: $\nabla G_0(\overline{R}, \overline{R}') = -\nabla' G_0(\overline{R}, \overline{R}')$, where ∇' denotes the gradient operator in the primed coordinate system and $\nabla' \cdot \left[\overline{J}(\overline{R}')G_0(\overline{R}, \overline{R}')\right] = G_0(\overline{R}, \overline{R}')\nabla' \cdot \overline{J}(\overline{R}') + \overline{J}(\overline{R}') \cdot \nabla' G_0(\overline{R}, \overline{R}')$.

C-3.4

Show that in the far-zone region the terms which are negligible in comparison with $\overline{E}$ and $\overline{H}$ given by (3.82) and (3.83) are of the order of $1/kR^2$.

Chapter 4

C-4.1

Find $\overline{\overline{G}}_{e2}$ and $\overline{\overline{G}}_{m1}$ for a half-space ($z \geq 0$). By definition,

$$\hat{z} \times \nabla \times \overline{\overline{G}}_{e2}(\overline{R}, \overline{R}') = 0$$
$$\hat{z} \times \overline{\overline{G}}_{m1}(\overline{R}, \overline{R}') = 0.$$

Prove that $\left[\overline{\overline{G}}_{m2}(\overline{R}', \overline{R})\right]^T = \overline{\overline{G}}_{m1}(\overline{R}, \overline{R}')$.

Hint: For three infinitesimal orthogonal current elements with current moments $c_j, j = 1, 2, 3$ placed at $\overline{R}'$, we have

$$\overline{J}(\overline{R}) = \sum_{j=1}^{3} c_j\hat{x}_j\delta(\overline{R} - \overline{R}'),$$

where

$$\overline{R}' = \sum_{j=1}^{3} x_j' \hat{x}_j.$$

Their images with respect to a magnetically perfectly conducting surface placed at $z = 0$ are described by

$$\overline{J}_i(\overline{R}) = (c_1\hat{x}_1 + c_2\hat{x}_2 - c_3\hat{x}_3)\,\delta\left(\overline{R} - \overline{R}_i'\right),$$

where

$$\overline{R}_i' = x_i'\hat{x}_1 + x_2'\hat{x}_2 - x_3'\hat{x}_3 \cdot$$

Answer:

$$\overline{\overline{G}}_{e2}(\overline{R}, \overline{R}') = \left(\overline{\overline{I}} - \frac{1}{k^2}\nabla\nabla'\right)(G_0 + G_i) - 2\hat{z}\hat{z}G_i$$

$$\overline{\overline{G}}_{m1}(\overline{R}, \overline{R}') = \nabla G_0 \times \overline{\overline{I}} + \nabla G_i \times \overline{\overline{I}}_i,$$

where

$$G_0 = \frac{e^{ik\left|\overline{R}-\overline{R}'\right|}}{4\pi\left|\overline{R} - \overline{R}'\right|}$$

$$G_i = \frac{e^{ik\left|\overline{R}-\overline{R}_i'\right|}}{4\pi\left|\overline{R} - \overline{R}_i'\right|}$$

$$\overline{\overline{I}}_i = \hat{x}\hat{x} + \hat{y}\hat{y} - \hat{z}\hat{z} \cdot$$

C-4.2

By means of the method of images find the electric dyadic Green function of the first kind for a right-angle conducting wedge $\left(\frac{\pi}{2} \geq \phi \geq 0\right)$. The axis of the wedge corresponds to the z-axis.

C-4.3

By definition, when the dyadic point source has only one component in the z-direction, the nonvanishing component of $\overline{\overline{G}}_{eo}$ is the vector component $\overline{G}_{eoz}$ defined by

$$\begin{aligned}\overline{G}_{eoz} &= \overline{\overline{G}}_{eo} \cdot \hat{z} \\ &= \hat{z}G_0 + \frac{1}{k^2}\frac{\partial}{\partial z}\nabla G_0,\end{aligned}$$

where

$$G_0 = \frac{e^{ik\left|\overline{R}-\overline{R}'\right|}}{4\pi\left|\overline{R} - \overline{R}'\right|}.$$

By means of the method of images find the electric vector Green function of the first kind, $\overline{G}_{e1z}$ for a 60° angle conducting wedge $\left(\frac{\pi}{3} \geq \phi \geq 0\right)$. The axis of the wedge corresponds to the z-axis.

Hint: There are five images of a z-directed electric dipole placed at $\overline{R}'$ inside the wedge.

C-4.4

Huygens' principle in free space can be cast in several different forms. One of them can be derived by using the vector-dyadic Green theorem of the second kind, (1.51), with $\overline{P} = \overline{E}(\overline{R}), \overline{\overline{Q}} = \overline{\overline{G}}_{eo}(\overline{R}, \overline{R}')$. The result yields

$$\overline{E}(\overline{R}) = -\oiint_S \Big\{ \left[\nabla \times \overline{\overline{G}}_{eo}(\overline{R}, \overline{R}') \right] \cdot \left[\hat{n}' \times \overline{E}(\overline{R}') \right] + i\omega\mu_0 \overline{\overline{G}}_{eo}(\overline{R}, \overline{R}') \cdot \left[\hat{n}' \times \overline{H}(\overline{R}') \right] \Big\} \, dS'.$$

It is assumed that the region of integration does not have a volume distribution of current.

You are required to fill in all the details of the derivation such as the use of the symmetrical relationships of $\overline{\overline{G}}_{eo}(\overline{R}, \overline{R}')$ and $\nabla \times \overline{\overline{G}}_{eo}(\overline{R}, \overline{R}')$ in arriving at the final result.

C-4.5

A formula due to Stratton and Chu [Stratton, 1941, p. 466] is another version of Huygens' principle. It was obtained by using the vector Green's theorem of the second kind, (1.18), with $\overline{P} = \overline{E}(\overline{R})$ and $\overline{Q} = G_0(\overline{R}, \overline{R}')\overline{a}$, where $G_0(\overline{R}, \overline{R}')$ is the three-dimensional free-space scalar Green function defined by (2.88) and $\overline{a}$ is an arbitrary constant. The result yields

$$\overline{E}(\overline{R}) = -\oiint_S \Big\{ \left[\hat{n}' \times \overline{E}(\overline{R}') \right] \times \nabla' G_0(\overline{R}, \overline{R}') + \left[\hat{n}' \cdot \overline{E}(\overline{R}') \right] \nabla' G_0(\overline{R}, \overline{R}') + i\omega\mu_0 \left[\hat{n}' \times \overline{H}(\overline{R}') \right] G_0(\overline{R}, \overline{R}') \Big\} \, dS'.$$

You are required to fill in the details of the derivation. It is observed that Stratton and Chu's formula involves not only the tangential components of $\overline{E}$ and $\overline{H}$ but also the normal component of $\overline{E}$.

If we let $\overline{P} = \overline{E}(\overline{R})$ and $\overline{Q} = \nabla G_0(\overline{R}, \overline{R}') \times \overline{a}$, an alternative formula for $\overline{E}(\overline{R})$ can be obtained. This formula is found in the works of Franz [1948] and Mentzer [1955, p. 147]. Derive this formula and compare it with the formula quoted in Exercise 4.4

C-4.6

According to the duality principle, for a fictitious magnetic current source the field equations have the form

$$\nabla \times \overline{E} = i\omega\mu_0 \overline{H} - \overline{J}_m$$
$$\nabla \times \overline{H} = -i\omega\epsilon_0 \overline{E}.$$

A pair of dyadic Green functions can then be introduced to integrate these two equations. They are

$$\nabla \times \overline{\overline{G}}_e = \overline{\overline{I}}\delta(\overline{R} - \overline{R}') + k^2\overline{\overline{G}}_m$$

$$\nabla \times \overline{\overline{G}}_m = \overline{\overline{G}}_e.$$

The function $\overline{\overline{G}}_e$ is now a solenoidal function while $\overline{\overline{G}}_m$ is not. With the aid of $\overline{\overline{G}}_m$ find the integral solution for $\overline{H}$. When the formulation is applied to problems involving an electrically perfectly conducting scattering body, show that

$$\begin{aligned}\overline{H}(\overline{R}) =& i\omega\epsilon_0 \iiint \overline{\overline{G}}_{m2}(\overline{R}, \overline{R}') \cdot \overline{J}_m(\overline{R}')\, dV' \\ &+ i\omega\epsilon_0 \oiint_{S_A} \overline{\overline{G}}_{m2}(\overline{R}, \overline{R}') \cdot \left[\hat{n}' \times \overline{E}(\overline{R}')\right] dS',\end{aligned}$$

where $\overline{\overline{G}}_{m2}$ obeys the boundary condition $\hat{n} \times \nabla \times \overline{\overline{G}}_{m2} = 0$ on S.

The volume integral can be converted to an integral of equivalent magnetization vector defined by

$$\overline{J}_m = -i\omega\mu_0\overline{M}.$$

Hint: Apply the second vector-dyadic Green's theorem, (1.51), to this problem with $\overline{P} = \overline{H}$ and $\overline{\overline{Q}} = \overline{\overline{G}}_m$. You may obtain the same answer with the help of the duality principle without repeating the analysis which leads to (4.181); however, you must be careful in coordinating the boundary conditions.

C-4.7

When we apply the vector Green's theorem of the second kind to derive the reciprocity theorems, the two cases treated in the text are Case 1, $\overline{P} = \overline{E}_a, \overline{Q} = \overline{E}_b$, and Case 2, $\overline{P} = \overline{E}_a, \overline{Q} = \overline{H}_b$. Discuss the case whereby $\overline{P} = \overline{H}_a, \overline{Q} = \overline{H}_b$.

C-4.8

Prove the symmetrical relationship of $\nabla \times \overline{\overline{G}}_e^{(ij)}$ stated by (4.205).

C-4.9

Verify the complementary reciprocity theorem for the transmission lines stated by (4.249).

Chapter 5

C-5.1

Verify the relationship $\nabla \times \overline{\overline{G}}_{e1} = \overline{\overline{G}}_{m2}$ with $\overline{\overline{G}}_{m2}$ and $\overline{\overline{G}}_{e1}$ given by (5.43) and (5.49).

Hint:

$$\nabla\times\left[-\frac{1}{k^2}\hat{z}\hat{z}\delta(\overline{R}-\overline{R}')\right]$$
$$=-\left[\frac{1}{k^2}\nabla\delta(\overline{R}-\overline{R}')\right]\times\hat{z}\hat{z}$$
$$=-\left[\frac{1}{k^2}\nabla_t\delta(\overline{R}-\overline{R}')\right]\times\hat{z}\hat{z}.$$

Decompose $\overline{N}_0(k_g)$ into $\overline{N}_{ot}(k_g)+\overline{N}_{oz}(k_g)$, where

$$\overline{N}_{ot}(k_g)=\frac{ik_g}{k}\nabla_t\left(S_xS_y\right)e^{ik_gz}$$
$$\overline{N}_{oz}(k_g)=\frac{k_c^2}{k}S_xS_ye^{ik_gz}\hat{z}.$$

Then, show

$$\nabla\times\sum_{m,n}C_{mn}\left[\overline{M}_e(k_g)\overline{M}_e'(-k_g)U(z-z')\right.$$
$$\left.+\overline{M}_e(-k_g)\overline{M}_e'(k_g)U(z'-z)\right]$$
$$=k\sum_{m,n}C_{mn}\left[\overline{N}_e(k_g)\overline{M}_e'(-k_g)U(z-z')\right.$$
$$\left.+\overline{N}_e(-k_g)\overline{M}_e'(k_g)U(z'-z)\right]$$

and

$$\nabla\times\sum_{m,n}c_{mn}\left[\overline{N}_0(k_g)\overline{N}_0'(-k_g)U(z-z')\right.$$
$$\left.+\overline{N}_0(-h_g)\overline{N}_0'(k_g)U(z'-z)\right]$$
$$=k\sum_{m,n}c_{mn}\left[\overline{M}_0(k_g)\overline{N}_0'(-k_g)U(z-z')\right.$$
$$\left.+\overline{M}_0(-k_g)\overline{N}_0'(k_g)U(z'-z)\right]$$
$$+\sum_{m,n}c_{mn}\hat{z}\delta(z-z')\times\left[\overline{N}_{0t}(k_g)\overline{N}_{0z}'(-k_g)\right.$$
$$\left.-\overline{N}_{0t}(-k_g)\overline{N}_{0z}'(k_g)\right].$$

The last term with the factor $\hat{z}\delta(z-z')$ can be transformed to

$$\overline{\overline{S}}=\sum_{m,n}\left(\frac{-2ik_gk_c^2}{k^2}\right)\nabla_t\left(S_xS_yS_{x'},S_{y'}\right)\times\hat{z}\hat{z}\delta(z-z')$$
$$=\nabla_t\delta(\overline{R}-\overline{R}')\times\hat{z}\hat{z}/k^2$$

with the aid of the two-dimensional expansion

$$\delta(x-x')\,\delta(y-y')=\sum_{m,n}\frac{4}{ab}S_xS_yS_{x'}S_{y'}.$$

The rest is straightforward.

C-5.2

Following a similar technique as hinted in the above exercise, show that

$$\nabla \cdot \bar{\bar{G}}_{e1}(\bar{R}, \bar{R}') = -\frac{1}{k^2} \nabla \delta(\bar{R} - \bar{R}'),$$

where $\bar{\bar{G}}_{e1}$ is given by (5.49).

C-5.3

Find $\bar{\bar{G}}_{e1}$ for a rectangular waveguide by using the vector functions defined by

$$\begin{aligned} \overline{m}_e &= \nabla \times \left(C_x C_y e^{ihz} \hat{z}\right) \\ \overline{n}_0 &= \hat{z} \times \overline{m}_0 = \hat{z} \times \nabla \times \left(S_x S_y e^{ihz} \hat{z}\right) \\ \bar{\ell}_0 &= k_c S_x S_y e^{ihz} \hat{z}, \end{aligned}$$

where the constants h and k_c and the functions C_x, C_y, S_x, and S_y have the same meaning as the ones contained in $\overline{M}_e$ and $\overline{N}_0$. It is observed that $\overline{m}_e$ is the same as $\overline{M}_e$ and $\overline{n}_0$ and $\bar{\ell}_0$, except some constants represent the transversal and the longitudinal components of $\overline{N}_0$

Hint: Shown first $\overline{m}_e, \overline{n}_0$, and $\bar{\ell}_0$ are orthogonal with the same normalization factor

$$\frac{1}{2}\left(1 + \delta_0\right) \pi ab k_c^2 \delta\left(h - h'\right) \cdot$$

Then derive the formula

$$\bar{\bar{I}}\delta(\bar{R} - \bar{R}') = \int_{-\infty}^{\infty} dh \sum_{m,n} C_{mn} \left(\overline{m}_e \overline{m}_e' + \overline{n}_0 \overline{n}_0' + \bar{\ell}_0 \bar{\ell}_0'\right),$$

with

$$C_{mn} = \frac{(2 - \delta_0)}{\pi ab k_c^2}.$$

The primed functions $\overline{m}_e'$, $\overline{n}_0'$, and $\bar{\ell}_0'$ are defined with respect to the primed variables and $-h$; that is,

$$\overline{m}_e' = \nabla' \times \left(C_{x'} C_{y'} e^{-ihz'} \hat{z}\right),$$

and so on. Then let

$$\begin{aligned} \bar{\bar{G}}_{e1}(\bar{R}, \bar{R}') = \int_{-\infty}^{\infty} dh \sum_{m,n} &\left[A \overline{m}_e \overline{m}_e' \right. \\ &\left. + \left(B_1 \overline{n}_0 + B_2 \bar{\ell}_0\right) \overline{n}_0' + \left(C_1 \bar{\ell}_0 + C_2 \overline{n}_0\right) \bar{\ell}_0'\right]. \end{aligned}$$

Show that the coefficients A_1, B_1, B_2, C_1, and C_2 are given by

$$
\begin{aligned}
A_1 &= \frac{1}{k^2\left(h^2-k_g^2\right)} \\
B_1 &= \frac{k_g^2}{k^2(h^2-k_g^2)} \\
B_2 &= -\frac{ihk_c}{k^2\left(h^2-k_g^2\right)} \\
C_1 &= -\frac{\left(h^2-k^2\right)}{k^2\left(h^2-k_g^2\right)} \\
C_2 &= -B_2.
\end{aligned}
$$

The coefficient C_1 can be split into two terms, namely,

$$C_1 = -\frac{1}{k^2} + \frac{k_c^2}{k^2\left(h^2-k_g^2\right)}.$$

The rest follows the same procedure described in the method of $\overline{\overline{G}}_e$; that is, extracting the singular term $-\hat{z}\hat{z}\delta(\overline{R}-\overline{R}')/k^2$ first and then regroup the remaining terms to obtain $\overline{\overline{G}}_{e1}$ in the form of (5.49).

C-5.4

Find the expressions for the electric field $\overline{E}(\overline{R})$ inside a rectangular waveguide excited by an infinitesimal electric dipole with current moment c located at $\overline{R}_0 = (x_0, y_0, z_0,)$ for the following cases:

$$
\begin{aligned}
&(a)\ \ \overline{J}(\overline{R}') = c\delta\left(\overline{R}'-\overline{R}_0\right)\hat{x} \\
&(b)\ \ \overline{J}(\overline{R}') = c\delta\left(\overline{R}'-\overline{R}_0\right)\hat{y} \\
&(c)\ \ \overline{J}(\overline{R}') = c\delta\left(\overline{R}'-\overline{R}_0\right)\hat{z}.
\end{aligned}
$$

Hint and the answer for (*c*):

In general,

$$\overline{E}(\overline{R}) = i\omega\mu_0 \iiint \overline{\overline{G}}_{e1}(\overline{R},\overline{R}')\cdot\overline{J}(\overline{R}')\,dV',$$

where

$$
\begin{aligned}
\overline{\overline{G}}_{e1}(\overline{R},\overline{R}') = &-\frac{1}{k^2}\hat{z}\hat{z}\delta(\overline{R}-\overline{R}') \\
&+\sum_{m,n} C_{mn}\left[\overline{M}_e\left(\pm k_g\right)\overline{M}'_e\left(\mp k_g\right)\right. \\
&\left.+\overline{N}_0\left(\pm k_g\right)\overline{N}'_0\left(\mp k_g\right)\right],\ z \gtrless z'
\end{aligned}
$$

$$C_{mn} = \frac{i\left(2-\delta_0\right)}{abk_c^2k_g}.$$

For Case (c), the answer is

$$\overline{E}(\overline{R}) = -\frac{1}{k^2} i\omega\mu_0 c\delta\left(\overline{R}-\overline{R}_0\right)\hat{z} + \sum_{m,n} D\left(\mp k_g\right)\overline{N}_0\left(\pm k_g\right), \; z \gtrless z',$$

where

$$D\left(\mp k_g\right) = -\frac{\omega\mu_0 c\left(2-\delta_0\right)}{kk_g ab}\sin k_x x_0 \sin k_y y_0 e^{\mp ik_g z_0}.$$

It should be mentioned that the classical method of treating this class of problems is based on the method of Hertzian potential [Chien, Infeld, Pound, Stevenson, and Synge, 1949] where different potential functions have to be found for different orientations of the dipole.

C-5.5

Find the expression for the electrical field inside a rectangular waveguide excited by an infinitesimal aperture field with field-moment f located on the wall of the waveguide. The cases to be treated are

$$(a)\;\; \overline{E}(\overline{R}') = f\delta\left(x'-x_0\right)\delta\left(z'-z_0\right)\hat{x},\; y'=b$$

$$(b)\;\; \overline{E}(\overline{R}') = f\delta\left(x'-x_0\right)\delta\left(z'-z_0\right)\hat{z},\; y'=b$$

$$(c)\;\; \overline{E}(\overline{R}') = f\delta\left(y'-y_0\right)\delta\left(z'-z_0\right)\hat{y},\; x'=a$$

$$(d)\;\; \overline{E}(\overline{R}') = f\delta\left(y'-y_0\right)\delta\left(z'-z_0\right)\hat{z},\; x'=a.$$

The dimension of the field-moment f is volt-meter.

Hint and the answer for (*c*):

In general, according to (4.183), we have

$$\overline{E}(\overline{R}) = -\oiint_{S_A}\left[\nabla\times\overline{\overline{G}}_{e2}(\overline{R},\overline{R}')\right]\cdot\left[\hat{n}'\times\overline{E}(\overline{R}')\right]dS'\cdot$$

For a rectangular waveguide

$$\nabla\times\overline{\overline{G}}_{e2}(\overline{R},\overline{R}') = \overline{\overline{G}}_{m1}(\overline{R},\overline{R}')$$

$$= \sum_{m,n} kC_{mn}\left[\overline{M}_e\left(\pm k_g\right)\overline{N}'_e\left(\mp k_g\right) + \overline{N}_0\left(\pm k_g\right)\overline{M}'_0\left(\mp k_g\right)\right], \; z \gtrless z',$$

where

$$C_{mn} = \frac{i\,(2-\delta_0)}{abk_c^2 k_g}.$$

For Case (c), the answer is

$$\overline{E}(\overline{R}) = \sum_{m,n} F\,(\mp k_g)\,\overline{M}_e\,(\pm k_g)\,, \quad z \gtrless z_0,$$

where

$$F\,(\mp k_g) = \frac{i(-1)^{m+1} f(2-\delta_0)}{k_g ab} \cos k_y y_0 e^{\mp i k_g z_0}.$$

C-5.6

Derive the expression for $\overline{\overline{G}}_{A1}(\overline{R}, \overline{R}')$ stated by (5.87).

C-5.7

Show that $\nabla \times \overline{\overline{G}}_e = \overline{\overline{G}}_m$ where the two functions are given by (5.157) and (5.158).

Hint:

$$\begin{aligned}
&\nabla \times \left[\hat{z}\hat{z}\delta(\overline{R}-\overline{R}')\right] = \nabla\delta(\overline{R}-\overline{R}') \times \hat{z}\hat{z} \\
&= \left(\hat{x}\frac{\partial}{\partial x} + \hat{y}\frac{\partial}{\partial y} + \hat{z}\frac{\partial}{\partial z}\right)\delta(x-x')\delta(y-y')\delta(z-z') \times \hat{z}\hat{z} \\
&= \left[-\frac{\partial}{\partial x}\delta(\bar{r}-\bar{r}')\hat{y}\hat{z} + \frac{\partial}{\partial y}\delta(\bar{r}-\bar{r}')\hat{x}\hat{z}\right]\delta(z-z'),
\end{aligned}$$

where

$$\delta(\bar{r}-\bar{r}') = \delta(x-x')\delta(y-y').$$

Derive a two-dimensional Fourier series representation of the delta function in the form

$$\delta(\bar{r}-\bar{r}') = \sum_{mn} \frac{4}{ab} S_x S_y S_x' S_y'.$$

In evaluating the curl of the residue series $\overline{\overline{S}}_{e1}$ in $\overline{\overline{G}}_{e1}$, take into consideration that the terms $\overline{N}_{0t}\overline{N}_{0z}'$ and $\overline{N}_{0z}\overline{N}_{0t}'$ are discontinuous at $z = z'$, where $\overline{N}_{0t}$ and $\overline{N}_{0z}$ denote the transversal and the longitudinal component of $\overline{N}_0$; that is, $\overline{N}_0 = \overline{N}_{0t} + \overline{N}_{0z}$. The contribution due to these terms cancel the terms resulting from $-\nabla \times \left[\hat{z}\hat{z}\delta(\overline{R}-\overline{R}')\right]/k^2$.

C-5.8

Complete the derivation of (5.167).

Hint: Split the term $\overline{N}_0\overline{N}_0'/(\kappa^2 - k^2)$ into two sums as follows:

$$\begin{aligned}
&\frac{1}{\kappa^2 - k^2}\overline{N}_0\overline{N}_0' \\
=&\frac{\kappa^2}{k^2(\kappa^2 - k^2)}\left(\frac{k^2 - k_c^2}{h^2}\overline{N}_{0t}\overline{N}_{ot}' + \overline{N}_{ot}\overline{N}_{0z}' + \overline{N}_{0z}\overline{N}_{0t}' + \overline{N}_{0z}\overline{N}_{0z}'\right) \\
&+ \frac{1}{k^2}\left(\frac{k_c^2}{h^2}\overline{N}_{0t}\overline{N}_{0t}' - \overline{N}_{0t}\overline{N}_{0z}' - \overline{N}_{0z}\overline{N}_{ot}' - \overline{N}_{oz}\overline{N}_{0z}'\right).
\end{aligned}$$

What is the main reason for such a splitting? See (5.75).

Chapter 6

C-6.1

A circular cylindrical waveguide of radius "a" is bifurcated by a conducting strip extended from the center of the guide to the wall. It is located at $\phi = 0$. Find the electric dyadic Green function of the first kind by the method of $\overline{\overline{G}}_m$. Which is the dominant mode of all the modes?

Hint: Half-order Bessel functions are involved in addition to Bessel functions of integral order. Figure 6-2 is a useful aid to do this exercise.

C-6.2

Find the electric dyadic Green function of the first kind for a semicircular waveguide with radius "a". The guide occupies the region $\pi \geq \phi \geq 0$.

C-6.3

A cylindrical waveguide with radius "a" is filled with air in region 1 ($0 > z > -\infty$) and with a dielectric of permittivity ϵ in region 2 ($\infty > z \geq 0$). Find the electric dyadic Green functions of the third kind $\overline{\overline{G}}_e^{(11)}$ and $\overline{\overline{G}}_e^{(21)}$ by the method of scattering superposition.

C-6.4

A semi-infinite circular waveguide occupies the region $\infty > z \geq 0$. It is terminated by a conducting plane at $z = 0$. Find $\overline{\overline{G}}_{e1}$ for this waveguide. When the guide is excited by an electric dipole with current moment c corresponding to a current distribution

$$\bar{J}(\overline{R}') = -\frac{1}{r'}c\delta(r' - a)\delta(\phi' - 0)\delta(z' - z_0)\hat{r}'$$

find the expression of the electromagnetic field, both $\overline{E}$ and $\overline{H}$, of the dominant mode (TE_{11}).

C-6.5

Verify the expression for $\overline{\overline{G}}_{E'1}(\overline{R}, \overline{R}')$ stated by (6.31).

Chapter 7

C-7.1

A plane wave in the form of

$$\overline{E}_i = E_0 \hat{y} e^{ik(x \sin\theta_0 + z\cos\theta_0)}$$

is incident upon a conducting cylinder with radius equal to "a". The axis of the cylinder coincides with the z-axis. Find the scattered field.

Hint: Consider a current source in the form of

$$\bar{J}(\overline{R}') = \frac{c\hat{y}\delta\left(R' - R_0\right)\delta(\theta' - \theta_0)\delta(\phi' - 0)}{R'^2 \sin\theta'}.$$

Determine the asymptotic expression for the scattered field when $kR_0 \gg 1$, following the technique described in Sec. 7-4.

C-7.2

Find the far-zone field of a half-wave dipole placed outside of a conducting cylinder with a current distribution described by

$$\begin{aligned} \bar{J}(\overline{R}') =& \frac{1}{b} I_0 \cos kz' \delta(r' - b)\delta(\phi' - 0)\hat{z} \\ & \frac{\lambda}{4} \geq z' \geq -\frac{\lambda}{4}. \end{aligned}$$

C-7.3

Find the function $\overline{\overline{G}}_{m1}(\overline{R}, \overline{R}')$ for a conducting cylinder of radius equal to "a" in free space. The cylinder is concentric with the z-axis. With the aid of $\overline{\overline{G}}_{m1}$ find the far-zone electric field of a longitudinal slot antenna cut on that cylinder with an aperture field described by

$$\begin{aligned} \overline{E}'(\overline{R}) &= E_0 \cos kz' \delta(\phi' - 0)\hat{\phi}', \; r' = a \\ & \frac{\lambda}{4} \geq z' \geq \frac{-\lambda}{4}. \end{aligned}$$

C-7.4

Find $\overline{\overline{G}}_{e1}$ for a semi-infinite conducting cylinder of radius "a" erected perpendicularly on a conducting ground plane. The region of the function corresponds to $\infty > r > a, 2\pi \geq \phi \geq 0, \infty > z \geq 0$.

Hint: Apply the method of images with respect to the conducting ground plane.

C-7.5

Find $\overline{\overline{G}}_{e1}$ for an infinitely long conducting half-cylinder with radius "a" placed on a conducting ground plane. The axis of the cylinder corresponds to the z-axis; the ground plane corresponds to $y = 0$ or the $x - z$ plane. The domain of the function occupies the region $\infty > r \geq a, \pi \geq \phi \geq 0, \infty > z > -\infty$.

C-7.6

A conducting cylinder of radius "a" is coated with a layer of dielectric with thickness $t = b - a$, where b denotes the outer radius of the dielectric layer. Being a problem involving two regions, $b \geq r \geq a$ and $r > b$, we are dealing with functions of the third kind. Outline the steps and the functions to be used to find $\overline{\overline{G}}_e^{(11)}$ and $\overline{\overline{G}}_e^{(21)}$, where region 1 corresponds to $b \geq r \geq a$ and region 2 to $r > b$. You are not required to find the complete answer, which is quite complicated.

Chapter 8

C-8.1

Find $\overline{\overline{G}}_{e1}$ for a semi-infinite conducting elliptical cylinder placed on a conducting ground plane. The ground plane corresponds to the $y - z$-plane and the axis of the elliptical cylinder coincides with the z-axis; the major and minor axes are denoted by "a" and "b".

C-8.2

Find the vector wave functions which can be used to construct the dyadic Green functions for an elliptical waveguide with major and minor axes denoted by "a" and "b".

C-8.3

Prove the normalization identities described by (8.24).

Chapter 9

C-9.1

Discuss the singular behavior of both $\overline{E}$ and $\overline{H}$ near the edge of a half-sheet with the aid of the series expansion of the Fresnel integrals given by (9.59). Consider the case where the incident field is perpendicular to the edge. Show that the result is compatible with Meixner's edge condition [Meixner, 1949].

C-9.2

Verify (9.23) by taking the partial derivative of the S-functions represented by (9.18) or (9.20).

C-9.3

Derive (9.32) by using (9.29) and (9.30) with the aid of the saddle-point method of integration.

C-9.4

An infinite line current source is placed inside a 90° conducting wedge at $x = x'$, $y = y'$, or $r = r'$ and $\phi = \phi'$. The scalar Green's function for E_z satisfies the differential equation

$$\left(\frac{\partial^2}{\partial x^2} + \frac{\partial^2}{\partial y^2} + k^2\right) g_1 = -\delta\left(x - x'\right)\delta\left(y - y'\right)$$

or

$$\frac{1}{r}\frac{\partial}{\partial r}\left(r\frac{\partial g_1}{\partial r}\right) + \frac{1}{r^2}\frac{\partial^2 g_1}{\partial \phi^2} = -\frac{1}{r}\delta\left(r - r'\right)\delta\left(\phi - \phi'\right).$$

The boundary conditions for g_1 are

$$g_1 = 0 \text{ at } x = 0 \text{ and } y = 0$$

or

$$g_1 = 0 \text{ at } \phi = 0 \text{ and } \phi = \frac{\pi}{2}.$$

The solution for g_1 can be obtained either by the method of eigenfunction expansion using a series of Bessel functions and Hankel functions of fractional order or by the method of images. The equivalence between these two solutions yields a relationship between a finite sum of Hankel functions of zeroth order and a series of Bessel and Hankel functions of fractional order. Find this identity.

Hint: The two-dimensional Green function in free space is given by (2.89).

Chapter 10

C-10.1

Find the dyadic Green functions $\overline{\overline{G}}_e^{(11)}$ and $\overline{\overline{G}}_e^{(21)}$ for a perfectly conducting sphere of radius "a" covered by a concentric layer of dielectric with permittivity ϵ and outer radius "b". Region 1 denotes the region inside the layer and region 2 the space outside.

C-10.2

The algebraic method of deriving $\overline{\overline{G}}_{eo}$ in terms of the spherical vector wave function does not provide for the singular term of this function, namely, $-\hat{R}\hat{R}\delta(\overline{R} - \overline{R}')/k^2$. However, if we apply the same method to expand $\nabla \times \left[\overline{\overline{I}}G_0(\overline{R}, \overline{R}')\right]$, then the singular term can be obtained by making use of the relationship $\overline{\overline{G}}_{eo} =$

$\left[\nabla \times \overline{\overline{G}}_{mo} - \overline{\overline{I}}\delta(\overline{R} - \overline{R}')\right] / k^2$. Sketch the steps that would give the complete answer for $\overline{\overline{G}}_{eo}$.

C-10.3

Find the expressions for the electromagnetic field due to a longitudinal slot or a transversal slot on a single cone [Bailin and Silver, 1956]. The error made in that work would have been avoided if the dyadic Green function technique were available then.

C-10.4

Find the expression for the back-scattering cross section for a conducting hemisphere resting on a conducting ground plane. Assume that the incident wave is polarized with the $\overline{E}$-vector parallel to the surface of the ground plane, but is impinging on the hemisphere at an oblique angle $\theta = \theta_0$ with the ground plane located at $\theta = \pi/2$. By definition, the back-scattering cross section is defined by

$$\sigma_b = \lim_{R\to\infty} \frac{4\pi R^2 \left|\overline{E}_s\right|^2}{\left|\overline{E}_i\right|^2},$$

where $\overline{E}_i$ denotes the incident plane wave and $\overline{E}_s$ the scattered field in the backward direction.

C-10.5

Derive Nomura's expression for $\overline{E}_s$ given by (10.70) by working out the details based on the boundary conditions.

C-10.6

Two infinitesimal electric dipoles with the same current moment are placed at the poles of a conducting sphere with radius equal to "a"; that is,

$$\overline{J}_1(\overline{R}') = \frac{c\delta\,(R' - a)\,\delta\,(\theta' - 0)\,\delta\,(\phi' - 0)\,\hat{R}'}{R'^2 \sin\theta'}$$

$$\overline{J}_2(\overline{R}') = \frac{c\delta\,(R' - a)\,\delta\,(\theta' - \pi)\,\delta\,(\phi' - 0)\,\hat{R}'}{R'^2 \sin\theta'}.$$

Find the expression for the electric field of this antenna, and the leading term of the field when $ka \ll 1$.

Hint: The asymptotic expression for the spherical Hankel functions and the series expansion of the spherical Bessel functions are found in Stratton [1941, p. 405].

Chapter 11

C-11.1

Find the expression for the electric field of a small loop (magnetic dipole) with a magnetic dipole moment $\overline{m}$ placed above a flat earth. Let the axis of the dipole lie (1) in the x-direction, $\overline{m} = m\hat{x}$ and (2) in the z-direction, $\overline{m} = m\hat{z}$.

Hint: It would be convenient to use the formula

$$\overline{E}(\overline{R}) = i\omega\mu_0 \iiint \left[\nabla \times \overline{\overline{G}}_e^{(11)}(\overline{R}, \overline{R}')\right] \cdot \overline{M}(\overline{R}')\, dV',$$

where $\overline{M}$ denotes the equivalent magnetization vector; that is,

$$\overline{J}_m = -i\omega\mu_0 \overline{M}$$

and $\overline{J}_m$, the fictitious magnetic current.

C-11.2

When the conductivity of a flat earth approaches infinity the earth becomes a perfectly conducting plane. Determine the eigenfunction expansion of $\overline{\overline{G}}_{e1}$ by finding the degenerate form of $\overline{\overline{G}}_e^{(11)}$ as σ approaches infinity.

C-11.3

Show that the field of a horizontal magnetic dipole placed above a flat earth can be derived by using one electric Hertzian potential function $\overline{\pi}_e$ and one magnetic potential function $\overline{\pi}_m$.

Hint: $\overline{E} = \nabla \times \pi_m + i\omega\left(\overline{\pi}_e + \frac{1}{k^2}\nabla\nabla \cdot \overline{\pi}_e\right)$.

C-11.4

Find the expression for $\overline{\overline{G}}_{m1}^{(12)}(\overline{R}, \overline{R}')$ by starting with the eigenfunction expansion for the free-space function $\overline{\overline{G}}_{m0}^{(2)}(\overline{R}, \overline{R}')$ in region 2 with constitutive constants μ_2 and ϵ_2 and then applying the method of scattering superposition to determine $\overline{\overline{G}}_{m1}^{(22)}(\overline{R}, \overline{R}')$ and $\overline{\overline{G}}_{m1}^{(12)}(\overline{R}, \overline{R}')$. The constitutive constants in region 1 are denoted by μ_1' and ϵ_1' for this model (model B). We consider the case $\mu_1 = \mu_2 = \mu_0$ in model A with electric constants ϵ_1 and ϵ_2; then

$$\begin{aligned} \frac{\mu_1'\mu_1}{\mu_2^2} &= \frac{\mu_1'}{\mu_0} \\ &= \frac{\epsilon_1}{\epsilon_2} = \left(\frac{k_1}{k_2}\right)^2. \end{aligned}$$

Hint:

$$\overline{\overline{G}}_{m0}^{(2)}(\overline{R},\overline{R}')=\int_0^\infty d\lambda\sum_n\frac{i(2-\delta_0)k_2}{4\pi\lambda h_2}\cdot$$

$$\cdot\left[\overline{N}(\pm h_2)\overline{M}'(\mp h_2)+\overline{M}(\pm h_2)\overline{N}'(\mp h_2)\right],\ z\gtrless z'.$$

Let

$$\overline{\overline{G}}_{m1}^{(22)}(\overline{R},\overline{R}')=\overline{\overline{G}}_{m0}^{(2)}(\overline{R},\overline{R}')+\overline{\overline{G}}_{ms}^{(22)}(\overline{R},\overline{R}'),$$

with

$$\begin{aligned}\overline{\overline{G}}_{ms}^{(22)}(\overline{R},\overline{R}')=&\int_0^\infty d\lambda\sum_n\frac{i(2-\delta_0)k_2}{4\pi\lambda h_2}\cdot\\&\cdot\left\{\overline{M}(h_2)\left[a_2^+\overline{N}'(h_2)+a_2^-\overline{N}'(-h_2)\right]\right.\\&+\overline{M}(-h_2)\left[b_2^+\overline{N}'(h_2)+b_2^-\overline{N}'(-h_2)\right]\\&+\overline{N}(h_2)\left[c_2^+\overline{M}'(h_2)+c_2^-\overline{M}'(-h_2)\right]\\&\left.+\overline{N}(-h_2)\left[d_2^+\overline{M}'(h_2)+d_2^-\overline{M}'(-h_2)\right]\right\}\\\overline{\overline{G}}_{m1}^{(12)}(\overline{R},\overline{R}')=&\int_0^\infty d\lambda\sum_n\frac{i(2-\delta_0)k_2}{4\pi\lambda h_2}\\&\cdot\left\{\overline{M}(h_1)\left[c_1^+\overline{N}'(h_2)+c_1^-\overline{N}'(-h_2)\right]\right.\\&\left.+\overline{N}(h_1)\left[d_1^+\overline{M}'(h_2)+d_1^-\overline{M}'(-h_2)\right]\right\}.\end{aligned}$$

The boundary conditions are

at $z=0$,

$$\hat{z}\times\overline{\overline{G}}_{m1}^{(22)}(\overline{R},\overline{R}')=0$$

at $z=d$,

$$\hat{z}\times\left[\frac{\overline{\overline{G}}_{m1}^{(22)}(\overline{R},\overline{R}')}{\mu_0}-\frac{\overline{\overline{G}}_{m1}^{(12)}(\overline{R},\overline{R}')}{\mu_1'}\right]=0$$

and

$$\hat{z}\times\left[\frac{\nabla\times\overline{\overline{G}}_{m1}^{(22)}}{k_2^2}-\nabla\times\frac{\overline{\overline{G}}_{m1}^{(12)}(\overline{R},\overline{R}')}{k_1^2}\right]=0.$$

The expression for $\overline{\overline{G}}_{m1}^{(22)}(\overline{R},\overline{R}')$ so obtained should be the same as (11.119).

Answer:

$$c_1^+=-c_1^-=-\left(\frac{k_1}{k_2}\right)^2\left(\frac{h_2}{h_1}\right)e^{i(\Delta_2-\Delta_1)}\frac{(1-\rho)}{D}$$

$$k_1^+=d_1^-=\left(\frac{k_1}{k_2}\right)^3\left(\frac{h_2}{h_1}\right)e^{i(\Delta_2-\Delta_1)}\frac{(1-\rho')}{D'}.$$

Chapter 12

C-12.1

Show that for the electromagnetically "subsonic" case ($n\beta < 1$), the Fourier transform of the dynamic scalar potential function $\varphi(\overline{R})$ is given by

$$\varphi(\overline{R}) = \frac{a^{-\frac{1}{2}}}{\epsilon} \iiint \frac{\left[\rho(\overline{R}') + \overline{\Omega} \cdot \overline{J}(\overline{R}')\right] e^{i\omega(\tau' + \tau_a)}}{4\pi R_a} dV'.$$

Determine $\varphi(\overline{R}, \tau)$ and then $\varphi(\overline{R}, t)$. Find the expression for $\overline{E}(\overline{R}, t)$ in terms of $\overline{A}\left(\overline{R}, t\right)$ and $\varphi\left(\overline{R}, t\right)$.

Hint: By taking the divergence of the equation for $\overline{A}(\overline{R})$, the Fourier transform of $\overline{A}\left(\overline{R}, \tau\right)$, one finds

$$\left(\nabla_a \cdot \nabla + k^2 a\right) \varphi(\overline{R}) = \frac{1}{\epsilon a} \left[\rho(\overline{R}) + \overline{\Omega} \cdot \overline{J}(\overline{R})\right].$$

Observe that

$$\nabla \cdot \overline{J}\left(\overline{R}, t\right) = -\frac{\partial \rho\left(\overline{R}, t\right)}{\partial t} = -\frac{\partial \rho(\overline{R}, \tau)}{\partial \tau}$$

$$\nabla \cdot \overline{J}\left(\overline{R}, t\right) = \nabla \cdot \overline{J}\left(\overline{R}, \tau\right) + \overline{\Omega} \cdot \frac{\partial \overline{J}\left(\overline{R}, \tau\right)}{\partial \tau}.$$

Hence

$$\nabla \cdot \overline{J}\left(\overline{R}, \tau\right) = -\frac{\partial}{\partial \tau} \left[\rho\left(\overline{R}, \tau\right) + \overline{\Omega} \cdot \overline{J}\left(\overline{R}, \tau\right)\right].$$

C-12.2

The Green function G_0 for the moving isotropic medium in the "supersonic" case satisfies the differential equation

$$\left(\frac{\partial^2}{\partial x^2} + \frac{\partial^2}{\partial y^2} - \frac{1}{|a|} \frac{\partial^2}{\partial z^2} - k^2 |a|\right) G_0(\overline{R}, \overline{R}')$$
$$= -\delta(\overline{R} - \overline{R}').$$

In the text, we identify the equation as a two-dimensional Klein-Gordon equation and quoted the work of M. H. Cohen for the answer of that equation, namely,

$$G_0(\overline{R}, \overline{R}') = \frac{|a|^{\frac{1}{2}} \cos k \, |a|^{\frac{1}{2}} R_a'}{4\pi R_a'},$$

where

$$R_a' = \left(|a| \, \xi^2 - r^2\right)^{\frac{1}{2}},$$

with

$$\xi = z - z', r = \left[\left(x - x'\right)^2 + \left(y - y'\right)^2\right]^{\frac{1}{2}}$$

for $|a| \, \xi > r$, and $G_0(\overline{R}, \overline{R}') = 0$ for $|a| \, \xi < r$.

Show that the solution can also be obtained by applying a Fourier and Hankel transform to that equation.

Hint: We define

$$g_0(\lambda, h) = \int_{-\infty}^{\infty} dz \int_0^{\infty} G_0(r, z) e^{-ihz} J_0(\lambda r) r\, dr$$

where $G_0(r, z)$ satisfies the equation

$$\begin{aligned}&\left[\frac{1}{r}\frac{\partial}{\partial r}\left(r\frac{\partial G_0}{\partial r}\right) - \frac{1}{|a|}\frac{\partial^2 G_0}{\partial z^2} - k^2 |a|\right] G_0(r, z)\\ &= -\frac{1}{2\pi}\delta(r-0)\delta(z-z');\end{aligned}$$

then

$$G_0(r, z) = \frac{1}{(2\pi)^2}\int_{-\infty}^{\infty} dh \int_0^{\infty} \frac{e^{ih(z-z')} J_0(\lambda r)}{\lambda^2 + k^2|a| - h^2|a|^{\frac{1}{2}}} \lambda\, d\lambda.$$

By evaluating the h-integration, one obtains

$$G_0(r, z) = \begin{cases} \frac{|a|}{2\pi}\int_0^{\infty} \frac{\sin h_1(z-z')}{h_1} J_0(\lambda r)\lambda\, d\lambda, & \text{for } z > z' \\ 0, \quad \text{for } z < z', \end{cases}$$

where

$$h_1 = \left(\lambda^2|a| + k^2|a|^2\right)^{\frac{1}{2}}.$$

Justify this step by invoking a priori condition for Cerenkov radiation. The Bessel integral is then carried out by making use of the Sonine-Gegenbauer formula [Watson, 1922, p. 415] by an appropriate change of the parameters. That formula states

$$\begin{aligned}&\int_0^{\infty} \frac{J_0(\lambda r)\sin\left[\left(\lambda^2 - b^2\right)^{\frac{1}{2}}\eta\right]}{\left(\lambda^2 - b^2\right)^{\frac{1}{2}}}\lambda\, d\lambda\\ &= \begin{cases} \dfrac{\cosh b\left(\eta^2 - r^2\right)^{\frac{1}{2}}}{\left(\eta^2 - r^2\right)^{\frac{1}{2}}}, \eta > r \\ 0, \quad \eta < r. \end{cases}\end{aligned}$$

C-12.3

Prove the symmetrical relationship as stated by (12.79), namely,

$$\left[\bar{\bar{b}} \cdot \bar{\bar{G}}_{eo}^{(b)}(\bar{R}', \bar{R})\right]^T = \bar{\bar{b}} \cdot \bar{\bar{G}}_{eo}^{(b)}(\bar{R}, \bar{R}').$$

C-12.4

Verify the identity stated by (12.86), namely,

$$\nabla \times \left[\bar{\bar{b}} \cdot \nabla \times \left(\bar{\bar{b}} \cdot \overline{\overline{G}}_{eo}^{(b)} \right) \right] = \frac{\left[\nabla_a \nabla \cdot \overline{\overline{G}}_{eo}^{(b)} - (\nabla_a \cdot \nabla) \overline{\overline{G}}_{eo}^{(b)} \right]}{a}.$$

C-12.5

Find the two-dimensional Fourier integral representation of the functions $\overline{\overline{G}}_e^{(11)}$ and $\overline{\overline{G}}_e^{(21)}$ for a moving semi-infinite medium. Region 1 is air and region 2 contains the moving medium with velocity $\overline{v} = v\hat{x}$. The interface corresponds to the $x - y$ plane or $z = 0$.

Remark: This is a relatively difficult exercise. It is a test of the mastery of some important techniques developed in the book. No hint will be provided for this exercise, which is a good candidate as a research topic to train a graduate student in analytic skills.

References

Books

Abramowitz, M., and I. A. Stegun, *Handbook of Mathematical Functions*, National Bureau of Standards, U. S. Government Printing Office, Washington, D.C., 1964.

Baker, B. B., and E. T. Copson, *The Mathematical Theory of Huygens' Principle*, Oxford University Press, London, 1950.

Baños, A., *Dipole Radiation in the Presence of a Conducting Half-plane*, Pergamon Press, New York, 1966.

Bladel, J. Van, *Singular Electromagnetic Fields and Sources*, Clarendon, Oxford, 1991.

Bowman, J. J., T. B. A. Senior, and P. L. E. Uslenghi, *Electromagnetic and Acoustic Scattering by Simple Shapes*, North-Holland, Amsterdam, 1969.

Collin, R. E., *Field Theory of Guided Waves*, IEEE Press, Piscataway, N.J., 1991.

Copson, E. T., *Theory of Functions of a Complex Variable*, Oxford University Press, London, 1948.

Courant, R., *Differential and Integral Calculus*, Vol. II, Interscience Publishers, New York, 1936.

Felsen, L. B., and N. Marcuvitz, *Radiation and Scattering of Waves*, Prentice Hall, Englewood Cliffs, N.J., 1973.

Feynberg, Y. L., *The Propagation of Radio Waves Along the Surface of the Earth*, English translation of 1961 Russian book published by the Foreign Technology Division, Wright-Patterson Air Force Base, Dayton, Ohio, 1961.

Fock, V. A., *Electromagnetic Diffraction and Propagation Programs*, Pergamon Press, New York, 1965.

Gelfand, I. M., and G. E. Shilov, *Generalized Functions*, Vols. 1–3, Academic Press, New York, 1964.

Ince, E. L., *Ordinary Differential Equations*, Dover Publications, New York, 1944.

King, R. W. P., T. T. Wu, and M. Owens, *Lateral Electromagnetic Waves*, Springer Verlag, New York, 1992.

Luneburg, R. K., *The Mathematical Theory of Optics*, Brown University Press, Providence, R.I., 1944.

Marcuvitz, N., *Waveguide Handbook*, Vol. 10 of M.I.T. Radiation Laboratory Series, McGraw-Hill, New York, 1951.

Maxwell, J. C., *Scientific Papers*, Dover Publications, New York, pp. 76–79, 1860.

Mentzer, J. R., *Scattering and Diffraction of Radio Waves*, Pergamon Press, New York, 1955.

Morse, P. M., and H. Feshbach, *Methods of Theoretical Physics*, Part II, McGraw-Hill, New York, 1953.

Nussbaumer, H. J., *Fast Fourier Transform and Convolution Algorithms*, Springer Verlag, New York, 1982.

Ramo, S., J. R. Whinnery, and T. Van Duzer, *Fields and Waves in Communication Electronics*, John Wiley & Sons, New York, 1965.

Sommerfeld, A., *Partial Differential Equations*, Academic Press, New York, 1949.

Sommerfeld, A., *Electrodynamics*, Academic Press, New York, 1952.

Sommerfeld, A., *Optics*, Academic Press, New York, 1954.

Stratton, J. A., *Electromagnetic Theory*, McGraw-Hill, New York, 1941.

Tai, C. T., *Dyadic Green's Functions in Electromagnetic Theory*, International Textbook Company, Scranton, Pa., 1971.

Tai, C. T., *Generalized Vector and Dyadic Analysis*, IEEE Press, Piscataway, N.J., 1992.

Wait, J. R., *Electromagnetic Radiation from Cylindrical Structures*, Pergamon Press, New York, 1959.

Wait, J. R., *Electromagnetic Waves in Stratified Media*, Pergamon Press, New York, 1962.

Watson, G. N. *Theory of Bessel Functions*, Cambridge University Press, Cambridge, 1922.

Whittaker, E. T., and G. N. Watson, *Modern Analysis*, Cambridge University Press, Cambridge, 1943.

Articles and Reports

Bailin, L. L., and S. Silver, "Exterior Electromagnetic Boundary Value Problems for Spheres and Cones," *IRE Trans., AP-4*, pp. 5–15, *AP-5*, p. 313 (correction), 1956.

Cheng, D. H. S., "On the Formulation of the Dyadic Green Function in a Layered Medium," *Electromagnetics*, Vol. 6, No. 2, pp. 171–182, 1986.

Chien, W. Z., L. Infeld, J. R. Pounder, A. F. Stevenson, and J. L. Synge, "Contributions to the Theory of Wave Guides," *Canadian J. Research*, Vol. 27, pp. 69–129, 1949.

Cohen, M., "Radiation in Plasma," Part I, *Phys. Rev.*, Vol. 123, pp. 711–721, 1961.

Compton, Jr., R. T., "Time-dependent Electromagnetic Field in a Moving Medium," *J. Math. Physics*, Vol. 7, pp. 2145–2152, 1966.

Dwight, H. B., "Tables of Roots for Natural Frequencies in Coaxial Type Cavities," *J. Math. Phys.*, Vol. 27, pp. 84–89, 1948.

Eaton, J. E., "On Spherically Symmetrical Lens," *IRE Trans. of Antennas and Propagation, A-P 4*, pp. 66–71, 1952.

Franz, W., "Zur Formulierung des Huygenschen Prinzips," *Z. Naturforchung*, Vol. 3a, pp. 500–506, 1948.

Hansen, W. W., "A New Type of Expansion in Radiation Problems," *Phys. Rev.*, Vol. 47, pp. 139–143, 1935.

Hansen, W. W., "Directional Characteristics of Any Antenna over a Plane Earth," *J. Appl. Phys.*, Vol. 7, pp. 460–465, 1936.

Hansen, W. W., "Transformations Useful in Certain Antenna Calculations," *J. Appl. Phys.*, Vol. 8, pp. 282–286, 1937.

Hargreaves, R., "A Diffraction Problem and an Asymptotic Theorem in Bessel Series," *Phil. Mag.*, Vol. 36, pp. 191–199: Supplementary Note, Vol. 36, pp. 317–319, 1918.

Harrington, R. F., "Current Element Near the Edge of a Conducting Half-Plane," *J. Appl. Phys.*, Vol. 24, pp. 547–550, 1953.

Jasik, H., "Electromagnetic Theory of Luneberg Lens," Dissertation, Department of Electrical Engineering, Polytechnic Institute of Brooklyn, New York, 1954.

Johnson, W. A., A. Q. Howard, and D. G. Dudley, "On the Irrotational Component of the Electric Green Dyadic," *Radio Science*, Vol. 14, pp. 961–967, 1979.

Lee, K. S. H., and C. H. Papas, "Electromagnetic Radiation in the Presence of Moving Simple Media," *J. Math. Phys.*, Vol. 5, pp. 1668–1672, 1964.

Lee, S. W., J. Boersma, C. L. Law, and G. A. Deschamps, "Singularity in Green's Function and Its Numerical Evaluation," *IEEE Trans. Antennas and Propagation*, Vol. 22, No. 3, May 1980.

Levine, H., and C. H. Papas, "Theory of Circular Diffraction Antenna," *J. Appl. Phys.*, Vol. 22, pp. 29–43, 1951.

Levine, H., and J. Schwinger, "On the Theory of Electromagnetic Wave Diffraction by an Aperture in an Infinite Plane Conducting Screen," *Comm. Pure and Appl. Math.*, Vol. III, pp. 355–391, 1950.

Meixner, J., "Die Kantenbedingung in der Theorie der Beugung elektromagnetischer Wellen an volkommen leitenden ebenen Schirmen," *Ann. Phys.*, Vol. 441, pp. 2–9, 1949.

Mie, G., "Beiträge zur Optik Trüber Medien, Speziell Killoidaler Metallösungen," *Ann. der Phys.*, Vol. 25, pp. 377–445, 1908.

Minkowski, H., "Die Grundgleichungen für die Elektromagnetischen Vorgange in Bewegten Korpern," *Gottingen Nachrichten*, pp. 53–116, 1908.

Nomura, Y., "On the Propagation of Electric Waves from a Horizontal Dipole over the Surface of the Earth Sphere," *Science Report of Tohoku University (Japan)*, Vols. 1–2, pp. 25–49, 1951.

Nomura, Y., and K. Takaku, "On the Propagation of the Electromagnetic Waves in an Inhomogeneous Atmosphere," *Science Report of Tohoku University (Japan)*, Vol. 7, pp. 107–148, 1955.

Pincherle, L., "Electromagnetic Waves in Metal Tubes Filled with Two Dielectrics," *Phys., Rev.*, Vol. 66, No. 5, pp. 118–130, 1944.

Rozenfeld, P., "The Electromagnetic Theory of Three-dimensional Inhomogeneous Lenses," *IEEE Trans. Antennas and Propagation*, Vol. 24, pp. 365–370, 1976.

Rozenfeld, P., "The Electromagnetic Theory of Three-dimensional Inhomogeneous Lenses and the Dyadic Green Functions for Cavities," Ph.D. Dissertation, Dept. of Electrical Engineering, The University of Michigan, Ann Arbor, Michigan, 1974.

Schelkunoff, S. A., "On the Teaching of the Undergraduate Electromagnetic Theory," *IEEE Trans. Educ.*, Vol. 15, pp. 15–25, 1972.

Senior, T. B. A., "The Diffraction of a Dipole Field by a Perfectly Conducting Half-plane," *Q.J. Mech. and Appl. Math.*, Vol. 6, pp. 101–114, 1953.

Stubenrauch, C. F., "Radiation from Source in the Presence of a Moving Dielectric Column," Ph.D. Dissertation, Department of Electrical Engineering, The University of Michigan, Ann Arbor, Michigan, 1972.

Stubenrauch, C., and C. T. Tai, "Dyadic Green Function for Cylindrical Waveguide with Moving Medium," *Appl. Sci.*, Vol. 25, pp. 281–289, 1971.

Tai, C. T., "On the Theory of Diffraction of Electromagnetic Waves by a Sphere," *Tech. Report no. 29*, Stanford Research Institute, Stanford, California, October 1952.

Tai, C. T., "Theory of the Cylindrical Luneberg Lens Excited by a Line Magnetic Current," Technical Report 678-3, Antenna Laboratory, The Ohio State University, Columbus, Ohio, 1956.

Tai, C. T., "Electromagnetic Theory of Spherical Luneberg Lens," *Appl. Sci. Res.*, Sec. B, Vol. 7, pp. 113–130, 1958a.

Tai, C. T., "Maxwell Fish-eye Treated by Maxwell Equations," *Nature*, Vol. 182, pp. 1600–1601, 1958b.

Tai, C. T., "Huygens' Principle in a Moving Isotropic, Homogeneous and Linear Medium," *Applied Optics*, Vol. 4, pp. 1347–1349, 1965a.

Tai, C. T., "The Dyadic Green Function for a Moving Isotropic Medium," *IEEE Trans., AP-13*, pp. 322–323, 1965b.

Tai, C. T., "Present Views on Electrodynamics of Moving Media," *Radio Science*, Vol. 2, pp. 245–248, 1967.

Tai, C. T., "On the Presentation of Maxwell's Theory," *Proc. IEEE*, Vol. 60, No. 8, pp. 936–945, 1972.

Tai, C. T., "On the Eigen-function Expansion of Dyadic Green Functions," *Proc. IEEE*, Vol. 61, p. 480, 1973; also "Math. Note," No. 28, Systems Laboratory, Kirkland Air Force Base, Albuquerque, New Mexico, 1973.

Tai, C. T., "Dyadic Green Functions for a Coaxial Line," *IEEE Trans. Antennas Propagation*, Vol. 31, pp. 355–358, 1983.

Tai, C. T., "Different Forms of Reciprocity Theorems," *Proc. Ann. Symp. Chinese Antenna Soc.* (in Chinese), Nanjing, China, Nov. 25–28, 1987.

Tai, C. T., "Dyadic Green Functions for a Rectangular Waveguide Filled with Two Dielectrics," *J. Electromagnetic Waves and Appl.*, Vol. 2, No. 3/4, pp. 245–253, 1988.

Tai, C. T., "Complementary Reciprocity Theorems in Electromagnetic Theory," *IEEE Trans. Antennas and Propagation*, Vol. 40, No. 6, pp. 675–681, 1992.

Tai, C. T., and P. Rozenfeld, "Different Representations of Dyadic Green's Functions for a Rectangular Cavity," *IEEE Trans. Microwave Theory and Technique*, Vol. 24, pp. 597–601, 1976.

Wilcox, C. H., "Electromagnetic Theory of Luneburg Lens," Technical Report No. MSD-1802, Lockheed Aircraft Corporation, Burbank, California, 1956.

Name Index

I

J

K

L

M

N

O

P

R

S

T

U

V

W

Subject Index

A

B

C

D

E

F

G

H

I

J

K

L

M

U

V

W